Multiscale Problems in Science and Technology

Springer
Berlin
Heidelberg
New York
Barcelona
Hong Kong
London
Milan
Paris
Tokyo

N. Antonić · C.J. van Duijn
W. Jäger · A. Mikelić (Eds.)

Multiscale Problems in Science and Technology

Challenges to Mathematical Analysis and Perspectives

Proceedings of the Conference on Multiscale Problems in Science and Technology, Dubrovnik, Croatia, 3-9 September 2000

With 52 Figures

Springer

Editors

Nenad Antonić

University of Zagreb
Department of Mathematics
Bijenička cesta 30
10000 Zagreb
Croatia

e-mail: nenad@math.hr

Willi Jäger

University of Heidelberg
IWR
Im Neuenheimer Feld 368
69120 Heidelberg
Germany

e-mail: jaeger@iwr.uni-heidelberg.de

C.J. van Duijn

CWI Amsterdam
Dept. of Applied Analysis
and Scientific Computing
P.O.Box 94079
Kruislaan 413
1090 GB Amsterdam
The Netherlands

e-mail: hans@cwi.nl

Andro Mikelić

Université Lyon 1
Analyse Numérique, Bât 101
U.F.R. Mathématiques
43, Bd. du 11 Novembre 1918
69622 Villeurbanne Cedex
France

e-mail: andro@lan.univ-lyon1.fr

Catalog-in-Publication Data applied for

Die Deutsche Bibliothek - CIP-Einheitsaufnahme

Multiscale problems in science and technology : challenges to mathematical analysis and perspectives / N. Antonić... - Berlin ; Heidelberg ; New York ; Barcelona ; Hong Kong ; London ; Milan ; Paris ; Tokyo : Springer, 2002
ISBN 3-540-43584-0

Mathematics Subject Classification (2000): 35-06 (primary), 35B37, 35R60, 49J45, 60H15, 60K37, 65N30, 74F10, 74Qxx, 74R10, 74S05, 76M50, 80M40, 85A05, 85A25

ISBN 3-540-43584-0 Springer-Verlag Berlin Heidelberg New York

Springer-Verlag Berlin Heidelberg New York
a member of BertelsmannSpringer Science+Business Media GmbH

http://www.springer.de

Typeset by the authors using a Springer TEX macro package
Cover production: *design & production* GmbH, Heidelberg

SPIN 10837661 46/3142db - 5 4 3 2 1 0 – Printed on acid-free paper

Preface

The International conference on *Multiscale problems in science and technology; Challenges to mathematical analysis and applications* brought together mathematicians working on multiscale techniques (homogenisation, singular perturbation) and specialists from applied sciences who use these techniques. Our idea was that mathematicians could contribute to solving problems in the emerging applied disciplines usually overlooked by them and that specialists from applied sciences could pose new challenges for multiscale problems.

Numerous problems in natural sciences contain multiple scales: flows in complex heterogeneous media, many particles systems, composite media, etc. Mathematically, we are led to study of singular homogenisation limits and the procedure is called upscaling or homogenisation. The processes to be upscaled are usually described by differential equations. For simple cases, when the differential equation is linear and the heterogeneities are periodic some progress has been made. However, most natural phenomena are described by nonlinear differential equations in a random nonhomogeneous medium and, despite an intensive development in recent years, there are many open problems.

The objective of the conference was to bring together leading specialists from Europe and the United States and to discuss new challenges in this quickly developing field. Topics of the conference were Nonlinear Partial Differential Equations and Applied Analysis, with direct applications to the modeling in Material Sciences, Petroleum Engineering and Hydrodynamics.

The Conference was co-organised by the Department of Mathematics of University of Zagreb, and the Inter-University Centre (IUC) in Dubrovnik, Croatia, where it took place from September 3rd to 9th, 2000. We hope that these Proceedings reflect the variety of subjects discussed; in order to provide a better overall view, we have reprinted the abstracts of those talks whose authors did not provide a full written account. Unfortunately, any attempt to convey the inspiring atmosphere of regular evening scientific discussions, or the beautiful surroundings of the ancient town and its neighbourhood, would be futile. We can only thank Mr. Zoran Jurović, a wine producer of *plavac mali* red wines from Postup, Orebić, for the contribution of his cellar to the atmosphere.

We would like to thank the members of the Local Organising Committee:

Mladen Jurak, University of Zagreb
Eduard Marušić-Paloka, University of Zagreb
Žarko Prnić, INA Oil Industry, Zagreb
Josip Tambača, University of Zagreb
Marko Vrdoljak, University of Zagreb

for all the work they did. The conference was organised, hotels booked, etc., while none of us was in Croatia. In Dubrovnik, the staff of IUC provided excellent support and warm reception. Their support was complemented by a number of graduate students who helped with the registration, provided directions for sightseeing and boat trip, etc. In particular, Martin Lazar was responsible for the local contacts in his home town, while Krešimir Burazin recorded Prof. Tartar's lectures.

Financial aid was generously provided by the Croatian Ministry of Science and Technology, INA Oil Industry, and the Open Society Foundation.

For these Proceedings, we would like to thank those authors who found time and energy to write their contributions. Professor Tartar's lectures were recorded, typed, and then the real work began. In spite of his other obligations (and no promises made), he found time to make several revisions, and a number of pages were added. Marko Vrdoljak generated all the figures for these lecture notes, and caught many misprints. Finally, Josip Tambača put all the contributions together, ironed the differences and provided the final version of the manuscript. To Springer, many thanks for constant encouragement and assistance.

Nenad Antonić was in charge of the Proceedings and other editors would like to thank him for his effort.

Zagreb, Amsterdam, Heidelberg, Lyon

February 2002

Nenad Antonić
C. J. van Duijn
Willi Jäger
Andro Mikelić

List of participants

1. Nenad Antonić: Department of Mathematics, University of Zagreb, Bijenička c. 30, 10000 Zagreb, Croatia,
   antonic@mis.mpg.de
2. Neven Balenović: Department of Mathematics, University of Zagreb, Bijenička c. 30, 10000 Zagreb, Croatia,
   neven@math.hr
3. Krešimir Burazin: Faculty of Electrical Engeneering and Computing, University of Zagreb, Unska c. bb, 10000 Zagreb,
   kreso@math.hr
4. Johan Byström: Luleå University of Technology, Luleå, Sweden,
   johanb@sm.luth.se
5. Pierre-Henri Chavanis: Université Paul Sabatier, Toulouse, France,
   chavanis@irsamcz.ups-tle.fr
6. Gregory A.Chechkin: Department of Differential Equations, Faculty of Mechanics and Mathematics, Lomonosov Moscow State University, Moscow 119899, Russia,
   chechkin@sci.lebedev.ru
7. Zvonimir Čerin: Department of Mathematics, University of Zagreb, Bijenička c. 30, Zagreb, Croatia,
   cerin@math.hr
8. Gianpietro Del Piero: Dipartimento di Ingegneria, via Saragat 1, 44100 Ferrara, Italia,
   gdpiero@ing12.unife.it
9. Zlatko Drmač: Department of Mathematics, University of Zagreb, Bijenička c. 30, Zagreb, Croatia,
   drmac@math.hr
10. C.J. van Duijn: CWI Amsterdam, The Netherlands,
    hansd@cwi.nl
11. Antonija Duvnjak: Faculty of Electrical Engeneering and Computing, University of Zagreb, Unska c. bb, 10000 Zagreb,
    antonija@msun.zpm.fer.hr
12. Bjorn Engquist: University of California, Los Angeles, USA,
    engquist@math.ucla.edu
13. Jose Luis Ferrin: Departamento de Matemática Aplicada, Universidad de Santiago de Compostela, 15706 Santiago de Compostela, Spain,
    maferrin@usc.es
14. Irene Fonseca: Department of Mathematical Sciences and Center for Nonlinear Analysis, Carnegie Mellon University, Pittsburgh, Pennsylvania 15231–3890, USA,
    fonseca@cmu.edu

15. Egon Gersch: Rafael, Inc., Israel,
e_gersch@netvision.net.il
16. Luka Grubišić: Department of Mathematics, University of Zagreb, Bijenička c. 30, 10000 Zagreb, Croatia,
luka@math.hr
17. Ignacio Guerra: CWI Amsterdam, The Netherlands,
ignaciog@cwi.nl
18. Flavius Guias: Institut für Angewandte Math., Universität Heidelberg, Im Neuenheimer Feld 294, 69120 Heidelberg, Germany,
Flavius.Guias@IWR.Uni-Heidelberg.de
19. Joost Hulshof: University of Leiden, The Netherlands,
hulshof@math.leidenuniv.nl
20. Willi Jäger: Universität Heidelberg, Germany,
jaeger@iwr.uni-heidelberg.de
21. Sanja Jovović: INA, Zagreb, Croatia,
sanja.jovovic@ina.hr
22. Mladen Jurak: Department of Mathematics, University of Zagreb, Bijenička c. 30, 10000 Zagreb, Croatia,
jurak@math.hr
23. Ivo Jurić: General hospital, Split,
ivo.juric1@st.tel.hr
24. Peter Knabner: Institute for Applied Mathematics, Universität Erlangen-Nürnberg, Germany,
knabner@am.uni-erlangen.de
25. Martin Lazar: Department of Mathematics, University of Zagreb, Bijenička c. 30, 10000 Zagreb, Croatia,
martin@math.hr
26. Stephan Luckhaus: Universität Leipzig, Germany,
luckhaus@mis.mpg.de
27. Eduard Marušić-Paloka: Department of Mathematics, University of Zagreb, Bijenička c. 30, 10000 Zagreb, Croatia,
emarusic@math.hr
28. Ana-Maria Matache: ETH Zürich Zentrum 8092 Zürich Switzerland,
amatache@math.ethz.ch
29. Andro Mikelić: Université Claude Bernard Lyon 1, Lyon, France,
andro@lan.univ-lyon1.fr
30. Nicolas Neuss: IWR, Universität Heidelberg, Im Neuenheimer Feld 368, 69120 Heidelberg, Germany,
Nicolas.Neuss@IWR.Uni-Heidelberg.de
31. Maria Neuss-Radu: Institut für Angewandte Math., Universität Heidelberg, Im Neuenheimer Feld 294, 69120 Heidelberg, Germany,
Maria.Neuss-Radu@IWR.Uni-Heidelberg.de
32. Mark Peletier: CWI Amsterdam, The Netherlands,
Mark.Peletier@cwi.nl

33. ANDREY PIATNITSKI: P.N. Lebedev Physical Inst. RAS, Leninski prosp. 53, Moscow 117333, Russia,
andrey@sci.lebedev.ru
34. ŽARKO PRNIĆ: INA, Zagreb, Croatia,
zarko.prnic@ina.hr
35. P.A.C. RAATS: Wageningen University, Roden, The Netherlans
36. ANDRIJA RAGUŽ: Department of Mathematics, University of Zagreb, Bijenička c. 30, 10000 Zagreb, Croatia,
andrija@math.hr
37. STEFAN RIEF: Universität Kaiserslautern, Germany,
rief@itwm.uni-kl.de
38. CHRISTOPH SCHWAB: Seminar für Angewandte Mathematik, ETH Zürich Zentrum 8092 Zürich, Switzerland,
schwab@math.ethz.ch
39. BEN SCHWEIZER: Institut fuer Angewandte Mathematik, INF 294 D-69120 Heidelberg, Germany,
Schweizer@iwr.uni-heidelberg.de
40. MAJA STARČEVIĆ: Department of Mathematics, University of Zagreb, Bijenička c. 30, 10000 Zagreb, Croatia,
mstarcev@math.hr
41. JOSIP TAMBAČA: Department of Mathematics, University of Zagreb, Bijenička c. 30, 10000 Zagreb, Croatia,
tambaca@math.hr
42. LUC TARTAR: Department of Mathematical Sciences, Carnegie Mellon University, Pittsburgh, PA 15231, USA,
tartar@cmu.edu
43. ZVONIMIR TUTEK: Department of Mathematics, University of Zagreb, Bijenička c. 30, 10000 Zagreb, Croatia,
tutek@math.hr
44. MARKO VRDOLJAK: Department of Mathematics, University of Zagreb, Bijenička c. 30, 10000 Zagreb, Croatia,
marko@math.hr
45. RAINER WEHRSE: Astrophysik, Heidelberg, Germany,
wehrse@foton.ita.uni-heidelberg.de
46. VASILIY V. ZHIKOV: Department of Mathematics, Vladimir State Pedagogical University, Prospect Stroiteley 11, 600024 Vladimir, Russia,
zhikov@vgpu.vladimir.ru

Contents

Mathematical Tools for Studying Oscillations and Concentrations: from YOUNG Measures to H-Measures and Their Variants

LUC TARTAR

Department of Mathematical Sciences, Carnegie Mellon University, Pittsburgh, Pennsylvania 15213–3890, U.S.A.
http://www.math.cmu.edu/public/people/fac/tartar.html

Foreword

The following set of notes was taken during the six lectures that I gave in Dubrovnik in September 2000. My purpose was to retrace the evolution of the subject over the last twenty years. A few years ago, I have thought that it should be helpful for young researchers to be told how and when new ideas were introduced, and what was known at the time when they appeared, so that they learn about the discovery process; in contrast, the traditional way of writing mathematical papers tends to suppress the human part of the research activity and only describes the results, which unfortunately are not always attributed to their real authors.

Teaching is like being a guide in a territory that one knows, for a group of travelers, some of them having already visited the area before, and some seeing it for the first time. I am not a very organised guide, and I rarely take two groups of visitors through exactly the same tour, and I hope that those who follow a few of these tours do learn a little at each time. I was not following a precise script in giving these lectures, and I had not thought about the optimal timing for presenting each piece, but I think that it came out as a nice tour among the particular area which I wanted to present, but some work was needed in order to transform into something readable what I had described in a slightly disorganised manner, and I am grateful to Nenad ANTONIĆ for having taken care of that.

Contents

1 LECTURE of Monday, September 4

1.1 A little history of the subject

I have to go back in the history in order to describe the origins of the problems I shall be talking about.

For me the history started back some thirty years ago, when I had been working with François MURAT in Paris (he actually started a little earlier to study these problems). In 1977 I gave a set of six lectures (Course Péccot), and my title was in French, of course, it was something like *Homogenisation in partial differential equations*, and I understood the term *homogenisation* in a much wider sense than what is usually meant now.

Just for historical purposes, let me stress that there are two aspects of homogenisation. One aspect is that of (linear) partial differential equations with oscillating coefficients, while the other is of nonlinear partial differential equations, with no oscillating coefficients there, but it could be oscillating initial data for an evolution equation or not. The question is to study the oscillations of the solutions and some of their partial derivatives. If there are no oscillations, then the problem is an easy one, and the question is to understand some type of averages that mathematically we describe by weak limits. And in some way find some kind of effective description that depends on the type of the equation.

In early seventies I understood how to use weak convergence and variants as a way to model what I called a relation between two levels, viz. *microscopic* and *macroscopic*. Since then I have learned that the term *microscopic* could be a little misleading for some people, as they use that term only to describe what happens at the level of atoms, while they would prefer the term *mesoscopic* for what I meant. Of course, that was what I meant, as we have to work with partial differential equations, and we start from principles like *Continuum Mechanics*, and in Continuum Mechanics you are not describing what happens at the level of atoms. The physicists tell us that you have to use *Quantum Mechanics* if you want to understand that.

Thus, when we are about to describe the effective properties of a mixture in conductivity or elasticity, we never go to the level of atoms, but we work from small pieces to actual size. One has to understand that this is a different model from the usual one I was taught, which has some probabilistic interpretation. In a way it has some advantages in applications, even though this is not the model that the physicists use. Unfortunately, if you do not talk their language, they are not able to understand what you are talking about. The point is that I am a mathematician, and that is why I prove theorems. Mathematics is something which is clear, but there are, however, different notions of what one could mean as *Applied Mathematics*, because in some parts of the world one does Mathematics with a strong look towards applications, while in other parts of the world one does formal expansions without caring whether they converge or not. I am of the school that does Mathematics, but I am trying to understand Continuum Mechanics and Physics.

What is important is to realise that in the initial work that I had started with François MURAT there was no relation to Physics. We were concerned with abstract academic problems. We discovered that there were already some results in the direction we wanted to pursue. Those results were mostly due to the Italian school, to Sergio SPAGNOLO [41,42] who benefited from the ideas of Ennio DE GIORGI.

At this level, we were only playing with Mathematics. What made a difference here is that we found some information in the work of Henri SANCHEZ-PALENCIA (for an overview consult his book [37]), who was studying some problems with a periodic structure. Part of his work was guessing an asymptotic expansion, then he would state some theorems. We realised that not only we could prove that SANCHEZ-PALENCIA was right, but that was how we understood that there was something behind those principles, since what we were doing for purely functional analysis reasons, or partial differential equations, had an interpretation as a model to understand something about Physics. And, of course, the point was that I had been taught about Continuum Mechanics and Physics before, but that was my new way to understand a few things that I had been taught, and therefore behind this was an idea that we were going to try to understand Physics from a purely mathematical point of view.

1.2 Suitable weak topologies

There is more than just weak convergence involved. In some way there is some kind of variant. *H-convergence* (see [34,59]) is a different variant of weak convergence. It is possible that this idea was well known, although I have not seen much of it so clearly exposed. Peter LAX once discussed problem, and said that Janos VON NEUMAN had followed a similar line of research at one time. But, except for that, I do think that people were only thinking in the terms of probabilistic and vague notions of averages when they were discussing the questions about Physics.

The basic problem that we were discussing was the conductivity of a mixture. Mathematically, in problem like that you have an equation of the form

$$-\mathsf{div}\,(a(\mathbf{x})\nabla u) = f\,.$$

This is a stationary equation, which should be supplemented by some boundary conditions. Here, a is something like *conductivity* (a scalar field) and u is something like *electric potential*. We have many different variants of the model. All those models share the need to express clearly the notion of the function a changing very rapidly in $\mathbf{x}$. This phenomenon is due to a having a domain which is made up of many small pieces. The value of a on one piece may be different from the value of a on the next. In that case we have to understand this equation in the sense of distributions. This idea goes back to Sergej L'vovič SOBOLEV [40], in the thirties. There were some works by Jean LERAY as well, but the distributions were not well defined before the time of Laurent SCHWARTZ (the interested reader might consult his book [39]).

One can avoid using distributions by writing down a weak formulation. It is important to understand that if we have derivatives of something discontinuous, we will be talking of the derivatives in the sense of distributions. The point is that, supposing you had something on the line and you would like to know what u looks like; so if you were to put u here, you would find that there will be a jump in the gradient of u each time one switches from one material to the other. Therefore there will be some kind of average u_0. But if you look at the gradient of u in a direction, then it is not like the gradient of u_0.

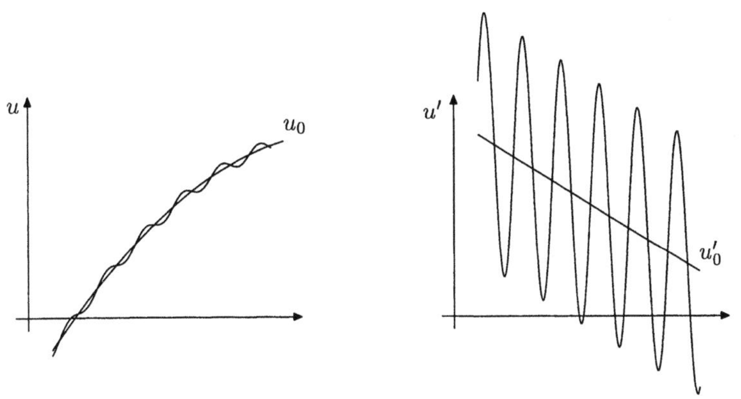

Figure 1. Functions having close values might have quite different derivatives.

The gradient of u does oscillate, so it does not seem to be near the gradient of u_0. It is significant that it is near in a different topology which is related to weak convergence.

However, we have to be a little careful; in Functional Analysis we have learned that the weak topology is not metrisable. Only if we restrict ourselves to bounded sets can we get a metrisable topology. This technical fact is important because in some situations, when for example one mixes materials which are insulators, you have to be very careful about the way in which you explain things, because the weak convergence may not be metrisable on unbounded sets and therefore the argument gets more technical.

It is important to realise that it is not only the weak convergence involved. It is natural to use the weak convergence on u, but it is not the weak convergence on a that matters. What you say is that u is very near u_0 and you would like to obtain some kind of expansion of u_0 plus a small term u_1 plus, for example, a periodic expansion that Henri SANCHEZ-PALENCIA was using.

The second term has some sort of oscillation, so it is small but its gradient is not. If you have a sequence which converges to a limit weakly, thus in the sense of distributions as well, then the derivatives converge to the derivatives in some weak sense, at least in the sense of distributions.

Therefore we have no problem understanding what this will be, if you have bounds. Sergej SOBOLEV had introduced the right spaces for that a long time ago. The problem is that we have to understand what happens to the product:

$$a^n \nabla u^n \; ?$$

We do have some bounds. The difficulty is that we were taught that weak convergence is not adapted to nonlinear problems, and that the product of two weakly converging sequences usually does not converge to the product of limits. So, what we rediscovered, was what Sergio SPAGNOLO had done. There was different type of convergence which we will call the *H-convergence*[1] and sometimes write

$$a^n \xrightarrow{H} a^{\text{eff}} .$$

We note that the solution satisfies an equation. We could show that

$$a^n \nabla u^n \rightharpoonup a^{\text{eff}} \nabla u^\infty ,$$

and we had a way to describe this kind of topology. This is not a weak convergence, but it is a natural thing if you understand the Physics. In 1974, when I was in Madison, I learned a lot from Joel ROBBIN. Mathematically there is something that you call differential forms. These are the objects that one integrates on manifolds. For these objects weak convergence is the right one. But, there are objects which are not differential forms. So, u is a function, while ∇u is actually a differential form of order 1. When one takes the divergence of a vector field one actually talks about a $N-1$ differential form. This a is something that transforms 1 forms into $N-1$ forms in a little different topology.

We were discussing which topology will be best adapted to that. But if we see that in more general terms, the question is that one does not know which topology to consider if one does not know what one wants to do.

---

[1] For $0 < \alpha \le \beta < \infty$, $M(\alpha,\beta;\Omega)$ will denote the set of $\mathbf{A} \in L^\infty(\Omega; \mathcal{L}(\mathbf{R}^N;\mathbf{R}^N))$ satisfying $\mathbf{A}(\mathbf{x})\boldsymbol{\xi}\cdot\boldsymbol{\xi} \ge \alpha|\boldsymbol{\xi}|^2$ and $\mathbf{A}(\mathbf{x})\boldsymbol{\xi}\cdot\boldsymbol{\xi} \ge \frac{1}{\beta}|\mathbf{A}(\mathbf{x})\boldsymbol{\xi}|^2$ for all $\boldsymbol{\xi} \in \mathbf{R}^N$, a.e. $\mathbf{x} \in \Omega$.

Equivalently $M(\alpha,\beta;\Omega)$ is the set of $\mathbf{A} \in L^\infty(\Omega; \mathcal{L}(\mathbf{R}^N;\mathbf{R}^N))$ satisfying $\mathbf{A}(\mathbf{x})\boldsymbol{\xi}\cdot\boldsymbol{\xi} \ge \alpha|\boldsymbol{\xi}|^2$ and $\mathbf{A}^{-1}(\mathbf{x})\boldsymbol{\xi}\cdot\boldsymbol{\xi} \ge \frac{1}{\beta}|\boldsymbol{\xi}|^2$, for all $\boldsymbol{\xi} \in \mathbf{R}^N$, a.e. $\mathbf{x} \in \Omega$.

We will say that a sequence $\mathbf{A}_n \in M(\alpha,\beta;\Omega)$ *H-converges* to $\mathbf{A}_\infty \in M(\alpha',\beta';\Omega)$ for some $0 < \alpha' \le \beta' < \infty$, if for every $f \in H^{-1}(\Omega)$, the sequence of solutions $u_n \in H^1_0(\Omega)$ of $-\text{div}\,(\mathbf{A}_n \nabla u_n) = f$ converges to u_∞ in $H^1_0(\Omega)$ weak and the sequence $\mathbf{A}_n \nabla u_n$ converges to $\mathbf{A}_\infty \nabla u_\infty$ in $L^2(\Omega;\mathbf{R}^N)$ weak, where u_∞ is therefore the solution of $-\text{div}\,(\mathbf{A}_\infty \nabla u_\infty) = f$ in Ω.

It will follow (compactness theorem) that one can always take $\alpha' = \alpha$ and $\beta' = \beta$.

If we only wanted to describe what is the limit u^∞ then the notion that Sergio SPAGNOLO had introduced [41], that of *G-convergence* (for a number of applications see [62]), is good. That is the convergence of GREEN functions. Then that is enough, it tells what does u^∞ satisfy and inside the proof, although he did not say much about it, they had the information which enabled them to identify what happens to $a^n \nabla u^n$ on the limit.

The main difference between G-convergence and H-convergence is twofold. When they were doing that, there was a *minimisation* problem behind. No minimisation context is necessary for H-convergence. That is one part.

On the other hand, one wants to understand oscillations and what has been called *correctors*. One wants to say that ∇u^n behaves like some operator applied to the limit

$$\nabla u^n = \mathbf{P}^n \nabla u^\infty .$$

Therefore we want to be able to say *macroscopically* (which means in term of weak limits) what is going on for functions of ∇u^n. We are going to use this u^∞ and its gradient, and we want to know if that is enough to understand where the oscillations are. So the limit is what I call the macroscopic variables, while the others are microscopic variables or mesoscopic ones. And one wants to be able to describe enough.

Once one has a theorem like that one, then one knows that the limit is described just by u^∞ and one has a different equation. The problem is that in some situations the form of the equation changes. In some way these are the more interesting cases, because these are the cases in which the mathematicians might be able to say something that the physicists have not yet discovered. Because, when you ask a physicist, some things are well known, but what does it mean to be well known in a Physics book?

1.3 On a formula of LANDAU and LIFSCHITZ

In 1974 I did not understand Physics so I looked in the book by Lev LANDAU and Evgenij LIFSCHITZ [27] and I found that there was a section (§9 in fact) on the conductivity of mixtures. Suppose you take two conductors and grind them to a fine powder and mix them in some proportion. Then the question is, what is the conductivity of the mixture?

Then LANDAU and LIFSCHITZ went and did some kind of strange computation where they added a sphere of the material imbedded in a uniform medium; everything was assumed to be isotropic. In the end they got a formula. When I had read that, I thought that the physicists did not know what they were talking about, because the conductivity of a mixture, except in one dimension, does not depend only on the proportion; that is the result of mathematical analysis.

If one takes an interpretation of these equations in the terms of electric conductivity, that would mean we are dealing with the equation of the form

$$\begin{cases} \mathsf{div}\,\mathsf{j} = \rho \\ \quad \mathsf{j} = \sigma \mathsf{E}\,. \end{cases}$$

The σ here is the conductivity and $1/\sigma$ is the resistivity. In one dimension we just have the wire and the current goes through. Meaning that we have put the resistances in series. When one has the resistances in series, that means that we add up resistivities, that is $1/\sigma$. When you are in two dimensions, the current can flow in various directions and there are different tubes and in each one you have the resistances in series, but then we put these tubes in parallels. When you have them in parallels, then you do the harmonic averages and when you have resistances in series, you add them — that is what we learn in physics. But if you have a mixture you do not know in advance where the current is going to flow. Therefore you cannot compute what the effective conductivity will be: it depends on the position of the pieces, the proportions are not enough.

It is easy to understand what happens if you understand what happens in a *layered material*. That is the case in which the coefficients depend only on one direction. If you believe in this analogy with Physics you are not surprised to learn that the effective conductivity will be different in the direction perpendicular to the layers in which you are going to do the harmonic average, because the layers are all in the series so you have to add $\frac{1}{a}$, from the perpendicular direction, where you have to compute the arithmetic average.

Figure 2. Layered material with layers perpendicular to x_1 axis.

You have the weak limits (in L^∞ weak $*$, after eventually passing to a subsequence) $a^n \overset{*}{\rightharpoonup} a_+$ and $\frac{1}{a^n} \overset{*}{\rightharpoonup} \frac{1}{a_-}$, but because we have weak limits we lose the relation that one is the inverse of the other. The difficulty then is that if you do not know about the geometry of the pieces you can only expect to have some bounds. The mathematical problem that we looked at was to try to compute all possible effective tensors $\mathbf{A}^{\text{eff}}$ corresponding to different proportions.

That was not easy, but if you imagine that you are working in two dimensions one gets Figure 3.

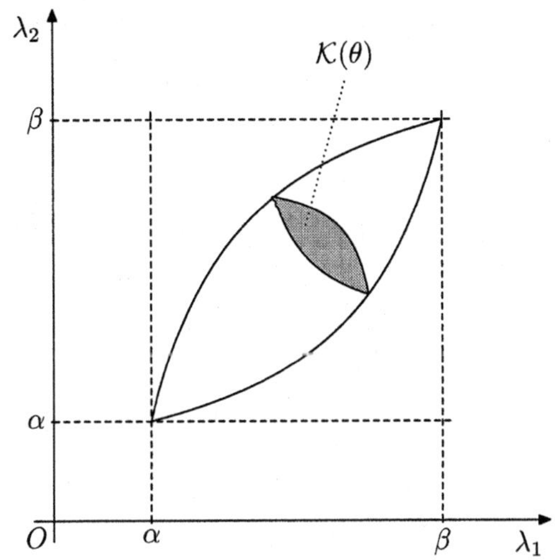

Figure 3. Effective coefficients (represented by their eigenvalues) obtained by mixing two isotropic materials in two-dimensional case.

What you put here are the eigenvalues of the matrix $\mathbf{A}^{\text{eff}}$. You find some kind of strange picture where the material that you can obtain by mixing some property has eigenvalues that depend on the set. Therefore, what you have found is that $\mathbf{A}^{\text{eff}}$ has eigenvalues $\lambda_1 \leq ... \leq \lambda_N$. This means that, if you start with two materials, you are looking at something like

$$\mathbf{A}^n = \alpha \mathbf{I} \chi_n + \beta \mathbf{I}(1 - \chi_n) \, .$$

The functions χ_n can be characteristic functions of arbitrary measurable sets. It is important to notice that we do not impose any technological constraint: *Can we recreate this mixture or not?* Then we can characterise that $(\lambda_1, ..., \lambda_n) \in \mathcal{K}(\theta)$. What is θ? It is the weak limit of χ_n, $\chi_n \overset{*}{\rightharpoonup} \theta$ in L^∞ weak $*$. So that is a mathematical way to say that this point in the average will use the proportion θ of material with conductivity α. So if I am looking just at the isotropic effective coefficients, there is not just one point there, there is an interval.

We did not know the precise interval when I read this argument. The argument of LANDAU and LIFSCHITZ seems to say that there exists just one value there which we have to compute, and that is not true. The point is that when you read something in Physics which is more Theoretical Physics, the authors do not even tell you if the formula is good in practice.

1.4 On HASHIN-SHTRIKMAN bounds

In the sixties (Was it 1962? I am not sure.) Zvi HASHIN and Shmuel SHTRIKMAN [21] have found an argument which was giving some bounds and the bounds were exactly these numbers that I am talking about. With the argument that we had, I proved that in 1980 [45]. From the mathematical point of view this was the first proof. It is difficult to accept HASHIN and SHTRIKMAN's bounds out of their proof. So the physicists' argument which I read in 1980 did not make any sense.

Later I have introduced something called *H-measures*. In some way with this HASHIN and SHTRIKMAN's proof made sense. Physicists often have an intuition about something which at a given moment does not make sense in a mathematical way. In the same way in which the DIRAC function did not make sense for the mathematicians until some time later. But it is very easy to understand what a point mass is, learn Physics and compute a lot of things. Paul Adrien Maurice DIRAC [14] even went further and took derivatives of the DIRAC function. That was rather bold, but it worked. The fact that it works is not the proof that you have done everything right. Until SCHWARTZ had defined distributions nobody understood what it meant to take the derivative of the DIRAC function. Here we have a situation which is similar.

Some arguments that you find in physics books, when you read them, they do not make any sense; and sometimes they will not make any sense because there is something which is wrong in a formula. The formula of LANDAU and LIFSCHITZ had to be wrong, because if you were mixing the two materials with 50% the formula was not symmetric in α and β. So it did not make sense. In the construction they used an explicit solution which was a sphere of one material embedded in another; therefore the two materials did not obey the same law. So you could say pretty reliably that they were assuming that there was a small proportion of material α. But they did not say what will happen if the pieces would have been squares, or obtuse. Therefore, nobody knew how good this formula was. They went a little further, which I had not seen at the beginning. They were saying that if the conductivity did not vary too much, i.e. $a(x) \approx \overline{a}$ (Note that in this lecture only the bar is used to denote the average of a function.), then they proposed the formula

$$a^{\text{eff}} = \overline{a} - \frac{1}{3\overline{a}}\overline{(a-\overline{a})^2}.$$

The number 3 was related to the dimension. Then they explained, that it looked like the formula $(a^{\text{eff}})^{\frac{1}{3}} \approx \overline{a^{\frac{1}{3}}}$. There are articles whose authors say that the effective conductivity in dimension three consists in doing that. If you read the argument, then you understand what it says. It just says that if you look at the TAYLOR expansion, and you look at what is the power of a which implies $(a^{\text{eff}})^\gamma = \overline{a^\gamma}$, then you have to identify some coefficients and you find that $\gamma = \frac{1}{3}$.

If they had been doing that in the dimension $N = 2$, then they would have found that it is $\ln a^{\mathrm{eff}} \approx \overline{\ln a}$. This time there are the formulas which say it is exactly that.

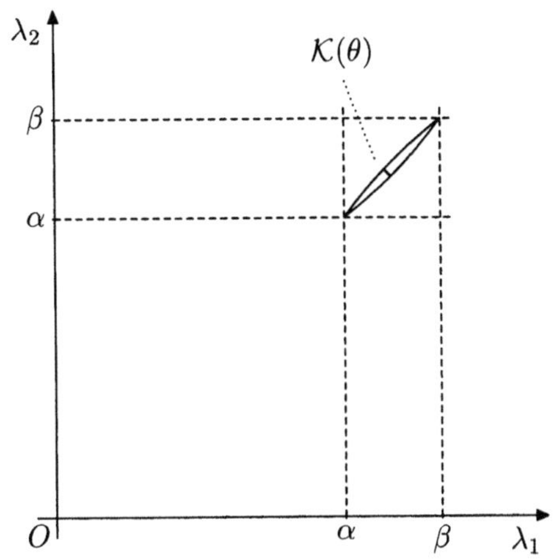

Figure 4. Effective coefficients obtained by mixing two isotropic materials such that $\beta - \alpha$ is small (two-dimensional case).

What is behind this argument? Behind this argument is the fact that the coefficients do not vary too much. Therefore in my picture I have to use that α and β are very close. That means that we have to consider this from a different point of view, which is that in the little picture denoted by $\mathcal{K}(\theta)$ this line becomes very flat. It says, suppose you admit that HASHIN-STRIKMANN bounds have been proved, and suppose you want to express that in terms of α and coefficient $\beta - \alpha$, which is *small*. The two bounds have the same expansion at order two in $\beta - \alpha$ and this is the formula which is given. In that formula you are allowed to mix as many materials you want, while here you are allowed to mix only two materials. In this formula, which I tried to prove in 1985, I found something very surprising, that an argument which really did not make any sense ended in proving a result which was very efficient. It does not explain why it is very efficient. Because in the development there is only an explicit computation for a sphere which has been used. However, you do not know if you can apply it if the pieces are not spherical and you do not know if you are allowed to apply it if the proportions are not small enough.

The fact that it works was proved using this new tool which I introduced, H-measures, and it is because of this computation that I thought of introducing it. I already had an idea about it, but I did not have the right motivation for introducing it. So the reason would be writing

$$\mathbf{A}^n = \mathbf{A}_* + \gamma b^n \mathbf{I}$$

with $b^n \xrightharpoonup{*} 0$, where γ is a *strength parameter*. If you look at the effective matrix, and SPAGNOLO had done that, this looks like $\beta - \alpha$ when you mix two materials. If you start with something that is analytic in γ, the limit is analytic in γ, so one has

$$\mathbf{A}^{\text{eff}} = \mathbf{A}_* + \gamma\mathbf{0} + \gamma^2\mathbf{M}_2 + \gamma^3\mathbf{M}_3 + \ldots.$$

The term multiplying γ^1 is zero, as $b^n \xrightharpoonup{*} 0$. What I had realised, when I tried to compute this using a new object which I called H-measure ('H' was for *Homogenisation*), is that you can express this formula in terms of this measure. Then there is a kind of miracle that appears, namely if the result is isotropic, i.e. $\mathbf{A}^{\text{eff}} = a^{\text{eff}}\mathbf{I}$, then one can compute c in $\mathbf{M}_2 = c\mathbf{I}$ using only proportions. Then we get something which corresponds more or less to this formula.

However, when I explained this at the meeting in Leesburg in 1988 [49], George PAPANICOLAOU was in the first row and he said that it was false. I knew why he said that. He was of that opinion because David BERGMAN had used extensively an idea of using cubic symmetry in order to compute the coefficients of γ^2 in terms of proportions [8]. Earlier this was the only case. And I had not said this, so he thought that it was not true. But the proof showed that it does not matter what the pieces are.

The miracle that appeared in LANDAU and LIFSCHITZ is that it does not matter what the geometry of the pieces is; if the coefficients are very near you can in some way compute the important part of it. Because it does not depend on geometry the computation with a sphere will give the right value. We do not know in advance that the result does not depend on the explicit construction. The approach of the mathematician is: we have to understand a more general question and having understood the more general question and introduced the new objects to do that we realise that there are some situations where the new objects are not necessary to express the limit.

In order to prove a theorem on polynomials you invent holomorphic functions and you do the proof on holomorphic functions and in the end you understand something about polynomials. That is the classical idea of Mathematics: you have to introduce new objects although in the case where the question uses only some level of the definition the proofs use something else and you have to create the objects that permit the answer to this question. It is not clear what the physicists wrote, and I had to guess what they meant, but I did not understand what they did. It could be some tricky idea the physicists know, but I did not know how they would have applied it to elasticity. I do not claim that this is easy; I actually made some mistake in the computation when I did it. But if you try to apply it to elasticity, there is a lot of Linear Algebra involved.

You can in some way predict what are the corrections in the elasticity coefficients (I am talking about *linearised elasticity*, of course), and there is a lot more Linear Algebra involved, because we have an object with four indices instead of two, and there is some difficulty there. However, the object was invented in order to explain a formula that physicists had written and in their example the formula was very good, but what came out we cannot consider a proof.

Before, there were only some other objects I shall describe in a moment, which I am going to call YOUNG measures. When I introduced them in the field of partial differential equations [44] I knew them under a different name which was *parametrised measures.*

1.5 Homogenisation and periodicity assumption

The term *homogenisation* I borrowed from the work of Ivo BABUŠKA [6]. BABUŠKA had done something similar to what SANCHEZ-PALENCIA was doing but he was doing that for a very important engineering question, which is that sometimes you have to compute things for periodic structures.

One of his examples was about heat conduction in a nuclear reactor. It is apparently made of hexagonal cells and uranium bars in the middle, which create a lot of heat and you want to take that heat away because you want to transform it into electricity. You want to compute many things as there are questions of safety which are important. Being a numerical analyst he said: *I will use some finite elements and I am going to have very small triangles and I can compute that, but that would be something huge.* He had the feeling that if he considered that the cells were made of some kind of homogeneous material and the problem here at some level was the same as computing something for a homogeneous material. For an homogeneous material you can take your mesh size to be relatively big. It is a fairly important question from the numerical analysis point of view. I used also the same idea. It is important to understand oscillations because if you understand oscillations then numerically you do not have to write the details of these oscillations. You will know which are the important macroscopic quantities and you will work on them.

This problem was with periodic structure and in some way it is important that in my work there is *no periodicity assumption.* What I have done in this problem has not been easy before even for a periodic structure. It is not difficult to show that for the effective coefficients a^{eff} you have some sort of expansion, and actually it is an expansion in $\beta - \alpha$, and there are explicit formulas for the effective coefficients which use the FOURIER coefficients of b.

The first time I heard of this kind of computation it was a computation done by Graeme MILTON [31]. When I constructed H-measures, Graeme MILTON told me that I had found a scale invariant part of the *two-point correlation* (i.e. that the H-measure sees a scale invariant part of two-point correlation). At the time I did not understand why he was saying that. I knew the formula, but it was not clear from my construction what it had to do with two point correlation. There is a formula for $\mathbf{M}_3$. You can write that in terms of two-point correlations for $\mathbf{M}_2$ and three-point correlations for $\mathbf{M}_3$, but it is a little bit more formal than this. There is no understanding at the moment of an object that will do the analogue in a non periodic situation for $\mathbf{M}_3$.

What one discovers by looking at the formula for H-measures, if earlier one had some experience with the formula in the periodic case, is that you are not surprised by the formula that is valid in the general case. This is more or less what HASHIN and SHTRIKMAN were doing. They were doing, in a general situation, some computations which had a very precise meaning in periodic case, or in constant coefficient case. The way they used it was a little strange. Everybody can see what the construction is in a periodic case, but at the moment I do not know how to transform that in a mathematical object for $\mathbf{M}_3$ which is valid in a non-periodic situation. Therefore, you have a difficult thing behind.

Suppose you have a periodic function a and suppose then that you look at $a(x/\varepsilon)$, where ε is small. What I am saying is that if you have a sequence a^n which is not periodic then I look here at the weak limit, so you can check that in the periodic case the weak limit will find the average. It is not easy to understand what is the general property that will do what we want in a general case. It is not by looking at the formula here that I invented H-measures. I invented them directly and then I checked whether they fit. I cannot do the same thing for the equivalent in *three-point correlation*. Knowing the precise formula in the periodic case is useful. But to understand what you must do in order to create a new object is something I cannot teach, as I do not understand it.

Therefore, *replacing the averages by weak limits* is in the same time easy and difficult. It is easy, because once you establish the convergence, the generalisation is mostly easy. If you see the H-measures, and you know the periodic case, that it is a natural generalisation. What is a natural generalisation of the formula for third correction? Therefore, there is something that you can consider most easy and natural, but it is only natural after you have done it. It is always more difficult to find a new proof of a mathematical theorem. When you have found it, then you can teach the proof to the students. But it takes years to find the proof. Here we are in a situation like that.

There are much more difficult problems ahead, but at least I can show you what were the ideas and natural things to do on some examples. How do we do the next step? It is only the tip of an iceberg and we have only understood the top. There is something below if you want to solve some of the problems that we were interested in if there are microstructures, so that is what I will describe in my lectures.

2 LECTURE of Tuesday, September 5

2.1 Study of oscillations

It is necessary to go back again in time to about 1975. That was the time when I introduced a method where I used the so called *div-curl lemma*, which I had proved with François MURAT [34]. It was a way to simplify our previous approach. That is the method which I called the *method of oscillating test functions* which I showed to Jacques-Louis LIONS and which he called the *energy method* [7].

This method applies to a lot of different questions. One thing I looked at immediately after introducing the method was the problem of conductivity with holes which is the same as the problem of conductivity coming to zero at the limit ($\alpha \longrightarrow 0$ in the notation used earlier); i.e. of using an insulator. Similarly, conductivity tending to infinity describes a perfect conductor.

I have studied the problem with periodic holes and I put either the DIRICHLET or the NEUMANN boundary condition. For the first, the solution tends to zero strongly $u_\varepsilon \longrightarrow 0$, and I did not think of showing that $\frac{u_\varepsilon}{\varepsilon^2} \rightharpoonup v$, which can be easily done by this type of argument. I also looked at the NEUMANN problem. I used the equation, with an interpretation for electrostatics, where it reads

$$\begin{cases} \mathsf{E} = \nabla u \\ \mathsf{D} = \mathsf{A}\mathsf{E}\,. \end{cases}$$

The point is that inside the holes nothing is defined, but there is something that could be naturally defined inside the hole. If you have the NEUMANN condition then you can extend D inside the hole by 0, and that would satisfy the equation $\operatorname{div} D = \rho$. In order to apply the div-curl lemma you have to extend E. Therefore, what I did was a natural thing, that you have a function $u \in \mathrm{H}^1$ here, and then you extend it inside the hole, but you have to be careful, you have to extend it in a way so that you have an extension of Pu and in the periodic case you do it for the unit cell. This means

$$||\nabla(Pu)||_{\text{inside}} \leq C(||u|| + ||\nabla u||)_{\text{outside}}.$$

However, one does not want this $||u||$ on the right hand side, because, when one rescales, different powers of ε will appear. Therefore, one needs to pay attention that when $u = 1$ then $\nabla u = 0$, and one extends by 1 inside. There is a small trick that one needs. When one gets rid of $||u||$, one gets the right type of information by rescaling.

Obviously, this trick can be used when the holes are different (i.e. in a nonperiodic setting), but one needs some kind of a security region around each hole and some kind of uniformity, so that we can bound the norm of the extension operator in an uniform way. This was actually extended later by Grégoire ALLAIRE and François MURAT [2]. In some way my technique needed the boundary of the holes to be LIPSCHITZ because, if you do not have a LIPSCHITZ boundary you may not have an extension operator. You do not really need the extension operator, because in the end it will disappear. It is just a trick. And you can avoid this trick and do something with slightly less regular holes.

It is important to have this sort of security region; if one has a domain and one tries to put the holes that almost touch, it is a way to cut the domain in two. This requires a different kind of scaling.

We are looking in some way at some kind of elliptic problem with degeneracy, as there are inclusions. The next thing I wanted to understand is a more general problem. Suppose you get the first order problem:

$$\mathsf{a} \cdot \nabla u + bu = 0 \ ,$$

which you can understand as a degenerate elliptic problem, where you just make some Laplacian disappear. Actually, you need some kind of general theorem from the sixties for this kind of problem. Unfortunately, for this problem the above method does not work. Here the a and b contain some kind of oscillations for which you can obtain some kind of estimate. That is very important, because in some way turbulence is related to this question. In the flow you have a transport operator

$$\frac{\partial}{\partial t} + \mathsf{u} \cdot \nabla_{\mathbf{x}} \ ,$$

and you transport different things. You can transport the *density* ρ, you can transport the components of the *momentum* ρu_k, and you can transport *energy* $\frac{1}{2}\rho||u||^2 + \rho e$, where you have also internal energy in the expression. What I have understood from discussing this with Joel ROBBIN was that the weak convergence is natural for *differential forms* (v. [59], Chapter II). When you look at conservation of mass, one finds

$$\frac{\partial \rho}{\partial t} + \operatorname{div} \mathsf{q} = 0 \ ,$$

where $\mathsf{q} = \rho \mathsf{u}$. Actually, q is momentum and ρ and q are the coefficients of a differential form, so the weak convergence is natural for them. Therefore, the weak convergence is natural for ρ and $\rho\mathsf{u}$. Defining u as the ratio of two things q/ρ, it is some average velocity for transporting mass.

It is a little strange what is this u around here. With this idea that each quantity has natural weak convergence adapted to it, like we had for the coefficients; it is not clear exactly how you should consider what this u does. If for a given sequence u^n the sequence φ^n solves an equation

$$\left(\frac{\partial}{\partial t} + \mathsf{u}^n \cdot \nabla_{\mathbf{x}}\right) \varphi^n = f \ ,$$

one would like to understand the limit of φ^n knowing what the oscillations of u^n are. Then you cannot expect to get the limiting problem of the same form. It is usually not true that

$$\left(\frac{\partial}{\partial t} + \mathsf{u}^n \cdot \nabla_{\mathbf{x}}\right)\varphi^n \longrightarrow \left(\frac{\partial}{\partial t} + \mathsf{u}^{\mathrm{eff}} \cdot \nabla_{\mathbf{x}}\right)\varphi^\infty .$$

I did not say what is true and in some way it is not known in general. What is actually well known is that turbulence is created by fluctuations in the velocity. The reason is that this type of the first order operator does not homogenise in something of the same form. The question is what should we do?

2.2 Problem modelled on the NAVIER-STOKES system

My first solution was to invent a simpler model than the first order equation for something that I wanted to model on; I started from the NAVIER-STOKES equations [46,47], and I considered

$$\begin{cases} -\nu\Delta\mathsf{u}^n + \mathsf{u}^n \times \mathrm{rot}\,(\mathsf{v}_0 + \mathsf{w}^n) + \nabla p^n = \mathsf{f} \\ \mathrm{div}\,\mathsf{u}^n = 0 . \end{cases}$$

There is a nonlinear term in the NAVIER-STOKES equation, which one can write as

$$\mathrm{rot}\,\mathsf{u} \times \mathsf{u} + \nabla\frac{|\mathsf{u}|^2}{2} .$$

That is the way I arrived at the equation here. I played with this and I decided to simplify it by not using time. So I said what would happen if I were looking at the problem where

$$\begin{cases} -\nu\Delta\mathsf{u} + [*] + \nabla p = \mathsf{f} \\ \mathrm{div}\,\mathsf{u} = 0 . \end{cases}$$

Here I was looking at the term [*] of the form $\mathsf{u} \times \mathsf{b}^n$, with some oscillating function b^n. This is the force, it looks a little like when you try to make the things mix.

I first did the computation with the function $\frac{1}{\varepsilon}b(\frac{\mathbf{x}}{\varepsilon})$, and I did the formal computation with Michel FORTIN, who shared my office that year. Later I proved what we had obtained by my method.

Let me explain the general case. Suppose we are looking at the problem

$$\begin{cases} -\nu\Delta\mathsf{u}^n + \mathsf{u}^n \times \mathrm{rot}\,(\mathsf{v}_0 + \lambda\mathsf{w}^n) + \nabla p^n = \mathsf{f} \\ \mathrm{div}\,\mathsf{u}^n = 0 , \end{cases}$$

with homogeneous DIRICHLET boundary condition, and one is looking at a stationary type of problem. One uses $\mathsf{w}^n \rightharpoonup 0$ and λ is a strength parameter. Now the theorem is obtained after extracting a subsequence; one gets a uniform bound for u^n, because $\mathsf{u}^n \times \mathsf{rot}\,(\mathsf{v}_0 + \lambda \mathsf{w}^n) \perp \mathsf{u}^n$. The limiting problem has the following form

$$-\nu\triangle\mathsf{u}^\infty + \mathsf{u}^\infty \times \mathsf{rot}\,\mathsf{v}_0 + \lambda^2\mathbf{M}\mathsf{u}^\infty + \nabla p^\infty = \mathsf{f}\ .$$

The $\mathbf{M}$ in the correction $\lambda^2\mathbf{M}u^\infty$ can be computed from w^n. What does it say? It says that you have got an extra term in the equation but you have a little more, that is

$$\nu|\nabla\mathsf{u}^n|^2 \rightharpoonup \nu|\nabla\mathsf{u}^\infty|^2 + \lambda^2\mathbf{M}\mathsf{u}^\infty\cdot\mathsf{u}^\infty.$$

We have a situation where the force [*] does no work. You have some very highly oscillating force which does no work directly but it creates oscillations for $\nabla\mathsf{u}$. Therefore it dissipates more than the limit. So

$$\nu||\nabla u^n||^2 \rightharpoonup \nu||\nabla u^\infty||^2 + [\star]\ ,$$

where [⋆] is something nonnegative which is usually not zero. The result which is important qualitatively is that the added quantity is not like a turbulent viscosity which is proportional to $||\nabla\mathsf{u}||^2$; the excess is proportional to u^2. The strange thing is that it is not the beginning of an expansion in λ, there is only the term in λ^2. When I have looked at the way in which I have computed this $\mathbf{M}$ I had the intuition of something else, the *H-measures* that I introduced a few years later. There is a formula for $\mathbf{M}$, which expressed it in terms of the H-measure of the sequence w^n. That is something that one sees on the Fourier transform of w^n, namely how it behaves at infinity.

My feeling is that I have cheated with the physical equation because I have decoupled u. I could not put the u for w because viscosity creates strong convergence for u^n. Thus in order to get some turbulent effect you have to let ν tend to zero. This was not done for the real problem, but my feeling was that if you wanted to understand what is going on in a real turbulent flow you would probably need more variables. At each point you may need u and an $\mathbf{M}$ and maybe u and $\mathbf{M}$ would satisfy something. Therefore there is a possibility that one needs more *macroscopic* variables which do not exist at a *microscopic level.* At the microscopic level you just have the usual things. If you changed a scale in order to take into account the oscillations you would want to say that because of the oscillations you need new variables and then we should have to write a system for describing that.

This was the first hint that there was another object which was really necessary. As I mentioned last time it was only when I tried to understand the small amplitude homogenisation that I created this object. This happened a few years later.

2.3 What are the H-measures?

Let us choose a scalar sequence u_n such that $u_n \rightharpoonup 0$ in $\mathrm{L}^2(\mathbf{R}^N)$ weakly. If you are working on $\Omega \subseteq \mathbf{R}^N$, then you extend the functions by zero outside Ω;

in such a way you get again a sequence converging weakly towards zero. Then we take a test function $\varphi \in \mathrm{C}_c(\mathbf{R}^N)$, and look at φu_n; one has supp $(\varphi u_n) \subseteq K$, where K is a fixed compact, and then you take the FOURIER transform of φu_n

$$\mathcal{F}(\varphi u_n)(\boldsymbol{\xi}) = \int_{\mathbf{R}^N} e^{-2\pi i \mathbf{x}\cdot\boldsymbol{\xi}} (\varphi u_n)(\mathbf{x})\, d\mathbf{x} \,.$$

As we are integrating only over the compact K, we have the bound

$$|\mathcal{F}(\varphi u_n)(\boldsymbol{\xi})| \leq C' \|\varphi u_n\|_{\mathrm{L}^2} \leq C \,,$$

but u_n goes to zero weakly, so actually $\mathcal{F}(\varphi u_n)(\boldsymbol{\xi}) \longrightarrow 0$. So we have pointwise convergence to zero, and it is uniformly bounded in $\boldsymbol{\xi}$. On the other hand, by PLANCHEREL'S theorem

$$\|\mathcal{F}(\varphi u_n)\|_{\mathrm{L}^2} = \|\varphi u_n\|_{\mathrm{L}^2} \,,$$

so it is bounded as well, but we cannot use the LEBESGUE theorem directly on $\mathbf{R}^N$ because the bound we have is not in L^2. On bounded sets the LEBESGUE dominated convergence theorem says that $\mathcal{F}(\varphi u_n) \longrightarrow 0$ strongly. So we say that $\mathcal{F}(\varphi u_n) \longrightarrow 0$ in $\mathrm{L}^2_{\mathrm{loc}}$ strong.

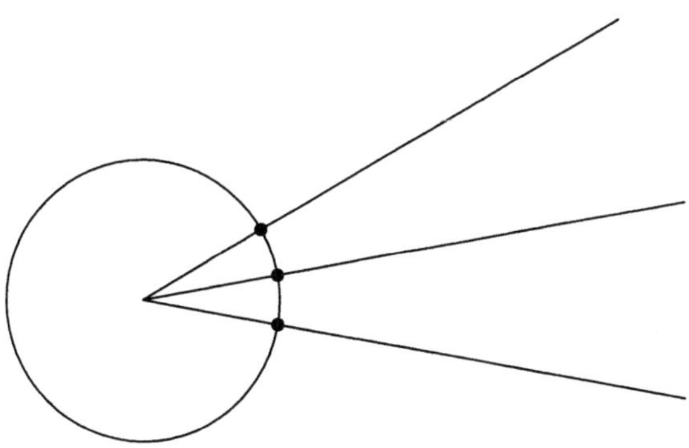

Figure 5. Graphical description of the definition of H-measures: Integrate the square of the Fourier transform along rays and put the results on the unit sphere.

Suppose that φu_n does not converge to zero in L^2 strong; then neither can $\mathcal{F}(\varphi u_n)$, as it has the same norm. So it converges to zero locally, but not in L^2. That means that there is some information that goes to infinity. So I have decided to look at how is this sequence going to infinity in different directions. I took the test function $\psi \in \mathrm{C}(S^{N-1})$ and I looked at the limit of the integral

$$\lim_n \int_{\mathbf{R}^N} \psi\left(\frac{\boldsymbol{\xi}}{|\boldsymbol{\xi}|}\right) |\mathcal{F}(\varphi u_n)|^2 d\boldsymbol{\xi} = \int_{S^{N-1}} \psi(\boldsymbol{\xi})\, d\nu_\varphi(\boldsymbol{\xi}).$$

The idea is that we want to integrate the square of the FOURIER transform on some rays and we want to put this information on the sphere, so we take all the square L^2 norm we have and project it on the sphere.

This limit, of course, is some kind of linear functional in ψ, so it is an integral on the sphere of some *nonnegative* RADON *measure* $d\nu_\varphi(\boldsymbol{\xi})$, but this measure depends on φ. It is just a way to say that if we look at the measures, we have a bounded sequence of measures, therefore a subsequence is going to converge vaguely to something. The critical thing to understand is how does ν_φ depend on φ? The theorem (cf. [49]) is:

Theorem. *There exists some* RADON *measure* $\mu = \mu(\mathbf{x}, \boldsymbol{\xi})$*, which I call the H-measure associated to a subsequence, such that this limit is actually an integral:*

$$\lim_n \int_{\mathbf{R}^N} \psi\left(\frac{\boldsymbol{\xi}}{|\boldsymbol{\xi}|}\right) |\mathcal{F}(\varphi u_n)|^2 d\boldsymbol{\xi} = \int_{\mathbf{R}^N \times S^{N-1}} |\varphi(\mathbf{x})|^2 \psi(\boldsymbol{\xi}) d\mu(\mathbf{x}, \boldsymbol{\xi}).$$

This explains how μ depends on φ, so behind ν_φ there is a measure in $\mathbf{x}$ and $\boldsymbol{\xi}$ which explains the limit. Then this measure has some interesting properties, because it can tell something that you cannot do with the weak convergence alone.

2.4 YOUNG measures and Compensated Compactness

You can look at the limit of u_n^2, or other functions of u_n: $g(u_n)$ goes to the limit l_g. If you do this pointwise, you have to be careful that g does not increase too fast, of course. These are related to so called YOUNG *measures*, which I have first learned of as *parameterised measures* in seminars in control theory, firstly from Alain GHOUILA-HOURI's work and later in seminars by Henri BERLIOCCHI and Jean Michel LASRY (all in French). But in this kind of problems, the variable is time. This is a form of control, and it is called *relaxation*, which is connected to the name of Jack WARGA (for more details cf. [61]). There was some kind of optimisation problem, but there were no partial differential equations.

What I wanted to do (see [44]) was to understand how to use the YOUNG measures in partial differential equations, not for the properties themselves, but as a way to make the difference in what happens when you have the information on the derivative and when you do not have. So at this time my idea of the Continuum Mechanics was usually *nonlinear* Continuum Mechanics of the following form: just some *nonlinear constitutive relations* between different quantities u^n, which could be expressed as a requirement that $\mathsf{u}^n(\mathbf{x}) \in K$, where K is some compact set in the range of u^n, and you have *linear balance equations*

$$\sum_{jk} A_{ijk} \frac{\partial u_j^n}{\partial x_k} = f_i \,,$$

which are linear partial differential equations with constant coefficients. If you do not have that kind of differential information and you only have the information that $\mathsf{u}^n \rightharpoonup 0$ in L^2 weak, then you want to know what kind of oscillating function can satisfy that, and then it would do to use YOUNG measures as they can predict the limit of functions of u^n. You have an arbitrary YOUNG measure which says what you can obtain.

Theorem. *If* U^n *is a bounded sequence in* $L^\infty(\Omega; \mathbf{R}^p)$, *taking (a.e.) its values in* K, *then there exists a subsequence* U^m *and a (weakly measurable) family of probability measures* $\nu_{\mathbf{x}}$ *on* $\mathbf{R}^p$, $\mathbf{x} \in \Omega$, *with support in* $\overline{K}$, *the closure of* K, *such that for every real continuous function* F *on* $\overline{K}$, $F(\mathsf{U}^m)$ *converges in* $L^\infty(\Omega)$ *weak* $\star$ *to the function* l_F *defined by* $l_F(\mathbf{x}) = \langle \nu_{\mathbf{x}}, F\rangle$ *a.e.* $\mathbf{x} \in \Omega$. *The family* $\nu_{\mathbf{x}}$, $\mathbf{x} \in \Omega$, *is called the* YOUNG *measure associated to the subsequence* U^m.

As soon as you have some linear balance relation you are putting a constraint on the type of oscillation. And here, in a way depending on the list that you have, sometimes you would get compactness, i.e. that the sequence u^n converges strongly, and sometimes you would not get anything. That was the *principle of the Compensated Compactness* [44] which I did with François MURAT, which was an extension of the *div-curl lemma*. The YOUNG measures are the best you can say if you do not have information of differential equations. If you have partial differential equations, the YOUNG measure is not the right thing to do. How will you express that there are other things?

2.5 H-measures and Compensated Compactness

The first step was to try out the ideas of Compensated Compactness in the new theory. What the H-measure does is to put it in a different type of framework, because it expresses the same thing with the aid of the measure μ, and sometimes this measure satisfies an equation. But here I am in the scalar case, and you cannot do much Continuum Mechanics with scalar functions only, as the equations are systems.

The next step is to look into vector valued functions. You assume that there is a sequence $\mathsf{u}^n \rightharpoonup 0$ in $L^2_{\rm loc}(\Omega; \mathbf{R}^p)$. You have to localise first. So you can do the same type of things as before, only now there are two test functions $\varphi_1, \varphi_2 \in C_c(\Omega)$, and you look at the limit with the other test function $\psi \in C(S^{N-1})$; actually, you look at the limit after extracting a subsequence, and you get some measure μ_{jk}:

$$\int_{\mathbf{R}^N} \psi\left(\frac{\boldsymbol{\xi}}{|\boldsymbol{\xi}|}\right) \mathcal{F}(\varphi_1 u_j^n)\overline{\mathcal{F}(\varphi_2 u_k^n)}\, d\boldsymbol{\xi} \longrightarrow \int_{\mathbf{R}^N \times S^{N-1}} \varphi_1(\mathbf{x})\overline{\varphi_2(\mathbf{x})}\psi(\boldsymbol{\xi})\, d\mu_{jk}(\mathbf{x}, \boldsymbol{\xi})\,.$$

This can be *complex*, because you may get complex functions when you do the FOURIER transform as before, except that now $\boldsymbol{\mu}$ is a p by p matrix of these measures, in fact it is Hermitian and nonnegative.

What is the relation with the Compensated Compactness theory? Firstly, I have to say what is the Compensated Compactness theory. You have the quadratic theorem. It says that if you have a quadratic function Q on $\mathbf{R}^p$ which has the property that it is positive on the particular set Λ associated to the equation:

$$\Lambda := \Big\{\boldsymbol{\lambda} \in \mathbf{R}^p : (\exists \boldsymbol{\xi} \in S^{N-1})(\forall i) \sum_{jk} A_{ijk}\lambda_j\xi_k = 0\Big\}$$

i.e. $Q(\boldsymbol{\lambda}) \geq 0$, for $\boldsymbol{\lambda} \in \Lambda$. So you just do the algebra to replace ∇ by $\boldsymbol{\xi}$.

If $Q(\boldsymbol{\lambda}) \geq 0$, and if $Q(\mathsf{u}^n) \rightharpoonup Q(\mathsf{u}^\infty) + r$, where the limit (in vague topology) may be a measure; then the remainder r is nonnegative. This could be a measure. Because you have a bound in L^2, it is assumed that $\mathsf{u}^n \rightharpoonup \mathsf{u}^\infty$ in L^2 weak. If you have no information on derivative, Λ would be all of $\mathbf{R}^p$, and Q being nonnegative will be convex. That would be a *convexity inequality*.

If you have here the list of all derivatives, the information on all $\partial_j u_k$ which are bounded, then the characteristic set Λ would be $\{0\}$, because here you have $\lambda_i\xi_k = 0$ for all ξ_k; and that would be a *compactness argument* for the convergence. Therefore, depending on the list, you may have an intermediate situation between convexity, which is what you get if you have no information on the derivative, and compactness if you have all the information on the derivatives. There are also intermediate cases, and the div-curl lemma is just one special case which is related to the *monotonicity argument*.

With this method, we have not paid attention to K, so let us try now to get more information what the constraint was on a function taking values in K, but also satisfying some restrictions on the derivatives. It is not a characterisation but in a way it explains some form of constraint that a YOUNG measure must satisfy. When you have an oscillating solution of some nonlinear partial differential equation, you know that some oscillations are forbidden. The H-measure says that this will come automatically from an information that if you have the equation you can transform some information of the measure.

If you look at the PLANCHEREL formula which was the case when $\psi = 1$, then you notice that the integral equals

$$\int_{\mathbf{R}^N} \mathcal{F}(\varphi_1 u_i^n)\overline{\mathcal{F}(\varphi_2 u_j^n)}\, d\boldsymbol{\xi} = \int_{\mathbf{R}^N} \varphi_1 u_i^n \overline{\varphi_2 u_j^n}\, d\mathbf{x}\,.$$

Therefore you see that it will tell you what is the limit of $u_j^n\overline{u_k^n}$. It is easy to see that when you integrate in $\boldsymbol{\xi}$, so this $\boldsymbol{\mu}$ knows the limit of all these quantities, but it does not know the limit of functions of u^n which are not of this form. Therefore the H-measure is not an improvement of the YOUNG measure. The YOUNG measure still sees something that the H-measure does not know about. But the H-measure sees the above type of information on derivatives very well, that is what I call a *localisation principle*. Suppose you just have one equation, in conservative form:

$$\operatorname{div}\left(\mathbf{B}(\mathbf{x})\mathsf{u}^n\right) \in \text{compact set in } \mathrm{H}^{-1} ,$$

where this $\mathbf{B}$ is a continuous matrix function. Then you have a (vector) equation

$$\boldsymbol{\xi}^\top \mathbf{B}(\mathbf{x})\boldsymbol{\mu} = 0 .$$

If you have a sequence which converges weakly you can introduce some measure and the measure sees something about the high frequencies in the FOURIER transform. Then you can localise in $\mathbf{x}$ and localise in $\boldsymbol{\xi}$. As ψ depends only on $\frac{\boldsymbol{\xi}}{|\boldsymbol{\xi}|}$, we are integrating on each ray, so we are looking at what happens at infinity, i.e. for high frequencies, but we do not seek the information on which frequency is used. It is not like in methods such as Geometrical Optics, where you pay attention to what happens for high frequency. There is no thing like a phase in my theory. The localisation property says, that each time you have a constraint like a partial differential equation, then the H-measure sees it and must satisfy some algebraic equation. This is a first type of constraint on the $\boldsymbol{\mu}$. For example, suppose you have the first order equation

$$\mathsf{b}(\mathbf{x}) \cdot \nabla u^n = f .$$

This is a scalar sequence u^n which is bounded in L^2, and the coefficients are C^1 functions. The localisation principle says that the H-measure is such that

$$\operatorname{supp} \boldsymbol{\mu} \subseteq \left\{(\mathbf{x}, \boldsymbol{\xi}) : P(\mathbf{x}, \boldsymbol{\xi}) = \mathsf{b}(\mathbf{x}) \cdot \boldsymbol{\xi} = 0\right\} ,$$

the zero set of the characteristic function P. Let us see this geometrically. Suppose that b is real; there is a vector field and you know then that there are characteristics.

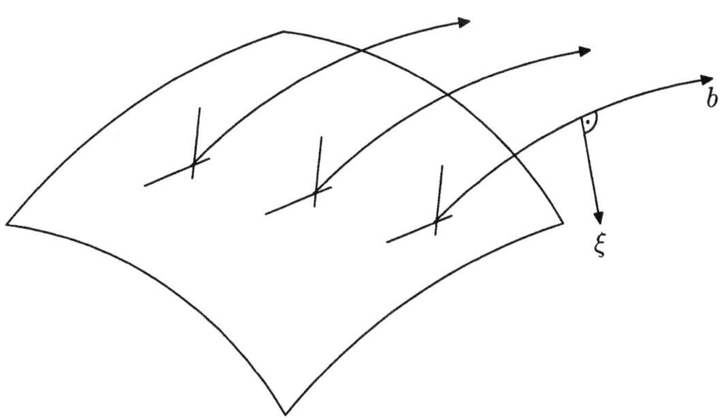

Figure 6. Characteristics determined by a real vector field b, and a noncharacteristic manifold where the data for u can be given.

You can prescribe data for u^n on a manifold which is not characteristic. If you take some oscillating data there then they are pushed along the curves. Therefore the solution here oscillates in some direction, but it cannot oscillate in the direction of b because the derivative in this direction is given by the equation. It can oscillate in a direction $\boldsymbol{\xi}$ which is perpendicular to b. That is exactly what the Localisation Principle says, $P(\mathbf{x}, \boldsymbol{\xi}) = 0$. The only possible directions where you can have oscillations are perpendicular to b.

It does not tell you exactly how are these oscillations transported along the way. You need to do a little more work. At this level one could have used complex coefficients b here. You have to be careful, because the first order system with complex coefficients may be elliptic. Being elliptic means actually that the H-measure is zero. When the H-measure is zero, you have strong convergence in L^2. Therefore in order to go a little further you want to realise that there are more precise equations which will tell you what the oscillations do. The idea is that in order to go a little further you would need some sort of partial differential equations for $\boldsymbol{\mu}$ which express the *propagation of oscillations*.

And, when I am talking about oscillations, I would like to say that it includes *concentration effects* as well.

2.6 Pseudodifferential calculus

In order to find the equation for the H-measure $\boldsymbol{\mu}$ we have to introduce some form of pseudodifferential calculus. The classical theory had some defects. Pseudodifferential operators, and FOURIER integral operators, were developed by mathematicians who had no interests in Continuum Mechanics and Physics (see [22], for example). Therefore they do not care if they have constraints like C^∞ coefficients, which is not satisfactory if one wants to do Continuum Mechanics where one may have two materials, and one wants to consider the *interface* between them. You want to know whether you are able to study such problems. It is important to look at the regularity of the coefficients of the equation.

I preferred just to reconstruct what I needed. Suppose you have $b \in \mathrm{L}^\infty(\mathbf{R}^N)$, and you define the operator M_b, *multiplication* by b, by

$$(M_b u)(\mathbf{x}) := b(\mathbf{x})u(\mathbf{x}) \ .$$

The operator M_b is in $\mathcal{L}(\mathrm{L}^2;\mathrm{L}^2)$, the space of linear continuous operators from L^2 to L^2, and $||M_b|| = ||b||_{\mathrm{L}^\infty}$. If $a \in \mathrm{L}^\infty(\mathbf{R}^N)$ then we use another operator P_a which is defined by

$$\mathcal{F}(P_a u)(\boldsymbol{\xi}) = a(\boldsymbol{\xi})\mathcal{F}u(\boldsymbol{\xi}) \ .$$

In other words $P_a = \mathcal{F}^{-1} M_a \mathcal{F}$, and you have the same expression for the norm, $||P_a|| = ||a||_\infty$. I do not need to have $a(\boldsymbol{\xi}/|\boldsymbol{\xi}|)$ here; in my theory this is what I have. The important thing is the following lemma:

Lemma. *If $b \in C_0(\mathbf{R}^N)$ and $a \in C(S^{N-1})$ then the commutator $[M_b, P_a] := M_b P_a - P_a M_b$ is a compact operator from L^2 to L^2.*

One needs some smoothness in order to do that. We can define this operator with just L^∞ regularity but the fact that the commutator is compact requires some kind of smoothness. We do not really need the continuity of b and we could choose $b \in L^\infty \cap \mathrm{VMO}$, but it is really not so important at this time (for details see [51] and [10]). This explains one step in the proof of existence. When you look at the integral

$$\int_{\mathbf{R}^N} \psi\left(\frac{\boldsymbol{\xi}}{|\boldsymbol{\xi}|}\right) \mathcal{F}(\varphi_1 u_j^n)\overline{\mathcal{F}(\varphi_2 u_j^n)} d\boldsymbol{\xi} ,$$

what you see here, it can be rewritten as

$$\int \mathcal{F}(P_\psi M_{\varphi_1} u_j^n)\overline{\mathcal{F}(M_{\varphi_2} u_j^n)} d\boldsymbol{\xi} .$$

By PLANCHEREL'S theorem you can suppress the FOURIER transform obtaining in that way

$$\int_{\mathbf{R}^N} (P_\psi M_{\varphi_1} u_j^n)\overline{M_{\varphi_2} u_k^n}\, d\mathbf{x} ,$$

and now you try to commute P_ψ and M_{φ_1}. So the commutator will appear and it is not zero, but it is compact so it will transform the weakly convergent sequence u_j^n into a sequence converging strongly to zero and it will not change the limit. Therefore all is the same and in the limit we get

$$\lim_n \int_{\mathbf{R}^N} M_{\varphi_1} P_\psi u_j^n M_{\overline{\varphi_2}} u_k^n d\mathbf{x} .$$

Therefore, when you get that, the limit depends only on ψ and $\varphi_1\overline{\varphi_2}$. Then you have to work a little in order to show that you indeed have a measure. Furthermore, you have localisation which is important, $\varphi_1\overline{\varphi_2}$ appears by the value at the same point. What is the constraint on this $\boldsymbol{\mu}$? The $\boldsymbol{\mu}$ gives the limit of the quantities that are obtained by pseudodifferential operators. We have the class of operators which have the symbol of the form

$$s(\mathbf{x}, \boldsymbol{\xi}) = \sum_m a_m(\boldsymbol{\xi}) b_m(\mathbf{x}) ,$$

where $a_m \in C(S^{N-1})$ and $b_m \in C_0(\mathbf{R}^N)$ such that

$$\sum_m ||a_m||_{L^\infty} ||b_m||_{L^\infty} < +\infty .$$

Then I define an operator which I call the *standard operator* which is just obtained by doing a sum

$$S := \sum_m P_{a_m} M_{b_m} .$$

It has the property that

$$\mathcal{F}(Sv)(\boldsymbol{\xi}) = \int_{\mathbf{R}^N} s\left(\mathbf{x}, \frac{\boldsymbol{\xi}}{|\boldsymbol{\xi}|}\right) v(\mathbf{x}) e^{-2\pi i \mathbf{x}\cdot\boldsymbol{\xi}} d\mathbf{x}\ .$$

If you look at the operator sum

$$L = \sum M_{b_m} P_{a_m}\ ,$$

then

$$Lv(\mathbf{x}) = \int s\left(\mathbf{x}, \frac{\boldsymbol{\xi}}{|\boldsymbol{\xi}|}\right) \mathcal{F}(v)(\boldsymbol{\xi}) e^{2\pi i \mathbf{x}\cdot\boldsymbol{\xi}} d\boldsymbol{\xi}\ .$$

People in the linear theory who play with pseudodifferential operators prefer this kind of formula (so called *standard*, or KOHN-NIRENBERG *quantisation*), while I prefer the other type (sometimes referred to as *adjoint quantisation*); but this makes sense as the difference is an operator which is compact. It does not matter which one you prefer. The first one is slightly more natural if you remember that if you have the function defined on the open set Ω you first have to cut it before using the FOURIER transform which is defined on $\mathbf{R}^N$. Therefore the localisation part in $\mathbf{x}$ should appear first before the operator that uses the FOURIER transform.

An operator T which is linear and continuous from L^2 to L^2 has symbol $s(\mathbf{x}, \boldsymbol{\xi})$ if $T - S$ is compact. If you take the operator and pass to the limit:

$$(T_1 u_j^n)\ \overline{(T_2 u_k^n)} \rightharpoonup \nu_{jk}(\mathbf{x})\ .$$

Because it is bounded in L^2 the result will be a measure. The H-measure says what $\boldsymbol{\nu}$ is, namely

$$\boldsymbol{\nu} = \int_{\boldsymbol{\xi}\in S^{N-1}} s_1(\mathbf{x}, \boldsymbol{\xi}) \overline{s_2(\mathbf{x}, \boldsymbol{\xi})} d\boldsymbol{\mu}(\mathbf{x}, \boldsymbol{\xi}).$$

The H-measure is exactly about knowing the limit of $(T_1 u_j^n)\ \overline{(T_2 u_k^n)}$; it tells this limit when you can use pseudodifferential operators. Pseudodifferential operators are little more general than the usual differential operators in the sense that there is some kind of continuity argument which is built in the theory. It simplifies a lot the Compensated Compactness theorem because it is a way to use the information for the all quadratic functions Q at the same time. From this you can prove a Compensated Compactness theory with variable coefficients. If you want to understand the propagation effects, then you need a *Second Commutation Lemma*.

It is sometimes useful to distinguish oscillations and concentration effects; for a sequence u_n converging weakly to u_∞ in $L^2(\Omega)$, one looks at the (weak $*$) limit ν of $|u_n|^2$, which may be a measure, and if $\nu = f\, d\mathbf{x}$ with $f \in L^1(\Omega)$ one says that the sequence only contains oscillations, while if ν is singular with respect to the LEBESGUE measure $d\mathbf{x}$ the sequence is said to contain concentration effects (a purely concentrating sequence corresponds to a nonzero measure ν which lives on a set of LEBESGUE measure 0). Although the particular questions of Homogenisation that I had considered were mostly concerned with oscillations, the Compensated Compactness theorems apply to general sequences and handle oscillation effects as well as concentration effects; I only paid more attention to concentration effects in the early eighties after the work of Pierre-Louis LIONS [28] and then of Ronald DIPERNA and Andrew MAJDA [13], but I thought that they were not doing it in the right way.

Pierre-Louis LIONS has chosen to call one of his methods the Concentration-Compactness Method in order to lure naive readers into error by the similarity in name with my Compensated Compactness Method, but there is an important distinction, which is that neither YOUNG measures (which are used for studying oscillations) nor the *measures of concentration* that Pierre-Louis LIONS had introduced (which had been used before by Paul LÉVY, as was pointed out by Srinivasa VARADHAN) need a differential structure, while the simpler Compensated Compactness approach or the more precise H-measures have a microlocal aspect without which one cannot really analyse well enough where oscillations and propagation effects are going; one can define H-measures on differentiable manifolds with a volume form, and it seems difficult to avoid completely the differential structure, while YOUNG measures can be defined on locally compact spaces endowed with a measure without atoms, and measures of concentration on locally compact metric spaces.

2.7 Examples of H-measures

One useful example where one can compute explicitly the H-measure of a sequence is that of a sequence which oscillates in a periodically modulated way, i.e.

$$u_n(\mathbf{x}) = v\left(\mathbf{x}, \frac{\mathbf{x}}{\varepsilon_n}\right),$$

where $\varepsilon_n \to 0$ and $v(\mathbf{x}, \mathbf{y})$ is a function which is periodic in $\mathbf{y}$ (and smooth enough) with average in $\mathbf{y}$ equal to 0 (so that u_n converges weakly to 0 in L^2_{loc}). If Y is the unit cell generated by basis vectors $\mathbf{e}_1, \ldots, \mathbf{e}_N$, and $g \in L^1(Y)$ is extended to $\mathbf{R}^N$ as a function G which has period $\mathbf{e}_j$ for $j = 1, \ldots, N$, then G has a FOURIER transform (according to Laurent SCHWARTZ's extension of FOURIER transform to his class of temperate distributions), and $\mathcal{F}G$ is a combination of DIRAC masses at points which have integer coefficients in the dual basis $\mathbf{e}^1, \ldots, \mathbf{e}^N$ (and the weight of each DIRAC mass is a corresponding coefficient in the FOURIER expansion of g); replacing $g(\mathbf{y})$ by $g(\frac{\mathbf{y}}{\varepsilon_n})$ corresponds to rescaling the positions of the DIRAC masses by a factor $\frac{1}{\varepsilon_n}$, and multiplying $g(\frac{\mathbf{y}}{\varepsilon_n})$ by a test function $\varphi \in C^\infty{}_c(\Omega)$ corresponds to regularising the DIRAC masses by convolution with $\mathcal{F}\varphi$ (which belongs to the SCHWARTZ space $\mathcal{S}$). Taking the period cell to be the unit cube (in order to avoid handling the dual lattice), one considers the decomposition of v in FOURIER series in variable $\mathbf{y}$,

$$v(\mathbf{x},\mathbf{y}) = \sum_{\mathbf{m}\in\mathbf{Z}^N} v_{\mathbf{m}}(\mathbf{x})e^{2\pi i\mathbf{m}\cdot\mathbf{y}}$$

(where $v_{\mathbf{0}} = 0$ by hypothesis); then, one can show that the whole sequence u_n defines an H-measure

$$\mu = \sum_{\mathbf{m}\in\mathbf{Z}^N\setminus\{\mathbf{0}\}} |v_{\mathbf{m}}|^2 \otimes \delta_{\frac{\mathbf{m}}{|\mathbf{m}|}}\,,$$

i.e.

$$\langle\mu,\Phi\rangle = \sum_{\mathbf{m}\in\mathbf{Z}^N\setminus\{\mathbf{0}\}} \int_{\mathbf{R}^N} \Phi\Big(\mathbf{x},\frac{\mathbf{m}}{|\mathbf{m}|}\Big)|v_{\mathbf{m}}(\mathbf{x})|^2\,d\mathbf{x}$$

for all continuous Φ with compact support in $\mathbf{R}^N \times S^{N-1}$. Each nonzero FOURIER coefficient indexed by $\mathbf{m} \in \mathbf{Z}^N \setminus \{\mathbf{0}\}$ creates a DIRAC mass in $\boldsymbol{\xi}$ at the point $\frac{\mathbf{m}}{|\mathbf{m}|}$, and one cannot in general recover the separate moduli of the FOURIER coefficients from the H-measure (notice that the set of all $\frac{\mathbf{m}}{|\mathbf{m}|}$ is dense in S^{N-1}).

One sees in this example why some computations made by physicists in a periodic setting may work well in general situations, as formulas computed using FOURIER coefficients may sometimes be interpreted in terms of H-measures, and the general result may be guessed, by having $\boldsymbol{\xi}$ to be an arbitrary point on S^{N-1} instead of a point in the direction of a nonzero point with integer coordinates.

Another useful example is to consider a sequence which concentrates at a point $\mathbf{z}$, i.e.

$$u_n(\mathbf{x}) = \varepsilon_n^{-N/2} f\Big(\frac{\mathbf{x}-\mathbf{z}}{\varepsilon_n}\Big)\,,$$

with $f \in \mathrm{L}^2(\mathbf{R}^N)$, and the scaling is such that the norm in $\mathrm{L}^2(\mathbf{R}^N)$ is independent of n; the sequence u_n converges weakly to 0 in $\mathrm{L}^2(\mathbf{R}^N)$ and $|u_n|^2$ converges vaguely (weakly $*$ in the space of RADON measures) to $C\,\delta_{\mathbf{z}}$, with $C = \int_{\mathbf{R}^N} |f(\mathbf{x})|^2\,d\mathbf{x}$. In that case the whole sequence u_n defines an H-measure $\mu = \delta_{\mathbf{z}} \otimes \sigma$, given by

$$\langle\mu,\Phi\rangle = \int_{\mathbf{R}^N} \Phi\Big(\mathbf{z},\frac{\boldsymbol{\xi}}{|\boldsymbol{\xi}|}\Big)|\mathcal{F}f(\boldsymbol{\xi})|^2\,d\boldsymbol{\xi}\,,$$

for all continuous Φ with compact support in $\mathbf{R}^N \times S^{N-1}$ (so that $\sigma \in \mathrm{L}^1(S^{N-1})$). Of course, one has also $C = \int_{\mathbf{R}^N} |\mathcal{F}f(\boldsymbol{\xi})|^2\,d\boldsymbol{\xi}$, and one sees that it is not the values taken by f but those taken by its FOURIER transform which are important for the H-measure; in particular the approach of using measures of concentration, which is valid for more general sequences than the preceding example of concentration at a point, will only see the value of C, and does not contain enough information for describing the propagation effects, which use the H-measure and therefore need the radial averages of $|\mathcal{F}f|^2$.

2.8 Localisation Principle

The Localisation Principle asserts that if a sequence u^n converging weakly to 0 in $L^2_{\text{loc}}(\Omega; R^p)$ is such that

$$\mathsf{div}\,(\mathbf{A}\mathsf{u}^n) \longrightarrow 0 \qquad \text{in } H^{-1}_{\text{loc}}(\Omega) \text{ strong,}$$

where the coefficients $\mathbf{A}$ are continuous, and if a subsequence u^m defines an H-measure $\boldsymbol{\mu}$, then $\boldsymbol{\mu}$ satisfies

$$\boldsymbol{\xi}^\top \mathbf{A}\boldsymbol{\mu} = 0\,.$$

This is a mere reformulation of what the quadratic Compensated Compactness theorem is about, but expressed in a much more precise way, which permits to deduce easily some applications which would have been quite difficult to obtain in the Compensated Compactness framework (where the coefficients were constant). Once the First Commutation Lemma has been proved and the class of symbols introduced, the Localisation Principle follows easily from some automatic calculus rules.

If for $j = 1, \ldots, N$, an operator L_j has symbol s_j, then it is easy to see that $v^m = \sum_{j=1}^p L_j u_j^m$ defines an H-measure $\mu_0 = \sum_{j,k=1}^p s_j \overline{s_k} \mu^{jk}$; one observes also that a sequence converges in $L^2_{\text{loc}}(\Omega)$ strong if and only if its H-measure is 0. Then one notices that $\mathsf{div}\,(\mathbf{A}\mathsf{u}^m)$ converges to 0 in $H^{-1}_{\text{loc}}(\Omega)$ strong if and only if $v^m = \sum_{j=1}^N \sum_{k=1}^p R_j A_{jk} u_k^m$ converges to 0 in $L^2_{\text{loc}}(\Omega)$ strong, where for $j = 1, \ldots, N$, the operator R_j is the RIESZ operator, which has symbol $\frac{i\xi_j}{|\boldsymbol{\xi}|}$ (so that $(-\Delta)^{1/2} R_j = \frac{\partial}{\partial x_j}$), and the preceding remark applies with $s_k(\mathbf{x}, \boldsymbol{\xi}) = \sum_{j=1}^N \frac{\xi_j}{|\boldsymbol{\xi}|} A_{jk}(\mathbf{x})$, and the condition $\sum_{k,l=1}^p s_k \overline{s_l} \mu^{kl} = 0$ is equivalent to $\sum_{k=1}^p s_k \mu^{kl} = 0$ for every l, by using the nonnegative Hermitian character of $\boldsymbol{\mu}$.

It is useful to use RADON-NIKODYM theorem and write that $\boldsymbol{\mu} = \mathbf{D}\mu_0$, where μ_0 is the nonnegative measure $\sum_{j=1}^p \mu^{jj}$ for example and $\mathbf{D}$ is μ_0-integrable (and μ_0-almost everywhere the matrix $\mathbf{D}(\mathbf{x})$ is Hermitian nonnegative, and has trace 1); the Localisation Principle states then that μ_0-almost everywhere one has $\boldsymbol{\xi}^\top \mathbf{A}\mathbf{D} = 0$.

I had chosen the term *Localisation Principle* because of the example of first order differential equations $\mathsf{b} \cdot \nabla u^n = f^n$, and in the case where the coefficients b are of class C^1 (and f^n converges strongly to 0 in H^{-1}_{loc} strong), one has $P(\mathbf{x}, \boldsymbol{\xi})\mu = 0$, where $P(\mathbf{x}, \boldsymbol{\xi}) = \mathsf{b}(\mathbf{x}) \cdot \boldsymbol{\xi}$, and therefore the support of μ is localised in the zero set of P (which may be empty if the coefficients b are complex).

Another typical example is the case of gradients, i.e. $\mathsf{rot}\, \mathsf{u}^n = 0$, and in that case one finds that $\boldsymbol{\mu} = \boldsymbol{\xi} \otimes \boldsymbol{\xi} \mu_0$ for a scalar nonnegative measure μ_0. Indeed, the condition $\frac{\partial u_j^n}{\partial x_k} - \frac{\partial u_k^n}{\partial x_j} = 0$ for $j, k = 1, \ldots, p$ implies that μ_0-almost everywhere one has $\xi_j D^{kl} = \xi_k D^{jl}$ for $j, k, l = 1, \ldots, p$, and one deduces that $D^{kl} = \xi_k e_l$ for $k, l = 1, \ldots, p$, with $e_l = \sum_{j=1}^p \xi_j D^{jl}$; then $\mathbf{D}(\mathbf{x})$ being Hermitian one has $\xi_k e_l = \xi_l \overline{e_k}$ for $k, l = 1, \ldots, p$, implying $\mathsf{e} = c\boldsymbol{\xi}$, with $c = \sum_{k=1}^p \xi_k \overline{e_k}$; finally the trace of $\mathbf{D}(\mathbf{x})$ being 1, one has $c = 1$. Notice that if $\mathsf{u}^n = \nabla \varphi_n$, then μ_0 is the H-measure associated to the sequence $(-\Delta)^{1/2} \varphi_n$.

3 LECTURE of Wednesday, September 6

3.1 Problem modelled on the NAVIER-STOKES system (revisited)

What are the H-measures?

We can get some intuition by studying a model constructed on NAVIER-STOKES system (for more details see [46,47]).

Physically, here I have a problem of trying to force the fluid to turn around, described by the equations

$$\begin{cases} -\nu\triangle \mathsf{u}^n + \mathsf{u}^n \times \operatorname{rot}(\mathsf{v}_0 + \mathsf{w}^n) + \nabla p^n = \mathsf{f} \\ \operatorname{div} \mathsf{u}^n = 0\,, \end{cases}$$

which should be supplemented by boundary conditions.

In homogenisation of an elliptic problem like this, it is important to realise that the questions are local; you actually do not need the boundary conditions, except to know that you have a bounded sequence in some space. Of course, then you have to know what are these boundary conditions satisfied for the variational formulation. However, in order to obtain the equation inside, usually you do not care whether they are DIRICHLET, NEUMANN, or anything, as long as you have the bounds.

Therefore, the boundary conditions are not important as long as you have u^n bounded in the SOBOLEV space $(H^1(\Omega))^3$, and as I have already said the theorem of homogenisation states that the weak limit of a subsequence satisfies an equation with an extra term in it:

$$\begin{cases} -\nu\triangle \mathsf{u}^\infty + \mathsf{u}^\infty \times \operatorname{rot} \mathsf{v}_0 + \mathbf{M}\mathsf{u}^\infty + \nabla p^\infty = \mathsf{f} \\ \operatorname{div} \mathsf{u}^\infty = 0\,, \end{cases}$$

and the method of oscillating test functions can be used to determine a formula for $\mathbf{M}$; so, in order to compute $\mathbf{M}$ you have to solve an extra problem for any $\mathsf{k} \in \mathbf{R}^3$

$$\begin{cases} -\nu\triangle \mathsf{z}^n + \mathsf{k} \times \operatorname{rot} \mathsf{w}^n = -\nabla r^n \\ \operatorname{div} \mathsf{z}^n = 0\,, \end{cases}$$

using boundary conditions such that $\mathsf{z}^n \rightharpoonup 0$ in H^1 (and it can be done in $W^{1,3}$), and you look at the limit $\lim \nu|\nabla \mathsf{z}^n|^2 = \mathbf{M}\mathsf{k}\cdot\mathsf{k}$.

All this has been obtained by using the standard tools: construct z^n and multiply the equation by it as a test function, then integrate by parts and pass to the limit obtaining the right terms. But now you see that for k given, you go from w^n to $\nabla \mathsf{z}^n$, applying the FOURIER transform on $\mathbf{R}^3$ we solve this problem on $\mathbf{R}^3$; actually, in order to do that we have first to extend the functions by zero to all of $\mathbf{R}^3$, but as we are only interested in what is happening inside Ω we do not care for the lack of regularity at the boundary of Ω. But you have to solve this on $\mathbf{R}^3$ by FOURIER transform, and you have to go through it explicitly.

What you find is that you go from one to the other by a pseudodifferential operator, and this one has to be standard (there are rational fractions involved), and then you can interpret this using H-measures. Therefore, when you look at this, you will see exactly what we were doing yesterday, that there is some integration along rays through the origin, for the FOURIER transform of $|\mathsf{w}^n|^2$, and this is just an interpretation.

You can now give an exact formula for $\mathbf{M}$. If w^n corresponds to H-measure $\boldsymbol{\mu}$, which is a 3×3 matrix of measures in $\mathbf{x}$ and $\boldsymbol{\xi}$, Hermitian nonnegative (so the nondiagonal terms could be complex), and the formula for M_{ij}, which I write as an integral is

$$M_{ij} = \int_{S^2} \xi_i \xi_j \Big(\sum_{k=1}^{3} d\mu^{kk} - \sum_{k,l=1}^{3} \xi_k \xi_l \, d\mu^{kl} \Big) d\boldsymbol{\xi} \, .$$

The part of the integrand in the parenthesis is nonnegative, as we can multiply it by $1 = |\boldsymbol{\xi}|^2$, and $\boldsymbol{\mu}$ is thus nonnegative.

If w^n satisfies $\mathsf{div}\, \mathsf{w}^n = 0$, then we get $\boldsymbol{\mu\xi} \cdot \boldsymbol{\xi} = 0$, and you always get this in problems like the NAVIER-STOKES, as the nonlinear term is

$$(\nabla \mathsf{u})\mathsf{u} = \mathsf{u} \times \mathsf{rot}\,(-\mathsf{u}) + \nabla \frac{|\mathsf{u}|^2}{2} \, .$$

The converse is not true in general, and from $\boldsymbol{\mu\xi} \cdot \boldsymbol{\xi} = 0$ one can only deduce that $\mathsf{div}\, \mathsf{w}^n \longrightarrow 0$ in $\mathrm{H}^{-1}_{\mathrm{loc}}(\Omega)$ strong, because the H-measure cannot see the difference between something that satisfies exactly the equation, and another sequence that satisfies it only modulo some regularising terms. So, in that case the term with moments of order four disappears, and you get only the moments of order two of the measure:

$$M_{ij} = \int_{S^2} \xi_i \xi_j \sum_{k=1}^{3} d\mu^{kk} \, .$$

3.2 Small amplitude Homogenisation

The reason why I have constructed this theory of H-measures, or rather the intuition behind it, was to create some small amplitude homogenisation theory, in order to explain why some formulae in LANDAU and LIFSCHITZ derived by physicists were in fact very efficient, although they were actually derived in a very strange way.

In Homogenisation, you have a problem with a sequence of equations:

$$-\mathsf{div}\,(\mathbf{A}^n \nabla u^n) = f \, ,$$

with corresponding differential operators being uniformly elliptic, i.e. for some $0 < \alpha < \beta < \infty$

$$\alpha|\boldsymbol{\lambda}|^2 \le \mathbf{A}^n(\mathbf{x})\boldsymbol{\lambda}\cdot\boldsymbol{\lambda} \le \beta|\boldsymbol{\lambda}|^2 ,$$

and we try to understand what happens with $\mathbf{A}^n$ when u_n tends to a limit u^∞. Actually, you want to understand the limit

$$\mathbf{A}^n\nabla u^n \rightharpoonup \mathbf{A}^{\text{eff}}\nabla u^\infty .$$

The effective $\mathbf{A}^{\text{eff}}$ is not, in general, a weak limit of $\mathbf{A}^n$. The symmetry of $\mathbf{A}^n$ is not necessary in the theory, but practically when you encounter such problem in Continuum Mechanics, the $\mathbf{A}^n$ are symmetric. For nonsymmetric $\mathbf{A}^n$ we have to write slightly different bounds.

In small amplitude Homogenisation you take $\mathbf{A}^n$ of the form

$$\mathbf{A}^n = \mathbf{A}_* + \gamma\mathbf{B}^n ,$$

where $\mathbf{A}_*$ is fixed, $\mathbf{B}^n \rightharpoonup \mathbf{0}$ (these are matrices), and γ is a small strength parameter.

Sergio SPAGNOLO has worked out this kind of G-convergence when you have a parameter, and it is not difficult to see that you then have $\mathbf{A}^{\text{eff}}$ analytic in γ, and you can write

$$\mathbf{A}^{\text{eff}} = \mathbf{A}_* + \gamma\mathbf{M}_1 + \gamma^2\mathbf{M}_2 + \cdots .$$

The term $\mathbf{M}_1$ with γ^1 is going to be $\mathbf{0}$ as we have zero weak limit, and the term with γ^2 will give us the first correction. Assuming γ small we assure that uniform ellipticity remains valid.

The H-measure will tell us what $\mathbf{M}_2$ is (we are going to suppress writing the subscript 2 in the sequel). How do we compute this $\mathbf{M}$?

What you do is to consider a sequence of problems

$$-\mathsf{div}\Big((\mathbf{A}_* + \gamma\mathbf{B}^n)\nabla u^n(\mathbf{x};\gamma)\Big) = f ,$$

where $\nabla u^n \rightharpoonup \boldsymbol{\lambda} \in \mathbf{R}^N$. The right hand side f has to be nice and analytic in γ; for simplicity take it to be zero. When we supply the boundary condition, the solution u^n will be analytic in γ as well:

$$\nabla u^n = \boldsymbol{\lambda} + \gamma\nabla w_1^n + \gamma^2\nabla w_2^n + \cdots ,$$

where each w_k^n converges weakly to zero.

Inserting this form of u^n back into the equation (here we need that ∇u^n is analytic in γ), and equating the terms with same powers of γ we obtain, first that $-\mathsf{div}\,(\mathbf{A}_*\boldsymbol{\lambda}) = 0$, and then with γ^1:

$$-\mathsf{div}\,(\mathbf{A}_*\nabla w_1^n + \mathbf{B}^n\boldsymbol{\lambda}) = 0 ,$$

while with γ^2 we get:

$$-\mathsf{div}\,(\mathbf{A}_* \nabla w_2^n + \mathbf{B}^n \nabla w_1^n) = 0\ .$$

Higher order terms could be obtained in the same way, but we do not need them.

Now we have to identify the limits of these terms. The limit of the first order one is zero (as $\mathbf{B}^n \rightharpoonup \mathbf{0}$, so by the continuity of the derivative for the convergence of distributions implied by the weak convergence we obtain the claim). For the second, the first part disappears (tends to zero as we commute div and $\mathbf{A}_*$), but the crucial term is the product of two weakly convergent sequences $\mathbf{B}^n \nabla w_1^n$.

As you transform $\mathbf{B}^n$ to ∇w_1^n by some pseudodifferential operator, the H-measure knows how to describe the weak limit of a product of such terms. Thus we have to reinterpret this and see what the formula would be.

However, the formula is somehow technical, as you should look at the H-measure of $\mathbf{B}^n$, which is a matrix of $\frac{N(N+1)}{2}$ times $\frac{N(N+1)}{2}$ elements (as $\mathbf{B}^n$ is a N by N symmetric matrix). And the H-measure is a matrix of measures. If you start with two indices for $\mathbf{B}^n$, you end up with four for $\boldsymbol{\mu}$. Therefore we look only at a simplified problem, where we consider scalar matrices

$$\mathbf{A}^n = \mathbf{A}_* + \gamma b_n \mathbf{I}\ ,$$

with $b_n \rightharpoonup 0$ defining the H-measure μ, which is a nonnegative (scalar) measure on $\Omega \times S^{N-1}$. And in that case the formula for $\mathbf{M}$ reads

$$\mathbf{M} = -\int_{S^2} \frac{\boldsymbol{\xi} \otimes \boldsymbol{\xi}}{(\mathbf{A}_* \boldsymbol{\xi} \cdot \boldsymbol{\xi})} d\mu(\mathbf{x}, \boldsymbol{\xi})\ ,$$

where the term is negative, as the effective conductivity is less than the average conductivity. As we integrate in $\boldsymbol{\xi}$, $\mathbf{M}$ depends only on $\mathbf{x}$.

If $\mathbf{A}_* = a_* \mathbf{I}$, then

$$\mathbf{M} = -\frac{1}{a_*} \int_{S^2} \boldsymbol{\xi} \otimes \boldsymbol{\xi} d\mu(\cdot, \boldsymbol{\xi})\ .$$

In particular, taking the trace of $\mathbf{M}$, instead of the moment of order two we get the moment of order zero

$$\mathsf{tr}\mathbf{M} = -\frac{1}{a_*} \int_{S^2} d\mu(\cdot, \boldsymbol{\xi}) = -\frac{\lim (b^n)^2}{a_*}\ ,$$

which is just the weak limit. Thus we can explain the trace of the correction using only the weak limit of $(b^n)^2$.

If $\mathbf{M} = m\mathbf{I}$ (which does not follow from earlier hypotheses), then

$$m = -\frac{1}{N a_*} \lim (b^n)^2\ .$$

And this explains a formula from LANDAU and LIPSCHITZ (see 1.3), when $N = 3$.

Therefore, there was some kind of miracle in that formula in LANDAU and LIPSCHITZ. In principle, you should obtain something depending on H-measure μ acting on some function in $\boldsymbol{\xi}$, while the correction actually depends on the moment of order two. On the other hand, if you could use the trace, it depends only on moments of order zero. But it was not clear in advance that this miracle will occur.

You should extend this to other problems: you can always do some small amplitude Homogenisation, because it is the question of analyticity and writing different questions and observing that some operators are actually pseudodifferential operators, so you can compute different quantities by using H-measures.

Unfortunately, we are not able to compute the next term, which will require the limit of term $\mathbf{B}^n \nabla w_2^n$, and w_2^n is obtained by a pseudodifferential operator applied to $\mathbf{B}^n \nabla w_1^n$, which now involves a trilinear quantity, and the theory does not tell you anything about that. The H-measure theory tells you only of the products of two pseudodifferential operators of order zero acting on the same thing.

Using this same idea, you can apply this to a wide variety of problems, as it is almost automatic. Then there is a lot of linear algebra needed in order to compute what are the moments of H-measures in the examples.

Let us consider the linearised elasticity problem; I take the isotropic coefficients in order to simplify the computations. The problem reads

$$\begin{aligned} \boldsymbol{\sigma}^n &= 2\mu^n \boldsymbol{\varepsilon}^n + \lambda^n \mathsf{tr} \boldsymbol{\varepsilon}^n \mathbf{I} \\ \mu^n &= \mu^* + \gamma \tilde{\mu}^n \\ \lambda^n &= \lambda^* + \gamma \tilde{\lambda}^n , \end{aligned}$$

where μ^n is the shear modulus, which together with the LAMÉ modulus λ^n oscillate, i.e. $\tilde{\mu}^n \rightharpoonup 0$ and $\tilde{\lambda}^n \rightharpoonup 0$. Of course, $\boldsymbol{\varepsilon}^n = (\nabla u^n + (\nabla u^n)^\top)/2$ is the symmetrised gradient of the displacement field u^n.

There is some 2 by 2 matrix of H-measures in $\mathbf{x}$ and $\boldsymbol{\xi}$ which appears, generated by the sequence (μ^n, λ^n)

$$\begin{bmatrix} \mu^{11} & \mu^{12} \\ \mu^{21} & \mu^{22} \end{bmatrix} .$$

While μ^{11} corresponds to the analysis of oscillations of μ^n, μ^{22} in a similar way to the analysis of oscillations of λ^n, the nondiagonal elements μ^{12} and μ^{21} correspond to the interaction of the oscillations in μ^n and λ^n.

The linear algebra is a lot more technical, but at the end one obtains that the effective coefficients are

$$C_{ijkl} = F_{ijkl} + \gamma^2 D_{ijkl} + \cdots .$$

The F_{ijkl} depend only on μ^* and λ^*, while the corrections D_{ijkl} use the moment of order 4 of μ^{11}, the moment of order 2 of μ^{12}, and the moment of order 0 of μ^{22}.

The dependence on the variation in shear modulus is much more sensitive than the variation in the Lamé parameters. Thus the shape of the pieces that you use for your mixture, whether they are squares, triangles, or something else, cannot be described by the Young measure, which sees only the proportion.

H-measure sees something about the geometry of microstructure, and it is only the moments of order zero which are related to the proportions. In the formula you even get the moment of order four. However, there is some kind of miracle on some quantity, that if the result is isotropic, some terms of order four disappear, and at the end you can express everything using the moments of order zero.

Which miracle, and on which quantity, it is not obvious until you do the computations. Thus, the formula in Landau and Lifschitz is now proved by saying that in the framework of a small amplitude Homogenisation, where you mix the materials with very similar conductivities, there is a theorem stating that we can express the effective conductivity using the H-measure at order two in the perturbation with a small parameter γ.

There are no a priori reasons for the effective conductivity to be isotropic, but if it is, then there is a formula expressing it in terms of moments of order zero, i.e. in the terms of proportions. However, there is no obvious reason to know in advance, for a given elliptic system, which are the quantities that will appear in the expression for the effective properties.

The H-measures transform the problem in one of Linear Algebra, which can be done automatically, although a complicated one. And you do not have to construct any special solutions, like the one describing the electric field around a conducting sphere in an infinite isotropic medium, or a similar thing in elasticity. You do not have to compute anything like that. Just do the linear algebra.

3.3 Propagation of oscillations and concentration effects

This is the third application of H-measures I wanted to check.

There is a theory which was started by Lars Hörmander (see [22]) in linear partial differential equations on propagation of singularities. It was clear from the beginning that this theory was not adapted to right questions stemming from Physics, but it was not immediately known what was exactly wrong in this regard in the theory. What is wrong is that this theory does not deal with the questions of propagation of singularities, but with the propagation of *microlocal regularity*. I shall describe in a minute what that means.

My motivation for studies in this direction was different, coming from Physics.

There are mathematical objects which satisfy some propagation rules. Can I use my measures to look at a real physical problem — the propagation of oscillations and concentration effects in the physical world, what HÖRMANDER did not have in mind?

Lars HÖRMANDER defined something called the *wave front set* $WF(u)$ of a distribution u; for simplicity let us consider scalar distributions.

A distribution could be equal to a very smooth function in a part of its domain. Laurent SCHWARTZ defined the notion of *singular support* of a distribution u similar to the definition of the support of a distribution. Recall that we define that a distribution u is zero on an open set ω if for any test function φ with $\operatorname{supp}\varphi \subseteq \omega$ it is necessary that $\langle u, \varphi \rangle = 0$. Then we define the support of u as the complement of the largest open set on which u is zero. Similarly, SCHWARTZ defined the singular support as the complement of the largest open set containing those points around which u is equal to a C^∞ function (the reader should convince her/himself that the definition makes sense).

HÖRMANDER was looking at something else; and there is indeed a little step to be made here. He defined the essential singular support, or the wave front set of a distribution u, which is not a set of points $\mathbf{x}$, but of the pairs of $(\mathbf{x}, \boldsymbol{\xi})$ ($\boldsymbol{\xi}$ is a direction) as follows: $(\mathbf{x}, \boldsymbol{\xi})$ is in the complement of the wave front set of u if there is a test function $\varphi \in C_c^\infty$ with $\varphi(\mathbf{x}) \neq 0$ and a conical neighbourhood around the direction $\boldsymbol{\xi}$ such that $\mathcal{F}(\varphi u)$ decays fast in that conical neighbourhood.

He was not interested that it decays fast in all directions (in which case φu would be smooth), but only in the conical directions around $(\mathbf{x}, \boldsymbol{\xi})$. Thus one says that u is *microlocally regular* at $(\mathbf{x}, \boldsymbol{\xi})$.

Then you have to work a little in order to show that this set is open, so its complement is closed. Now, we do not know what happens in the complement, because we do not go there. What the theory says is of the form:

If a distribution u satisfies some linear hyperbolic partial differential equation, then one can deduce some information on its wave front set.

As an example we take a first order equation

$$\mathsf{b} \cdot \nabla u + cu = f \ ,$$

(for hyperbolicity the coefficients b_j are assumed to be real), and there are the characteristics, and we define

$$P(\mathbf{x}, \boldsymbol{\xi}) := \mathsf{b}(\mathbf{x}) \cdot \boldsymbol{\xi} \ .$$

Of course, HÖRMANDER took both b and c to be C^∞. He proved two things; first, that

$$WF(u) \subseteq \{(\mathbf{x}, \boldsymbol{\xi}) : P(\mathbf{x}, \boldsymbol{\xi}) = 0\} \ ,$$

by using the *stationary phase method;* and then, asuming the coefficients b_j to be real, he proved that if $(\mathbf{x}, \boldsymbol{\xi}) \notin WF(u)$ then the bicharacteristic rays (or bicharacteristic) through $(\mathbf{x}, \boldsymbol{\xi})$ do not intersect $WF(u)$.

What is a bicharacteristic ray? *Bicharacteristics* are the integral curves of the system:

$$\begin{cases} \dfrac{d\mathbf{x}}{d\tau} = \nabla_{\boldsymbol{\xi}} P \\ \dfrac{d\boldsymbol{\xi}}{d\tau} = -\nabla_{\mathbf{x}} P \,. \end{cases}$$

Notice that $P(\mathbf{x}, \boldsymbol{\xi})$ is conserved on such a bicharacteristic curve. This is a way to fibrate the set in $\mathbf{x}, \boldsymbol{\xi}$ space. The first equation corresponds to the one determining the characteristic curves. So, when you solve the first order characteristic equation, you know that there are the characteristic curves. What you actually do is to take the initial data on a surface which is transversal to the characteristic curves, and you push the value to the required point.

But if you want to study how a nearby characteristic curve winds around this one, then you might discover that it is useful to introduce these bicharacteristic curves. If you start with a little line being perpendicular to $\boldsymbol{\xi}$, the little line turns around, and it will be perpendicular to something that satisfies the equations of bicharacteristics. What the result says is that if you have a distribution, the coefficients are C^∞, and b being real, then if it is microlocally regular at $(\mathbf{x}, \boldsymbol{\xi})$, it is microlocally regular along all bicharacteristics through $(\mathbf{x}, \boldsymbol{\xi})$ as well.

As a consequence, the wave front set is a union of bicharacteristics. Therefor, it is the set where we do not know what happens. This could be done for other equations, viz. the wave equation, and you get the light rays which show up.

What I say is that this is not the right question from the physical viewpoint, as the regularity is not important in that context. The important question for the wave equation is where does the energy go? And not where does the regularity go. What I checked is that there are similar properties using H-measures. What is the difference? The difference is that we are no longer talking about one distribution, with arbitrary regularity. We are talking about L^2 sequences. The difference is that u_n is a sequence of solutions in L^2 of some partial differential equation; it is not a distribution with arbitrary regularity, it is a sequence in L^2 which converges weakly to zero. Therefore it defines (possibly after passing to a subsequence) an H-measure μ. Furthermore, we assume that

$$\mathsf{b} \cdot \nabla u_n + c u_n = f_n \,,$$

where $f_n \longrightarrow 0$ in $H^{-1}_{\text{loc}}(\mathbf{R}^N)$ strong.

Then the Localisation principle will say that

$$P(\mathbf{x}, \boldsymbol{\xi})\mu = \Big(\mathsf{b}(\mathbf{x}) \cdot \boldsymbol{\xi}\Big)\mu = 0 \,,$$

where b is in C^1. That is the analogue of HÖRMANDER's theorem (mentioned above) that the wave front set $WF(u)$ is contained in the set where $P = 0$.

Similarly, the oscillations and concentrations live in the zero set of P. However, you have to understand that we are not talking about the same thing. We need for b just to be C^1, because I have not written the equation in conservative form. If I had written it in conservative form, I would have needed only continuous coefficients. And now, because of the hyperbolicity we assume that b_j are real as well, and μ satisfies a partial differential equation in $\mathbf{x}$ and $\boldsymbol{\xi}$, which is of the form that for any test function φ in $\mathbf{x}$ and $\boldsymbol{\xi}$:

$$\int \{\varphi, P\}\, d\mu + \cdots = [\text{source term}]\,.$$

The POISSON bracket will have derivatives in $\mathbf{x}$ and $\boldsymbol{\xi}$, and this will just be a way to write a first order partial differential equation for μ, while the characteristic curves for the equation are the ones described earlier (bicharacteristics). So, there is something going on there.

Not at all the same thing as in HÖRMANDER's theory. We are on the set where something is going on. Lars HÖRMANDER studies what happens away from that set, where the singularities are supposed to be. So, the usage of the term *propagation of singularities* by HÖRMANDER is not exactly right. Suppose that there are two characteristics that go through the same point, but with different directions. It is not clear whether they transfer information or not. Here they cannot, because it is a differential equation in $\mathbf{x}$ and $\boldsymbol{\xi}$, and we have different points in $\mathbf{x}, \boldsymbol{\xi}$ space. But what is important is that if you look at the wave equation, it is one thing to know that the regularity propagates along the light rays, and another that the energy propagates along light rays. Because we are not talking about the same things. And my claim is that the Physics is related to propagation of energy, and not regularity.

But in the definition I used something similar to the definition of wave front set: localise in $\mathbf{x}$, look at the FOURIER transform, and how does it behave at infinity, whether it decays and what is the integral of the square of FOURIER transform. What we are doing here is having a sequence instead of one distribution, trying to describe a new object. This is a standard procedure in Mathematics, like describing real numbers using sequences of rationals.

In the theory of distributions, as it was introduced by Laurent SCHWARTZ, when you do not know how to do something, like how to take a derivative, you can take a sequence of C^∞ functions approaching the distribution, and the derivative of the distribution will be the limit of derivatives of the functions. Because the definitions introduced by SCHWARTZ were the right ones (i.e. the derivative is a continuous operator on distributions) this procedure works well.

So, the H-measures are some kind of closure, and you need a sequence to define an H-measure. You cannot define it by just one function, as it is not related to just one function. On the other hand, the wave front set is related to only one distribution; it is a property of just one distribution.

Let me now look more closely at the definition of the POISSON bracket of two functions f and g in $\mathbf{x}$ and $\boldsymbol{\xi}$:

$$\{f,g\} = \nabla_{\boldsymbol{\xi}} f \cdot \nabla_{\mathbf{x}} g - \nabla_{\mathbf{x}} f \cdot \nabla_{\boldsymbol{\xi}} g \,.$$

This is a natural way to describe how a quantity varies when one plays with a Hamiltonian system. Actually, the POISSON bracket was not introduced by Siméon Denis POISSON, but by Joseph Louis LAGRANGE.

Why does the POISSON bracket appear? The point is that we need the second commutation lemma. Why do we need it?

Recall that we have a, a function of $\boldsymbol{\xi}$ on the unit sphere, and b, a function of $\mathbf{x}$. Then, the commutator of P_a and M_b, $[P_a, M_b]$ is compact. This is true if a is continuous, and b continuous tending to zero at infinity.

Now we need a little more, that it is regularising and transforms L^2 into H^1. Therefore, we want to say: if a and b are smooth, then this commutator maps L^2 into H^1, and $\frac{\partial}{\partial x_k}[P_a, M_b]$ has a symbol $\xi_k \nabla_{\boldsymbol{\xi}} a \cdot \nabla_{\mathbf{x}} b = \xi_k\{a,b\}$. As a is homogeneous of degree zero, when you take the derivative, and multiply it by $\boldsymbol{\xi}$, it is homogeneous of degree zero as well. Now, the point is that this is what $\xi_k\{a,b\}$ is, because a has no derivative in $\mathbf{x}$, and b has no derivative in $\boldsymbol{\xi}$.

There is some technicality about what kind of regularity you need of a and b, but I shall discuss that later.

Why do we need that? Here, for the propagation, I do not only need b real, but $f_n \rightharpoonup 0$ in L^2 weak in the equation. And I did not say exactly what the source term is. I need it in order to assert that there is an H-measure associated to the sequence (u_n, f_n), that plays a rôle. The μ I was discussing is what I call now μ^{11}, while the mixed one appears as the source term for the propagation. There are cases where f_n is obtained by a pseudodifferential operator acting on u_n, like $f_n = cu_n$. We can throw the term cu_n inside that f_n, and that makes no difference.

Let us consider the case $c = 0$. After multiplying the equation by u_n, we immediately see that we have an equation

$$\mathsf{b} \cdot \nabla \frac{(u_n)^2}{2} = f_n u_n \,.$$

However, it is not exactly this property that we need for the equation. We need a conservation law for the modulus of u_n squared, when u_n is complex. Even though I usually think of u_n being real.

Multiply the equation by the complex conjugate of u_n. So, therefore you would have

$$\mathsf{b} \cdot \nabla u_n \bar{u}_n = f_n \bar{u}_n \,,$$

and similarly

$$\bar{\mathsf{b}} \cdot \nabla \bar{u}_n u_n = \bar{f}_n u_n \,.$$

We need b to be real; if it were not, the sum of the terms on the left would not be $\mathsf{b} \cdot \nabla |u|^2$. Adding these two equalities together, we can get the real part of $2 f_n \bar{u}_n$.

Why do I need to assume that u_n might be complex? I am going to apply the equation not only to u_n, but to some operator applied to u_n as well, like $P_a u_n$, and we will need

$$\mathsf{b} \cdot \nabla_{\mathbf{x}} (P_a u_n) = P_a (\mathsf{b} \cdot \nabla_{\mathbf{x}} u) + [P_a, M_{\mathsf{b}}] u \, .$$

And the commutator is exactly of the above type. Then we need the limit of this operator times $\overline{u}_n$, and we need the limit of u_n times $\overline{u}_n$, and f_n times $\overline{u}_n$. The first one comes from the μ^{12}, and the second one from the properties of H-measures, as the limit of an operator acting on u_n multiplied by $\overline{u}_n$. Then you get something for every test function a, but you can use a test function depending on $\mathbf{x}$ as well. If you realise that the POISSON bracket is a simplifying thing, then out of this you get the right equation.

So, what we have deduced is that we have quadratic, or sesquilinear conservation law for complex solutions. That is also true for the wave equation.

Indeed if you do it for the wave equation, you have the same type of property. You can start with the equation

$$\rho u^n_{tt} - \mathsf{div} \, (\mathbf{A} \nabla u^n) = f_n \, .$$

For the first part, which is the localisation principle, one uses continuity of ρ and $\mathbf{A}$. And now you have a new symbol, which is

$$Q(\mathbf{x}, \boldsymbol{\xi}) = \rho(\mathbf{x}) \xi_0^2 - \mathbf{A}(\mathbf{x}) \boldsymbol{\xi} \cdot \boldsymbol{\xi} \, ,$$

where ξ_0 is the dual variable of time denoted by x_0. And then if ρ and $\mathbf{A}$ are C^1, and positive, you can take the same thing, and you find an equation where the POISSON bracket appears (Φ is a test function in variables $\mathbf{x}$ and $\boldsymbol{\xi}$):

$$\langle \mu, \{Q, \Phi\} \rangle = 0 \, ,$$

and that is exactly the theorem which says that the energy propagates along the light rays.

Question: Could a trilinear quantity be treated by H-measures, even in a very special case? Are there any known examples?

Answer: I do think that we need a new idea for that. It is possible that there will be an abstract definition, but a miracle as in LANDAU and LIFSCHITZ may occur.

4 LECTURE of Thursday, September 7

I want to mention something which I should have mentioned at the beginning, that Patrick GÉRARD, for completely different reasons, has independently introduced the same objects as I did, which I called H-measures, to be reminded of Homogenisation, but he gave them a different name: *microlocal defect measures* (see [18]).

While my motivation was from the Homogenisation theory, his motivation was from microlocal analysis. I always mention that this was not a good name, coming from HÖRMANDER's philosophy that microlocal regularity, or rather the lack of microlocal regularity is an important question. I claim that this is not an important question from the physical viewpoint, as this is a different type of problem. GÉRARD arrived at H-measures independently, a little later than me, but for different reasons.

4.1 Propagation for the wave equation

Today, I want to describe the properties of propagation which I described yesterday very quickly.

If you work with the wave equation

$$\begin{cases} \rho u^n_{tt} - \mathsf{div}\,(\mathbf{A}\nabla u^n) = 0 \\ u^n(0,\cdot) = v^n \\ u^n_t(0,\cdot) = w^n \,, \end{cases}$$

where $\rho > 0$ and $\mathbf{A}$ elliptic (i.e. $\alpha|\boldsymbol{\lambda}|^2 \leq \mathbf{A}(\mathbf{x})\boldsymbol{\lambda}\cdot\boldsymbol{\lambda} \leq \beta|\boldsymbol{\lambda}|^2$) are in $\mathrm{L}^\infty(\mathbf{R}^N)$, while $v^n \in \mathrm{H}^1(\mathbf{R}^N)$ and $w^n \in \mathrm{L}^2(\mathbf{R}^N)$ converge both to zero weakly in corresponding spaces. Then for each n the problem is well posed, and we have the conservation of energy. The density of the energy is

$$\frac{1}{2}\rho|u^n_t|^2 + \frac{1}{2}\mathbf{A}\nabla u^n\cdot\nabla u^n \rightharpoonup e \geq 0\,.$$

If you take a sequence of initial data which converges weakly to zero, then, because the problem is linear, and the coefficients are fixed, the solutions tend to zero. Even though they converge to zero, it is a weak convergence, and the density of energy will tend (vaguely, in weak $*$ topology for bounded Radon measures, and it might be necessary to pass to a subsequence in order to get the convergence) to a limit e which is nonnegative (and could be strictly positive).

Therefore you have the conservation of energy, and no problem with the existence of solutions, but you have the problem to know what is the energy that you get. For example, in the applications you sometimes have explosions, and you want to know what is going on.

This is related to the case when v^n and w^n concentrate at a point. So, at the time $t = 0$ all the energy is concentrated in one point. Of course, the classical theory, which is formal, tells you that the information is going to propagate on the wave cone (that is the case when the coefficients are constant), and the problem is that we have no way of using weak convergence to say how much of this energy goes in one direction. Because, in order to know that, you will need to know the variable $\boldsymbol{\xi}$, which is a direction, and therefore what the H-measures do is to do the analysis, with respect to $\boldsymbol{\xi}$, so that you can really decide what is going on in that direction.

But I did not work with the initial data for this problem, and I considered the scalar case. The question for initial data was done by Gilles FRANCFORT and François MURAT [15], with some technical advice from Patrick GÉRARD (see also [19]).

What I was describing yesterday was an equation in the domain, and I did not take care of the initial condition, what is the information about the initial data of this equation. What the H-measure says is exactly what the physicists know. But their meaning of knowing is different from ours. So, it is important to see that what we are talking about here is something like *Geometrical Optics*.

There is something interesting what we can say using only classical tools, that the difference of kinetic and potential energy (the action) tends to zero. I will call this the equipartition of energy. If you put in some energy of very high frequency, and of course it goes somewhere, but you do not know where, as there is no variable $\boldsymbol{\xi}$ to know the direction of propagation. But you know that wherever it will be, you look at the energy, and half of it will be in kinetic form, and half of it in potential form. This is just an application of the div-curl lemma to the equation.

However, without the introduction of the variable $\boldsymbol{\xi}$ there is no way to know where this energy will go. Geometrical Optics is some formal theory, which is more or less that you look at the solution in the form $Ae^{i\varphi}$, where A is the amplitude, while φ is the phase, and you have a large coefficient which is the frequency. You insert that in the equation, and you get the *eikonal equation* for the phase

$$|\varphi_t|^2 = |\nabla_{\mathbf{x}}\varphi|^2$$

(in the case of $u_{tt} - \Delta u = 0$). This is a HAMILTON-JACOBI problem, and you have a problem with caustics. Knowing φ, one then finds that the amplitude A satisfies a transport equation, but you want to know the direction of $\frac{\nabla\varphi}{|\nabla\varphi|}$, as it appears in the equation.

Now, this is slightly formal, and if you do it correctly, what you have proved here is that there exists some oscillating solution of the wave equation whose energy follows light rays.

What the H-measure theory says is something different: Firstly, there is no phase. What we have done in the FOURIER transform space was the integration along the rays. And the frequency is lost by doing that. So the H-measure loses information on the phase. Therefore this makes no sense.

But the H-measure $\boldsymbol{\mu}$ satisfies a partial differential equation. It is a partial differential equation in $(t, \mathbf{x})$ and dual variables $(\tau, \boldsymbol{\xi})$. And $\boldsymbol{\xi}$ plays the rôle of $\frac{\nabla\varphi}{|\nabla\varphi|}$. Therefore then this equation does not care much about caustics, it is a weak form of a partial differential equation, and the caustics are only trouble if you look at the regularity of the density of this measure. If you want to know that this measure depends on $\mathbf{x}$ and $\boldsymbol{\xi}$ smoothly, then you get the problem with caustics, with respect to the smoothness of the density.

But the measure exists. And one reason that the caustics are not so bad is that we have a measure in $\mathbf{x}$ and $\boldsymbol{\xi}$, so we may put different DIRAC masses at different points $\boldsymbol{\xi}$. That would correspond to a different wave, with different directions going through the same point $\mathbf{x}$. And the problem with caustics is that when you have a reflection, in some way you have waves of slightly different directions which arrive near such points. Therefore they cannot be very well described by something that looks like a plane wave.

However, at the end, all oscillating solutions of the wave equation satisfy the rules of Geometric Optics, which is that the energy will follow the light rays. Therefore, there is a big difference in principle between the two theories. They are not really talking about the same thing. Here we follow some kind of physicist's approach. You do the computation in the special case, you learn of the general principle, and then apply it to a different situation. This is a little bit like the computation in LANDAU and LIFSCHITZ, where they did the computation for the sphere, and then they say it is true for any mixture of two materials. But they have not proved that, as they had no way to prove that.

On the other hand, we have proved in fact that Geometric Optics works. But we did not follow the traditional approach. It would have been a little difficult to do, and the reason I could do it was that I introduced the H-measures for something else. And once I had introduced them, I looked whether they might be useful for propagation problems.

Of course, you have to be careful with the precise hypotheses, and they are not only that $\mathbf{A}$ is real, but it is also C^1. I did not mention much about the regularity needed in the second commutation lemma, but it is a little technical, and my initial result required $a \in C^1(\mathbf{R}^N)$ and $b \in X^1$, which is a technical space meaning that $\nabla b \in \mathcal{F}(L^1(\mathbf{R}^N))$, being a little more regular than C^1. But Pierre-Louis LIONS mentioned that if I might use the result by Guy DAVID and Jean-Lin JOURNÉ in Harmonic Analysis, in order to improve my hypotheses. Looking a little more in that direction I found that Alberto CALDERÓN [10] had already done the commutation lemma for singular integrals, and just by applying one of his results I could have a a little smoother, and b in C^1.

I looked at the paper of CALDERÓN, but I did not check the details; actually he did not really need b in C^1, but only Lipschitz, or even in $W^{1,p}$ in his theory, therefore he was able to do the commutation lemma with much less regularity than I could. Putting more regularity on a is not a problem, as it appears as a test function, so we can even take it in C^∞. But b corresponds to the coefficients in the equation, so it is useful to know that it does not need to be so regular.

However, the coefficients b must have the derivatives at least in L^p. That is important, as you could not deduce anything about the refraction effects, which is related to a discontinuity in the coefficients. So, we are talking about the rules of Geometrical Optics applied to any medium if the coefficients are smooth enough, which we take to mean C^1, but Patrick GÉRARD noted that the solution of the equation for the measure must be unique (and this is the case when coefficients are C^2).

Therefore, what I have proved is that when I look at H-measure, the H-measure in this case is not related to the solution u, but to its gradient ∇u. As u is bounded in $H^1(\mathbf{R}^N)$, so the derivatives are bounded in $L^2(\mathbf{R}^N)$, and when I have a gradient, the H-measure is actually coming from a scalar measure, it is $(\boldsymbol{\xi} \otimes \boldsymbol{\xi})\mu$.

4.2 Geometric Theory of Diffraction

The intuition there is that the H-measures are not the right object for studying refraction, because the result is supposed to be frequency dependent, and I said that H-measures have no phase, because when we applied the FOURIER transform, we were integrating $|\mathcal{F}u^n|^2$ on each ray. The FOURIER transform sees the frequency, but integration loses it and H-measures do not see it.

Why would that be important for refraction? There is an improvement of the Geometrical Optics method due to Joseph KELLER (see [23]). It is formal too, and called *Geometric Theory of Diffraction.* However it gives you something you can compute: if you send a wave on an obstacle, you can compute how much energy comes back. It is good for the wave equation, for MAXWELL's system, radar applications, acoustics, etc. And, of course, the first thing is that the law you expect is some kind of reflection, a geometric condition on a perfectly reflecting body. On the other hand, in the theory of Joseph KELLER there are also rays that turn around and leave the body, and the way they lose their energy you have to integrate and then take the exponential; the law is

$$e^{-\int k^{\frac{1}{3}}},$$

where k is the wave number. Therefore, the effect is very much frequency dependent. This has something to do with computations with AIRY's function, and I think George AIRY was interested in the very problem of the shade of the light coming from infinitely distant source, behind an obstacle, where there is still some light (looking at the solution of the wave equation). There is some light in what is supposed to be the shadow. And in order to compute how much light there is, you must understand what happens with the solution, and find that something is going on, so that there is actually some light there.

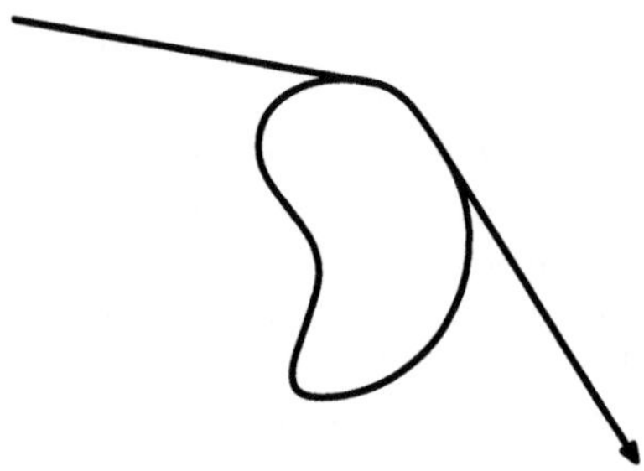

Figure 7. A light ray turns around an obstacle, so that there is some light even in the shade.

And I think that is the same type of thing that Joseph KELLER in some way guessed from special computations, and the precise theory is that light is going to follow geodesics on the boundary of the body, which are very complicated things. And this could be very efficient. But not on caustics. The theory there gives a wrong answer. But if we believe this analysis, then we expect that the question of refraction, which is caused either by the difference of two materials and discontinuity, or the boundary, is a frequency dependent effect, and therefore you have no hope of doing that with H-measures. We need a more precise microlocal object which can see the frequency. And we shall see a variant of that later on.

4.3 Quantum Mechanics and Electrodynamics

My feeling is that what I was doing here has some very interesting qualitative application in an attempt to understand what the physicists do when they study Quantum Mechanics.

They start from a Hamiltonian system, which is an ordinary differential equation, and they construct a partial differential equation, following the approach of Erwin SCHRÖDINGER, but it is important to notice that there are examples which are different, as the DIRAC equation.

What I am doing is the opposite: I start with a partial differential equation. So, what the H-measures do is the reverse. We get an oscillating solution of the equation, and you can describe the important features of oscillating solutions by some measure, and this measure satisfies a partial differential equation of first order. Such an equation is like an ordinary differential equation, and you can interpret that as propagation of some objects, which we can call particles. They do not really look like classical particles. When you learn Quantum Mechanics, you learn that the particles can behave like waves in some situations. Here we have a different point of view. There are no particles, there are only the waves, but in the high frequency part of the oscillations the waves are well described by a simpler system, which you can interpret as particles moving on.

However, the theory is not yet in the situation where you can apply it to real Physics, because the equation has to be linear. I mentioned yesterday the first order problem that I could have here

$$\mathsf{b} \cdot \nabla u^n = f_n ,$$

and this f_n could be nonlinear in u^n. Then the H-measure of u^n is not enough; you need the H-measure of the pair (u, f), namely

$$\begin{bmatrix} \mu^{11} & \mu^{12} \\ \mu^{21} & \mu^{22} \end{bmatrix} ,$$

and the equation for this one uses as the source term the real part of μ^{12}. So if f_n is a nonlinear function of u^n, then we do not know what μ^{12} is in terms of μ^{11}. Therefore it will be important to understand what are the ways to deal with H-measures, in the case when there are nonlinear relations. Of course, we need another object, and maybe that object will be better adapted to this task.

DIRAC's equation is a semilinear hyperbolic system, which has only one characteristic speed — the speed of light, and that is important. That is the reason why I was not working with the SCHRÖDINGER equation, which is not hyperbolic. In order to work with it, we have to take a small parameter tending to zero, and to use some kind of scale, and that is what I will describe when I shall talk about a variant of H-measure, with one characteristic length.

What do we get for the MAXWELL equations? Let us take the classical MAXWELL's equations:

$$\begin{cases} \operatorname{div} \mathsf{B} = 0 \\ \mathsf{B}' + \operatorname{rot} \mathsf{E} = 0 , \end{cases}$$

and

$$\begin{cases} \operatorname{div} \mathsf{D} = \rho \\ -\mathsf{D}' + \operatorname{rot} \mathsf{H} = \mathsf{j} , \end{cases}$$

where ρ is the density of charge and j the current density. If you look at the form of the equations, the first part of the system describes the propagation of light, while the second describes its interaction with matter.

Paul Adrien Maurice DIRAC introduced a function ψ, taking the values in $\mathbf{C}^4$, and his equation has the linear part in ψ, and then a bilinear term which is linear in ψ, and it does not use the electric or magnetic field, but scalar and vector potentials. So, if you freeze that, i.e. suppose that the electromagnetic field is fixed, you get a linear equation in ψ. And my conjecture was that the equation would be like the equation for the electrons.

If you do that for MAXWELL's system, where ρ and j are given, there are linear relations, of course, what you see is not scalar light, but polarised light. When you study the linear algebra corresponding to the equation, it is a little more difficult than for the wave equation, what you will get is what corresponds to the polarised light.

I expected DIRAC'S equation to show effects related to electrons, and Nenad ANTONIĆ [5] worked a little on that kind of problem, but a much better way to say that the work is in the right direction was done by Patrick GÉRARD [17], using what I shall describe later as *semiclassical measures*. This is the case where DIRAC corrected a little the equation, by adding another term, which has the mass of the electron multiplied by ψ. And that breaks some kind of symmetry.

Knowing the mass of the electron gives us some kind of a length scale. If we denote this length scale corresponding to the electron by ε_n, and we use the analysis that Patrick GÉRARD introduced at this length scale, then we will find something that will look like the transport equation for the H-measure. However the transport equation tells two things: it is not only for electrons, but also for positrons. And it is relativistic. The result of these computations is that in some way what we will find at the end will describe the oscillations of this system, but as seen using the characteristic length ε_n, corresponding to the movement of two types of particles, with the charges $+e$ and $-e$, both with mass m_0, and which satisfy the LORENTZ force law, the force is $\mathsf{E} + \mathsf{v} \times \mathsf{B}$, and the mass is relativistic, depending on the velocity. So this computation is a way to say that the DIRAC system does a very good job by using the same type of ideas in a different way. And the equation we guess has this kind of property. But E and B are not given in a real problem.

This is a difficulty in Physics. When you learn Quantum Mechanics, you are always taught that the equations are linear. But that is not true. There are nonlinear effects, and Quantum Field Theory is supposed to be good in dealing with nonlinear effects. But I do not know of any good theory that will give estimates so that you can do something quantitative. But certainly we already know that the H-measure is not the right object. Probably it will be some kind of variant, but it is not H-measure itself, which is the answer to the problem. Thus it is important to understand the problems corresponding to one characteristic length, and even more than one.

4.4 Compactness by integration

I want to tell you what was the reason why Patrick GÉRARD [18] introduced the same objects as H-measures.

I tried to solve this same problem, but I could not. It is something I learned from Benoit PERTHAME, who was working with Rémi SENTIS, related to some transport equation

$$\frac{\partial f}{\partial t} + \mathsf{v} \cdot \nabla f + \cdots = 0 .$$

We define the function $\rho(t,\mathbf{x})$ by integrating f in v, actually we take a test function φ smooth enough and let

$$\rho(t,\mathbf{x}) := \int f(t,\mathbf{x},\mathsf{v})\varphi(\mathsf{v})\,d\mathsf{v}\ ,$$

and the result from the bounds is that ρ converges strongly even though f converges only weakly. This is compactification by integration in the velocity variable.

My idea was to try to start with a function f satisfying some partial differential equation, and try to deduce what the H-measure of ρ would be, in terms of the H-measure of f. I could not find any relation.

Patrick GÉRARD solved this problem differently. He introduced something slightly different, H-measures with values in a HILBERT space. I called them H-measures, but he used a different word for them (microlocal defect measures). Measures valued in the HILBERT space, because it was L^2 in the velocity variable, and therefore he had to be slightly more precise. He used HÖRMANDER's theory of FOURIER integral operators, as that was the theory he was trained in, and at the end he obtained the same object as H-measure, but in a slightly more general setting. The same thing, as with the values in $\mathbf{R}^N$ could be done, but you should be more careful in the proofs. However, the answer is not an expression for the H-measure of ρ in the terms of H-measure of f.

4.5 Variants of H-measures: One characteristic length

The first variant is a variant with one characteristic length ε_n.

My idea, which I am going to explain on an example, was that when you have a characteristic length ε_n, you have a way to include it into the framework.

In order to do that, I introduce another variable x_{N+1} and consider the H-measure with one more variable. More precisely, you have a sequence $u_n(\mathbf{x})$ converging weakly to zero in $\mathrm{L}^2_{\mathrm{loc}}$ and you know that there will be some characteristic scale ε_n. Define the following function

$$U_n(\mathbf{x},x_{N+1}) := u_n(\mathbf{x})\cos\frac{x_{N+1}}{\varepsilon_n}\ ,$$

where ε_n is the chosen characteristic lenght tending to zero. Here we have added one space variable, and the cosine will provide the oscillations.

When you do the FOURIER transform of something like that, it is going to make two copies of $\mathcal{F}u_n$, but in the perpendicular direction the scale will be $\frac{1}{\varepsilon_n}$. Just by looking at the FOURIER transform of cosine, it is two DIRAC masses, so you take two copies of the same; then the H-measure will look at everything and map it back to the unit sphere. Therefore, it will look like something that will be at the right size of distance in $O(\frac{1}{\varepsilon_n})$.

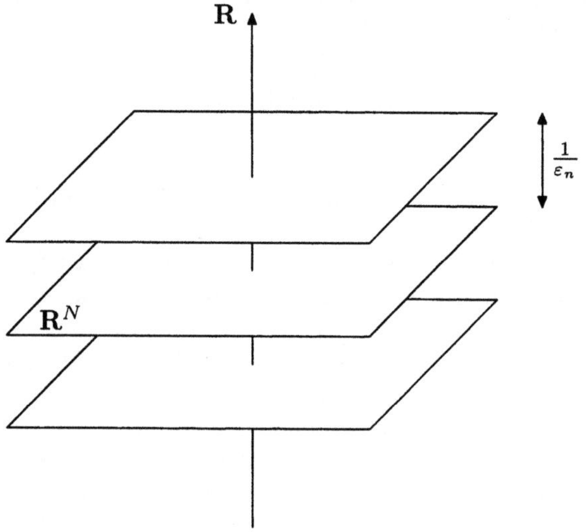

Figure 8. The FOURIER transform of a function periodic with period at scale ε_n in x_{N+1} supported on parallel hyperplanes separated at scale $\frac{1}{\varepsilon_n}$.

As U_n converges weakly to 0 in $L^2_{loc}(\mathbf{R}^N \times \mathbf{R})$, one can extract a subsequence corresponding to an H-measure μ, living on $(\mathbf{R}^N \times \mathbf{R}) \times S^N$, but looking at the definition one sees easily that if one stays away from $\xi_{N+1} = 0$, then that measure is independent of x_{N+1}, and its projection on $\mathbf{R}^N \times (S^N \setminus S^{N-1})$ is essentially the same measure on $\mathbf{R}^N \times \mathbf{R}^N$ which Patrick GÉRARD [17] had defined (considering the point $(\boldsymbol{\xi}, 1)$ in each equivalence class instead of the point in S^N). However, Patrick GÉRARD's definition is more easy to handle, and he thought of more general situations than the ones I had in mind. Nevertheless, it had some defect.

What he did was the following. The idea was to look at the sequence $u_n(\mathbf{x}) = v(\mathbf{x}, \frac{\mathbf{x}}{\varepsilon_n})$. What is the way to do that? When you do the FOURIER transform of u_n, this is mostly for $\boldsymbol{\xi} = O(\frac{1}{\varepsilon_n})$, so in the FOURIER space you expect that the information is mostly contained very far from the origin.

The H-measure integrates along the ray, and does not know how far it is. But now we decide to take a closer look at it, so what you want to look at is the integral after we perform a rescaling. Thus we compute (in the scalar case)

$$\int_{\mathbf{R}^N} |\mathcal{F}(\varphi u_n)(\boldsymbol{\xi})|^2 \psi(\varepsilon_n \boldsymbol{\xi})\, d\boldsymbol{\xi}\,,$$

where GÉRARD has chosen the test function $\psi \in \mathcal{S}$ (the SCHWARTZ space) and $\phi \in C^\infty_c$ as before. He did not impose that u_n converge weakly to 0, and he wanted to consider all the possible limit points because of some different situations that he had in mind. The above integral will tend to a limit, which will behave like

$$\int_{\mathbf{R}^N \times \mathbf{R}^N} |\varphi(\mathbf{x})|^2 \psi(\boldsymbol{\xi})\, d\mu_{sc}(\mathbf{x}, \boldsymbol{\xi}) = \left\langle \mu_{sc}\,,\, |\phi|^2 \otimes \psi \right\rangle ,$$

where μ_{sc} is the semi-classical measure. It is important to note that $\boldsymbol{\xi}$ is taking values in all of $\mathbf{R}^N$, and not only on the unit sphere.

Actually, I made a small technical mistake from the geometrical point of view, that $\boldsymbol{\xi}$ does not live on unit sphere, but on the quotient space of $\mathbf{R}^N$ by the equivalence relation where we identify two points that are on the same ray.

After the H-measures were introduced, I gave a talk in Los Angeles, and Gregory ESKIN mentioned that he had seen some measures in $\mathbf{x}$ and $\boldsymbol{\xi}$ used before, by Alexander SCHNIRELMAN [38], and it was used later also by Yves COLIN DE VERDIÈRE [11]. This had something to do with the eigenfunctions of the Laplacian. If you consider the eigenproblem for the Laplacian

$$-\Delta u_n = \lambda_n u_n$$

on the rectangle, where eigenfunctions can be expressed as the products of cosines and sines, then you get something that looks like the reflection of light. The idea was to do that in general, and the oscillations in the eigenfunction tend to follow the rules of Geometrical Optics, with reflection at the boundary.

SCHNIRELMAN has introduced a measure in $\mathbf{x}$ and $\boldsymbol{\xi}$ for that purpose, and $\boldsymbol{\xi}$ was on the unit sphere. But this measure was not an H-measure. His measure was a *semiclassical measure*, which was forced to live on the unit sphere by the equation (as a result of the localisation principle).

So, it is not exactly the same type of object. This was also worked on by Patrick GÉRARD and Eric LEICHTNAM [20], and it has some similarities with the theory of Joseph KELLER [23] I mentioned earlier, trying to understand something about what was going on at the boundary.

That is the definition. The difference between the H-measure and the semiclassical one is that one has to do slightly different variant of the earlier described H-measure theory. When we do the H-measure theory, we said that to a function $b(\mathbf{x})$ we associate the operator M_b, and to $a(\boldsymbol{\xi})$ the operator P_a. We wanted to know that the commutator $P_a M_b - M_b P_a$ is compact. And that requires some regularity on a and b. The a was continuous on the unit sphere, and extended to $\mathbf{R}^N$ (without the origin, of course) by the homogeneity of order zero. However, there are situations where we have to consider functions which are not homogeneous of order zero.

For semiclassical measures you have also a function $a(\boldsymbol{\xi})$, but defined on all of $\mathbf{R}^N$, and what you do is to look at the operator $\overline{\mathcal{F}} a(\varepsilon_n \boldsymbol{\xi}) \mathcal{F}$. Therefore, the operator is not associated to one a, but to a sequence of its rescaled version, and we have a sequence of operators P_{a_n}. So, we are not in the same situation as before. We consider the commutator

$$[P_{a_n}, M_b] = P_{a_n} M_b - M_b P_{a_n} ,$$

and it will not be enough that each of these operators is compact, as we do not have the same operator each time. The reason why we are interested in compactness is to have an operator which will transform weakly convergent sequences into strongly convergent ones, which we can then discard. Here we have a sequence of operators, applied to a sequence (u_n) that converges weakly to zero, and we would like that the norm

$$||(P_{a_n} M_b - M_b P_{a_n}) u_n||_{L^2(\mathbf{R}^N)} \longrightarrow 0 .$$

Being compact is not the right thing any more. We have to estimate the norm of this sequence of the operators. Thus, the problem is slightly different.

It will appear in a moment that the choice of smooth functions ($\psi \in \mathcal{S}$) has some advantages, but also some defects. The defect it has is that it is continuous at the origin, and tends to zero at infinity. Let us see what is not good here.

Suppose $u_n(\mathbf{x}) := v(\mathbf{x}, \frac{\mathbf{x}}{\delta_n})\varphi(\mathbf{x})$, where φ is smooth and v is periodic in the second variable. We distinguish three different cases:

1. $\delta_n \ll \varepsilon_n$,
2. $\delta_n = O(\varepsilon_n)$,
3. $\delta_n \gg \varepsilon_n$, but δ_n still tends to 0.

So, you are looking at something that has oscillations, but perhaps you are looking at it with a wrong wavelength.

In the first case, you apply the Fourier transform, and the Fourier transform of u_n is going to be at the distance of order $O(\frac{1}{\delta_n})$. If you take ε_n which is too big, after rescaling you see that the information goes to infinity. As ψ goes to zero at infinity, you lose all the information. This is the defect of taking the test function tending to zero at infinity.

The second case is good.

In the third case, all the information is pushed towards zero, and the information in all directions get mixed up at zero. As ψ is continuous at zero, all the information you get is related to all possible directions. It is not really lost, but mixed up, and you cannot use it.

There is one way to avoid that. Take a function ψ that behaves like $\psi_\infty(\frac{\boldsymbol{\xi}}{|\boldsymbol{\xi}|})$ around infinity and like $\psi_0(\frac{\boldsymbol{\xi}}{|\boldsymbol{\xi}|})$ around zero, and then you will be able to preserve the information.

Tomorrow we are going to see how we can use this object. It is useful, but cannot do everything. Notably, it cannot recover the H-measure, but can be used in some situations where H-measures cannot.

It is a variant of H-measures.

5 LECTURE of Friday, September 8

5.1 Semi-classical measures: One characteristic length

Semi-classical measures were introduced by Patrick GÉRARD [17].

We have already seen how does the sequence (we can take vector valued u_n here as well)

$$U_n(\mathbf{x}, x_{N+1}) = u_n(\mathbf{x}) \cos\left(\frac{x_{N+1}}{\varepsilon_n}\right)$$

generate an H-measure in $\mathbf{R}^{N+1}$.

GÉRARD passed to the limit

$$\int_{\mathbf{R}^N} \mathcal{F}(\varphi_1 U_n)\overline{\mathcal{F}(\varphi_2 U_n)}\psi(\varepsilon_n \boldsymbol{\xi})\, d\boldsymbol{\xi} \longrightarrow \int_{\mathbf{R}^N \times \mathbf{R}^N} \varphi_1(\mathbf{x})\bar{\varphi}_2(\mathbf{x})\psi(\boldsymbol{\xi})\, d\mu(\mathbf{x}, \boldsymbol{\xi})\,.$$

As a test function he took $\psi \in \mathcal{S}(\mathbf{R}^N)$. It would be better to take $\psi \sim \psi_0\left(\frac{\boldsymbol{\xi}}{|\boldsymbol{\xi}|}\right)$ around zero, and $\psi \sim \psi_\infty\left(\frac{\boldsymbol{\xi}}{|\boldsymbol{\xi}|}\right)$, as explained yesterday. The function ψ should be bounded and uniformly continuous.

This is different from two-scale convergence, as introduced by Gabriel NGUETSENG [36] and further studied by Grégoire ALLAIRE [1]. In two-scale convergence theory one considers a sequence u_n and one wants to identify a function U in variables $\mathbf{x}$ and $\mathbf{y}$, such that

$$\int_\Omega u_n \varphi\left(\mathbf{x}, \frac{\mathbf{x}}{\varepsilon_n}\right) d\mathbf{x} \longrightarrow \int_{\Omega \times Y} U(\mathbf{x}, \mathbf{y})\varphi(\mathbf{x}, \mathbf{y})\, d\mathbf{x}\, d\mathbf{y}$$

for any test function φ which is Y-periodic in the second variable (and has compact support in $\mathbf{x}$); this definition is obviously dependent upon what the unit cell Y is, so it is not only related to the choice of ε_n. Two-scale convergence involves resonance effects, so if one considers one-dimensional case, takes $Y = \langle 0, 1\rangle$ and $u_n = v\left(\frac{\kappa x}{\varepsilon_n}\right)$ with v having period 1, then if κ is irrational one finds that U is constant and equal to the average of v, while for $\kappa = 1$ one finds $U(x, y) = v(y)$. There is no such discontinuity in κ for semi-classical measures.

Without going too much into the details, the proof for H-measures is based on the First Commutation Lemma which asserts that a commutator is compact, while the proof for semi-classical measures requires estimating the norm of a similar commutator and showing that it tends to 0; the reason is that to a function $a(\boldsymbol{\xi})$ defined on $\mathbf{R}^N$ (or $\mathbf{R}^N \setminus \{\mathbf{0}\}$), one associates now an operator $P_{a(\varepsilon_n \cdot)}$ which depends upon n, and in order to assert that it transforms a weakly converging sequence into a strongly converging sequence, one imposes that its norm tends to 0.

In the proof of the First Commutation Lemma, one uses the fact that a limit in norm of a sequence of compact operators is compact, and one approaches b uniformly by a sequence b_k such that $\mathcal{F}b_k \in C_c(\mathbf{R}^N)$, and one proves the lemma for b_k instead; the same step works for the case of semi-classical measures, because one makes a small error in norm by replacing b by some b_k and one must show that the norm of the commutator with b_k tends to 0.

In the case of H-measures, one has

$$\mathcal{F}\big([P_a, M_{b_k}]v\big)(\boldsymbol{\xi}) = \int_{\mathbf{R}^N} (a(\boldsymbol{\xi}) - a(\boldsymbol{\eta}))\mathcal{F}b_k(\boldsymbol{\xi} - \boldsymbol{\eta})\mathcal{F}v(\boldsymbol{\eta})\, d\boldsymbol{\eta}\,,$$

and the integral only takes into account $|\boldsymbol{\xi} - \boldsymbol{\eta}| \leq r_k$, thus one can cut the integral in two pieces. The first one is for large $|\boldsymbol{\xi}|$ and large $|\boldsymbol{\eta}|$, which is small because of the continuity of a (which is homogeneous of degree 0 in this computation). The second one is for bounded $|\boldsymbol{\xi}|$ and bounded $|\boldsymbol{\eta}|$ which comes from a HILBERT-SCHMIDT operator, which is compact (the lemma does not need a to be homogeneous of degree 0, and some variants do require a different behaviour for a and the same proof of the lemma applies). In the case of semi-classical measures one must estimate the norm of an operator T_n such that

$$\mathcal{F}(T_n v)(\boldsymbol{\xi}) = \int_{\mathbf{R}^N} (a(\varepsilon_n\boldsymbol{\xi}) - a(\varepsilon_n\boldsymbol{\eta}))\mathcal{F}b_k(\boldsymbol{\xi} - \boldsymbol{\eta})\mathcal{F}v(\boldsymbol{\eta})\, d\boldsymbol{\eta}\,,$$

and now a is defined on all of $\mathbf{R}^N$, but if $|\boldsymbol{\xi} - \boldsymbol{\eta}| \leq r_k$ then $|\varepsilon_n\boldsymbol{\xi} - \varepsilon_n\boldsymbol{\eta}| \leq \varepsilon_n r_k$, and one can bound $|a(\varepsilon_n\boldsymbol{\xi}) - a(\varepsilon_n\boldsymbol{\eta}))|$ by a small number if a is uniformly continuous, and then one has

$$\int_{\mathbf{R}^N} |\mathcal{F}b_k(\boldsymbol{\xi} - \boldsymbol{\eta})|\,|\mathcal{F}v(\boldsymbol{\eta})|\, d\boldsymbol{\eta} = \Big(|\mathcal{F}b_k| * |\mathcal{F}v|\Big)(\boldsymbol{\xi})\,,$$

a convolution of $|\mathcal{F}b_k| \in \mathrm{L}^1(\mathbf{R}^N)$ and $|\mathcal{F}v| \in \mathrm{L}^2(\mathbf{R}^N)$, thus in $\mathrm{L}^2(\mathbf{R}^N)$.

Patrick GÉRARD's definition has two defects; the first one comes from having ψ continuous at the origin, and the information corresponding to wavelengths tending to 0 but much longer than ε_n is mixed for all directions (and with the information coming from the nonzero limit of the sequence, as he had not started with a sequence converging weakly to 0); the second one comes from having ψ tend to 0 at infinity, the information corresponding to wavelengths much smaller than ε_n is completely lost. For that reason Patrick GÉRARD has introduced two definitions to name those sequences where no information is lost, and without these precautions one cannot in general recover the H-measure from the semi-classical measure, contrary to what Pierre-Louis LIONS and Thierry PAUL have written [29].

For sequences converging weakly to 0, the two defects of Patrick GÉRARD's definition can be easily fixed by considering ψ to be of the form $\psi_0\big(\frac{\boldsymbol{\xi}}{|\boldsymbol{\xi}|}\big)$ near the origin and either the same condition near infinity or a more general one like requiring ψ to be bounded and uniformly continuous at infinity; the first choice consists in compactifying $\mathbf{R}^N \setminus \{\mathbf{0}\}$ by a sphere at origin and a sphere at infinity, while in the second case the compactification at infinity is more subtle; only after a compactification like one of these can one expect to recover the H-measure from the *generalised* semi-classical measure in all cases.

In the localisation theorem for semi-classical measures of Patrick GÉRARD there are terms like $(\varepsilon_n \partial_x)$, or $(\varepsilon_n \partial_{x_1})(\varepsilon_n \partial_{x_2})$, and in the H-measure there is some kind of homogeneity being used —- this works when $t \sim x$, but not otherwise. Thus, it is successful for $u_{tt} - u_{xx}$, but it is not successful for $u_t - u_{xx}$, or $u_t - \varepsilon_n^2 u_{xx}$.

5.2 WIGNER transform and two-point correlations

Pierre-Louis LIONS and Thierry PAUL [29] have introduced essentially the same objects, but in a more complicated way. They were motivated by the work of some physicists, and they called WIGNER measures what GÉRARD called semi-classical measures.

For a function u from a good enough subset of $L^2(\mathbf{R}^N)$, it is possible to define its WIGNER transform

$$W(\mathbf{x}, \boldsymbol{\xi}) := \int_{\mathbf{R}^N} u(\mathbf{x} + \mathbf{y}/2)\overline{u(\mathbf{x} - \mathbf{y}/2)} e^{-2\pi i \mathbf{y}\cdot\boldsymbol{\xi}} d\mathbf{y} \ ,$$

and it is not difficult to show that $W \in C_0(\mathbf{R}^N, \mathbf{R}^N)$.

I have heard of that transform from George PAPANICOLAOU in the late seventies when I had mentioned to him my idea of splitting YOUNG measures in $\boldsymbol{\xi}$ (an idea which I had to abandon in order to introduce H-measures); he had stressed that the interest of WIGNER transform was that it could see both u and its FOURIER transform: indeed, allowing for a little more regularity for u, one has

$$\int_{\mathbf{R}^N} W(\mathbf{x}, \boldsymbol{\xi})\, d\mathbf{x} = |\mathcal{F}u(\xi)|^2 \ ,$$

for $u \in L^2(\mathbf{R}^N) \cap L^1(\mathbf{R}^N)$, while for $u \in L^2(\mathbf{R}^N) \cap \mathcal{F}L^1(\mathbf{R}^N)$

$$\int_{\mathbf{R}^N} W(\mathbf{x}, \xi)\, d\xi = |u(x)|^2 \ .$$

I had not seen how George PAPANICOLAOU's idea of using WIGNER transform could help for my purpose, but he himself had pursued his idea, so that recently, in collaboration with Joseph KELLER and their student Leonid RYZHIK, they were able to obtain results for propagation of waves in random media [25,26].

Eugene WIGNER worked on the SCHRÖDINGER equation without potential

$$iu_t - \Delta u = 0 \ .$$

After applying the WIGNER transform, one gets a transport equation for W

$$W_t + \boldsymbol{\xi} \cdot \nabla W = 0 \ ,$$

where $\boldsymbol{\xi}$ played the role of velocity. This W cannot be given an interpretation of a density of particles having velocity $\boldsymbol{\xi}$, as it is not nonnegative. However, if you take a convolution in $\boldsymbol{\xi}$ with $e^{-|\boldsymbol{\xi}|^2}$, then you obtain a nonnegative function, which was known to WIGNER. This composition of the WIGNER transform and convolution with Gaussian Pierre-Louis LIONS and Thierry PAUL called the HUSIMI transform and use its positivity as a crucial observation in order to show that the limit is a nonnegative measure.

Pierre-Louis LIONS and Thierry PAUL have shown that if you take

$$W_n(\mathbf{x},\boldsymbol{\xi}) := \int_{\mathbf{R}^N} u(\mathbf{x}+\varepsilon_n\mathbf{y}/2)\overline{u(\mathbf{x}-\varepsilon_n\mathbf{y}/2)}e^{-2\pi i\mathbf{y}\cdot\boldsymbol{\xi}}d\mathbf{y}\,,$$

then the sequence W_n converges vaguely to the semi-classical measure μ_{sc} of GÉRARD. The main difficulty in their proof consisted in proving directly that the limit was a nonnegative measure.

Together with Patrick GÉRARD I have simplified their proof, and noticed that it can be directly linked to two-point correlations. Recall that the YOUNG measures can handle only one-point correlations.

In a probabilistic framework one may work with n-point correlations, without using any characteristic lengths. In the periodic case, one uses the characteristic length ε for the definition of correlations.

What shall we do in general?

For *two-point correlation* consider a bounded sequence u_n in L^2, and look at the a limit (on a subsequence, vaguely in the sense of measures)

$$u_n(\mathbf{x}+\varepsilon_n\mathbf{y}^1)\overline{u_n(\mathbf{x}+\varepsilon_n\mathbf{y}^2)} \rightharpoonup G_2(\mathbf{x};\mathbf{y}^1,\mathbf{y}^2)\,.$$

(By using continuous test functions with compact support one can take care of the case where $\mathbf{x}\in\Omega\subseteq\mathbf{R}^N$, as on the support one has $\mathbf{x}+\varepsilon_n\mathbf{y}$ and $\mathbf{x}+\varepsilon_n\mathbf{z}$ in Ω for n large enough.)

Similarly, for *three-point correlation* (note that there is no complex conjugate here), for a bounded sequence u^n in L^3:

$$u_n(\mathbf{x}+\varepsilon_n\mathbf{y}^1)u_n(\mathbf{x}+\varepsilon_n\mathbf{y}^2)u_n(\mathbf{x}+\varepsilon_n\mathbf{y}^3) \rightharpoonup G_3(\mathbf{x};\mathbf{y}^1,\mathbf{y}^2,\mathbf{y}^3)\,.$$

As G_2 is invariant by the addition of the same vector to both $\mathbf{y}^1$ and $\mathbf{y}^2$, it can be written as

$$G_2(\mathbf{x};\mathbf{y}^1,\mathbf{y}^2) = \Gamma_2(\mathbf{x};\mathbf{y}^1-\mathbf{y}^2)\,,$$

Γ_2 being called the *correlation function*. If you take $\boldsymbol{\lambda}\in\mathbf{C}^m$, and look at the linear combination

$$\begin{aligned}\lim\left|\sum_{j=1}^m \lambda_j u_n(\mathbf{x}+\varepsilon_n\mathbf{y}^j)\right|^2 &= \sum_{j,k=1}^m \lambda_j\bar\lambda_k G_2(\mathbf{x};\mathbf{y}^j,\mathbf{y}^k)\\ &= \sum_{j,k=1}^m \lambda_j\bar\lambda_k \Gamma_2(\mathbf{x};\mathbf{y}^j-\mathbf{y}^k)\geq 0\,.\end{aligned}$$

Thus we can apply the BÔCHNER-SCHWARTZ theorem on the characterisation of FOURIER transforms of nonnegative measures, to conclude that $\Gamma_2(x,\cdot) = \mathcal{F}\mu(x,\cdot)$, where μ is just the semi-classical measure defined by GÉRARD.

It is not clear why we have to use the FOURIER transform here (it works, but that seems an obstacle for further extensions).

5.3 Examples with one characteristic length

I knew that for some problems I was going to need a characteristic length, and I was thinking about diffusion equations with a small diffusion coefficient like

$$u_t - \varepsilon^2 \mathsf{div}\,(\mathbf{A}\nabla u) + \mathsf{b}(\mathbf{x},t)\cdot\nabla u + c(\mathbf{x},t)\,u = f,$$

with initial data generalising $e^{-\varepsilon \mathbf{D}\mathbf{x}\cdot\mathbf{x}}$ for which I knew a more direct approach. Taking b affine in $\mathbf{x}$, and c a second degree polynomial in $\mathbf{x}$, there are explicit solutions of exponential type, having a function quadratic in $\mathbf{x}$ with coefficients depending upon t in the exponent, i.e. of the form:

$$u(\mathbf{x},t) = a(t)e^{-\mathbf{B}(t)(\mathbf{x}-\mathbf{c}(t))\cdot(\mathbf{x}-\mathbf{c}(t))}\ .$$

Substitution leads to differential equations for $\mathbf{B}$, $\mathbf{c}$ and a.

Together with Patrick GÉRARD I have neglected the lower order terms, and for simplicity looked at the equation (in one space dimension):

$$u_t^n - \varepsilon_n^2 u_{xx}^n = 0\ .$$

In order to avoid some technical difficulties we additionally assume that $\varepsilon_n u_x^n$ is bounded in L^∞.

We would like to determine

$$\frac{\partial}{\partial t}[u^n(x,t)u^n(x+\varepsilon_n y,t)]\ .$$

Multiplying the above equation by $\overline{u^n(x+\varepsilon_n y,t)}$, and combining the terms, while using

$$u_t^n(x+\varepsilon_n y,t) - \varepsilon_n^2 u_{xx}^n(x+\varepsilon_n y,t) = 0\ ,$$

and neglecting the complex conjugate, we obtain

$$\varepsilon_n^2 u_{xx}^n(x,t)u^n(x+\varepsilon_n y,t) = \varepsilon_n^2\Big(u_x^n(x,t)u^n(x+\varepsilon_n y,t)\Big)_x - \varepsilon_n^2\Big(u_x^n(x,t)u_x^n(x+\varepsilon_n y,t)\Big)\ .$$

The sequence $\varepsilon_n u_x^n(x,t)$ is bounded.

After we note that $\varepsilon_n u_x^n(x+\varepsilon_n y,t) = \big(u^n(x+\varepsilon_n y,t)\big)_y$, we get

$$\varepsilon_n\Big(u^n(x,t)u^n(x+\varepsilon_n y,t)\Big)_{xy}$$

and we can apply the same idea a second time

Finally, we obtain a partial differential equation for Γ_2:

$$\frac{\partial}{\partial t}\Gamma_2(x,y,t) + \frac{\partial^2}{\partial x \partial y}\Gamma_2(x,y,t) + \cdots .$$

At this point we could have done the same with three-point correlation.

From the approximation point of view, it may be preferable to approach correlations, which are more classical objects to handle, and although one has not yet defined suitable microlocal objects that could describe trilinear or more general multilinear effects, one can always define correlations if the corresponding L^p bounds are available, and obtain equations that they satisfy, as in the following computations, done with Patrick GÉRARD. If the coefficients $b_k, k = 1, \ldots, N$, are of class C^1 and real and if $u^{(m)}$ satisfies the equation

$$\frac{\partial u^{(m)}}{\partial t} + \sum_{k=1}^{N} b_k \frac{\partial u^{(m)}}{\partial x_k} + c\,u^{(m)} = 0 \text{ in } \Omega \times \langle 0,T\rangle,$$

and $u^{(m)}$ defines the semiclassical measure μ in $\Omega \times \langle 0,T\rangle$, one can deduce an equation satisfied by μ from the equation satisfied by the correlation function Γ: one considers the equation evaluated at $\mathbf{x} + \varepsilon_m \mathbf{z}$ and multiplied by $\overline{u^{(m)}}$ evaluated at $\mathbf{x}$, and one adds the complex conjugate of the equation evaluated at $\mathbf{x}$ and multiplied by $u^{(m)}$ evaluated at $\mathbf{x} + \varepsilon_m \mathbf{z}$, and letting ε_m tend to 0, one finds that Γ satisfies the equation

$$\frac{\partial \Gamma}{\partial t} + \sum_{k=1}^{N} b_k \frac{\partial \Gamma}{\partial x_k} + \sum_{k,l=1}^{N} \frac{\partial b_k}{\partial x_l} z_l \frac{\partial \Gamma}{\partial z_k} + 2\mathsf{Re}\,(c)\Gamma = 0 \text{ in } \Omega \times \langle 0,T\rangle \times \mathbf{R}^N,$$

and therefore μ, its FOURIER transform in the variable $\mathbf{z}$, satisfies the equation

$$\frac{\partial \mu}{\partial t} + \sum_{k=1}^{N} b_k \frac{\partial \mu}{\partial x_k} - \sum_{k,l=1}^{N} \frac{\partial b_k}{\partial x_l} \frac{\partial(\xi_k \mu)}{\partial \xi_l} + 2\mathsf{Re}\,(c)\mu = 0 \text{ in } \Omega \times \langle 0,T\rangle \times \mathbf{R}^N,$$

and if one denotes $P(\mathbf{x},\boldsymbol{\xi}) = \sum_{k=1}^{N} b_k \xi_k$, and one identifies the POISSON bracket $\{P,\mu\}$ among the terms, one finds the same equation that I had derived for H-measures

$$\frac{\partial \mu}{\partial t} + \{P,\mu\} + (2\mathsf{Re}\,(c) - \mathsf{div}\,\mathsf{b})\mu = 0 \text{ in } \Omega \times \langle 0,T\rangle \times \mathbf{R}^N.$$

At this point, I should warn of a dangerous trap: as Patrick GÉRARD has noticed, although all the semiclassical measures satisfy the same equation independently of what sequence ε_n has been chosen, it does not prove that the H-measures do satisfy that equation; indeed there are situations where for every sequence ε_n the information is either lost at zero or at infinity, and therefore H-measures cannot even be deduced from the knowledge of all semiclassical measures, for all sequences tending to zero.

If in the equation for $u^{(m)}$ one then adds a term $-\varepsilon_m^2 \frac{\partial}{\partial x_i}\Big(A_{ij}\frac{\partial u^{(m)}}{\partial x_j}\Big)$ with $\mathbf{A}$ Hermitian and continuous, the equation for Γ will contain a new term $2\sum_{i,j=1}^{N} A_{ij}\frac{\partial^2 \Gamma}{\partial z_i \partial z_j}$, and the equation for the semiclassical measure μ will contain a new term $8\pi^2\Big(\sum_{i,j=1}^{N} A_{ij}\xi_i\xi_j\Big)\mu$. It seems easier then to approximate the two-point correlation Γ, and obtain a discrete version of the equation that it satisfies, than to approximate the semiclassical measure μ itself; actually for what concerns three-point correlations, one can define an analogue of Γ but one does not know how to define an analogue of μ: if $u^{(m)}$ tends to 0 weakly in $L^3_{loc}(\Omega)$, then one can extract a subsequence such that

$$u^{(m)}(\mathbf{x}+\varepsilon_m\mathbf{z}_1)u^{(m)}(\mathbf{x}+\varepsilon_m\mathbf{z}_2)u^{(m)}(\mathbf{x}+\varepsilon_m\mathbf{z}_3) \rightharpoonup G_3(\mathbf{x};\mathbf{z}_1,\mathbf{z}_2,\mathbf{z}_3)$$

vaguely in $\mathcal{M}(\Omega\times\mathbf{R}^N\times\mathbf{R}^N\times\mathbf{R}^N)$, and G_3 satisfies

$$\sum_{j=1}^{3}\frac{\partial G_3}{\partial \mathbf{z}_j}=0, \text{ i.e. } G_3(\mathbf{x};\mathbf{z}_1+\mathsf{h},\mathbf{z}_2+\mathsf{h},\mathbf{z}_3+\mathsf{h}) \text{ is independent of } \mathsf{h}.$$

If $u^{(m)}$ satisfies

$$\frac{\partial u^{(m)}}{\partial t} - \varepsilon_m^2\triangle\, u^{(m)} = 0 \text{ in } \Omega\times\langle 0,T\rangle\,,$$

then G_3 satisfies

$$\frac{\partial G_3}{\partial t} - \sum_{i\neq j}\frac{\partial^2 G_3}{\partial \mathbf{z}_i\partial \mathbf{z}_j} = 0 \text{ in } \Omega\times\mathbf{R}^N\times\mathbf{R}^N\times\mathbf{R}^N.$$

Let us now consider another example: the wave equation in N space dimensions, which reads:

$$u^n_{tt} - \Delta u^n = 0\,.$$

For any sequence ε_n converging to zero from above, the semi-classical measure μ_{sc} satisfies a partial differential equation in $(\mathbf{x},t)$ and $(\boldsymbol{\xi},\tau)$. We would like to deduce that the H-measure satisfies the same equation.

Unfortunately, this cannot be done. Patrick GÉRARD has shown that the knowledge of the semi-classical measure corresponding to arbitrary sequences $\varepsilon_n \searrow 0$ is not enough to determine the H-measure.

With the semi-classical measure it is not clear that we may find a sequence ε_n such that the information will not go either to zero or to infinity.

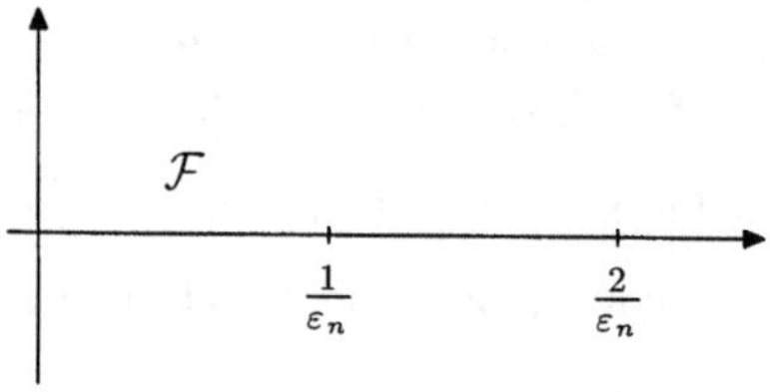

Figure 9. In FOURIER space, it may happen that for every sequence ε_n the integral of $|\mathcal{F}u_n|^2$ integrated from $\frac{1}{\varepsilon_n}$ to $\frac{2}{\varepsilon_n}$ tends to 0, and semiclassical measures are not the right tool in that case.

Patrick GÉRARD and Eric LEIGHTMANN [20] have studied the sequence u_n of normalised eigenfunctions of the Laplacian:

$$-\Delta u_n = \lambda_n u_n ,$$

with ε_n related to the eigenvalues λ_n by $\frac{1}{\varepsilon_n^2} = \lambda_n$.

For a more complicated problem it is not clear what kind of object will appear.

5.4 Semi-classical measures: Two characteristic lengths

The motivation for considering two characteristic lengths comes from the study of domains in micromagnetism, where it is not known in advance what the characteristic lengths are. They are related to the thickness of domain walls and the size of domain. One faces the problem of its determination.

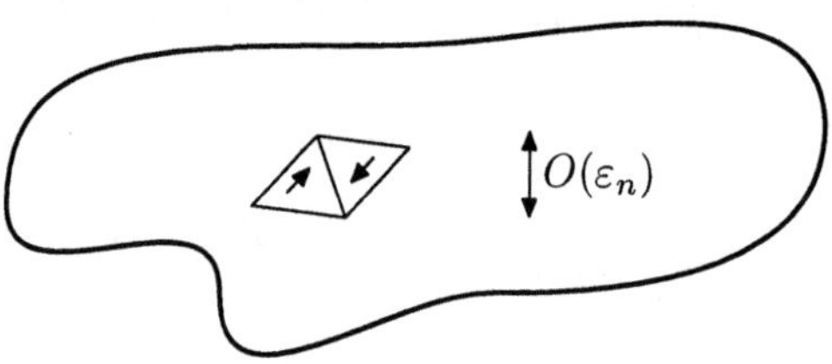

Figure 10. A wall of characteristic size ε_n may separate two regions having a different orientation of the macroscopic spin m.

The fundamental equations are:

$$\begin{cases} \mathsf{B} = \mathsf{H} + \mathsf{m} \\ |\mathsf{m}| = 1 \\ \operatorname{div} \mathsf{B} = 0 \,, \end{cases}$$

where the magnetisation m is zero outside of the material occupying the domain Ω.

The energy functional reads

$$\int_{\mathbf{R}^3} \frac{|\mathsf{H}|^2}{2} d\mathbf{x} + \int_\Omega \varphi(\mathbf{x}, \mathsf{m})\, d\mathbf{x} + \varepsilon_n^2 \int_\Omega |\nabla \mathsf{m}|^2 d\mathbf{x}\,,$$

and ε_n is related to the size of domain walls.

Physicists usually take small samples, with only one domain (for example, in magnetic tapes).

As a model, consider the one-dimensional problem of minimisation of

$$\int_0^L \Big((1-(u')^2)^2 + \frac{u^2}{a^2} \Big) dx\,,$$

over $u \in W_0^{1,4}([0, L])$ (i.e. $u(0) = u(L) = 0$).

There is no minimiser for this problem in the above class, but only for the relaxed problem (in YOUNG measures).

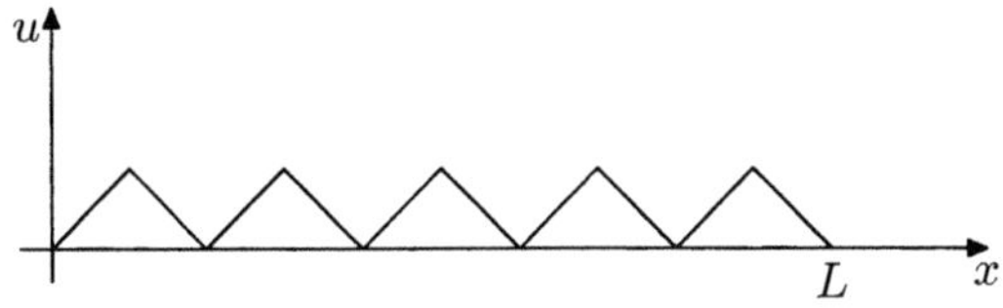

Figure 11. Sawtooth function as an approximate minimiser for the above nonconvex minimisation problem.

We have two length scales: a and L, neither of which is small.

John BALL has asked what will happen if we regularise the problem by adding another term $\varepsilon^2 (u'')^2$ to the integrand? This introduces a new scale ε. I learned of this question by Nick OWEN and later Stefan MÜLLER [32] provided an answer concerning the periodicity of the minimiser. For each jump of u' we pay a price.

What happens for $v = u'$?

$$\min \int_{-\infty}^{\infty} (\varepsilon^2 (v')^2 + (1 - v^2)^2)\, dx\,,$$

where by scaling we can replace ε by 1, and have that $v(\pm\infty) = \pm 1$; then the minimum is $K\,\varepsilon$, where K is the minimum of $\int_{-\infty}^{+\infty} \big((v')^2 + (1-v^2)^2\big)\, dx$ when $v(\pm\infty) = \pm 1$, obtained for $v = \tanh$. Then one looks at

$$\min \int_0^L \frac{u^2}{a^2} dx + NK\varepsilon\,,$$

$u' = \pm 1$, and there are N jumps with $\varepsilon^{\frac{1}{3}} a^{\frac{2}{3}}$ being the size of the domains.

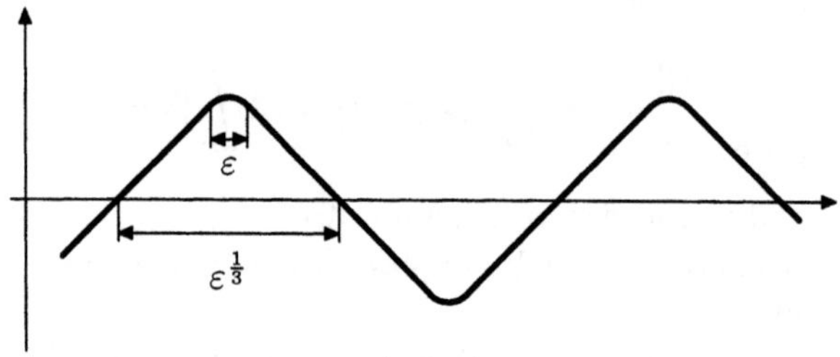

Figure 12. Regularisation of sawtooth function with precise asymptotics of walls and domains.

Semi-classical measures are not good for this problem, as they cannot handle more than one characteristic length.

6 LECTURE of Saturday, September 9

6.1 Variants of H-measures: Two characteristic lengths

We continue the discussion of semiclassical measures from the last time. These measures were also called WIGNER measures by Pierre-Louis LIONS and Thierry PAUL [29].

So, yesterday we devoted most of the time to the problems with one characteristic length, where we were looking at the sequences of the form $\mathcal{F}u(\mathbf{x}, \frac{\mathbf{x}}{\varepsilon_n})$. Today we shall describe a problem where you have two characteristic lengths.

In the two-phase transition problems (linked to the names of John CAHN and John HILLIARD), where you get some domain of the size $\varepsilon^{\frac{1}{3}}$ and some walls, where the phase changes, which are of the size ε. Knowing that, what will happen if you use the approach of semiclassical measures in order to find ε_n.

And the fact is, in some more complicated problems, like electromagnetism, you do not really know what you should expect for the size of the characteristic length.

Realistic problems often contain more than one characteristic length, and some may contain an infinity of them. Therefore it is useful to see how the preceding approach fails, by looking at the following computation, which I did with Patrick GÉRARD, and where we took a simplified example. On the domain $[0,1]$ consider a sequence of periodic functions u_n of period $\frac{1}{n}$

$$u_n(x) = \begin{cases} \sqrt{n} & \text{, if } \frac{k}{n} \le x < \frac{k}{n} + \frac{1}{n^2} \\ 0 & \text{, otherwise} \end{cases}$$

$(k = 0, \ldots, n-1)$. However, there is another characteristic length $\frac{1}{n^2}$ here, and the height of function is $\sqrt{n}$; where, of course, $n \longrightarrow \infty$.

If you look at u_n^2, you have n rectangles of the size n times $\frac{1}{n^2}$, thus the integral is 1, and the sequence converges in the sense of distributions (or vaguely as bounded RADON measures, taking continuous test functions) to 1. On the other hand, this convergence is not valid in L^1 weak (i.e. for bounded measurable test functions). This is a typical counter-example when you have to show the difference between a sequence which is bounded in L^1, converging when multiplied by a continuous test function, but not converging when multiplied by a L^∞ test function (it does not satisfy the equi-integrability property of DUNFORD-PETTIS's theorem).

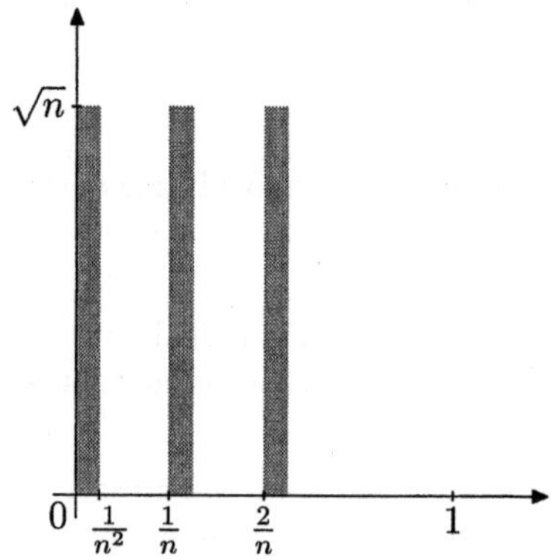

Figure 13. Graph of periodic function u_n, with $||u_n||_{L^2([0,1])} = 1$. The sequence u_n contains two scales, $\frac{1}{n}$ and $\frac{1}{n^2}$.

On the other hand u_n is bounded in L^2, and it converges weakly to zero in L^2. However this is not the point of interest here, but the fact that this sequence obviously contains two scales $1/n^2$ and $1/n$, and the question is to guess what the semiclassical measure would be, depending upon the choice of ε_n. With the intuition behind the definition of semiclassical measures, we expected to observe the following five cases.

1. If $\frac{1}{n} \ll \varepsilon_n$, the information is lost at infinity, and we are not able to analyse this sequence.
2. If $\varepsilon_n = O(\frac{1}{n})$, we expect to lose something at infinity, and find a nonzero semiclassical measure
3. If $O(\frac{1}{n^2}) \ll \varepsilon_n \ll O(\frac{1}{n})$, we expect that some of the information will be lost at zero and some of the information will be lost at infinity.
4. If $\varepsilon_n = O(\frac{1}{n^2})$, we expect to find a nonzero semiclassical measure and lose some information at zero.
5. If $\varepsilon_n \ll O(\frac{1}{n^2})$, we expect to lose all information at zero.

This is a guess based on the intuition gathered from a previous analysis. But it is not true. This is not what really happens. When we computed the various semiclassical measures, we observed the following:

For 1 and 5, the picture is correct.

The true picture in the second and the third case is that everything is lost at infinity.

For 4, there is nothing lost at zero.

What is wrong is that the thing you expect to see is what corresponds to the scale $\frac{1}{n}$, but it is not there. It is not at the place you expect. But this is a piece of a function which has the period $\frac{1}{n}$, and it is difficult to claim that $\frac{1}{n}$ is not one of the length scales of the sequence. What happens is the following: we will expect in this picture that if you look at the FOURIER transform of φu_n, in order to localise it:

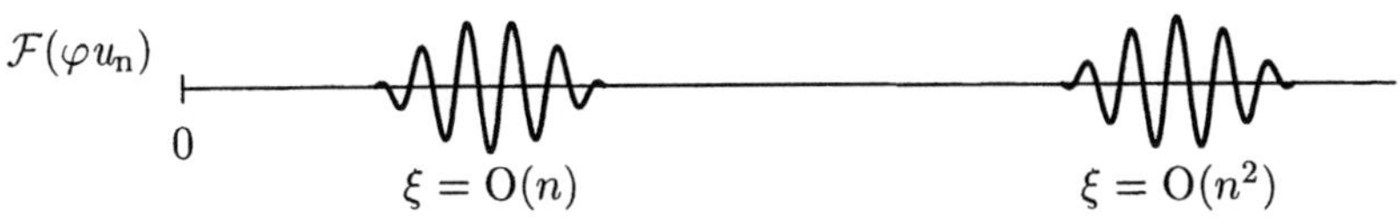

Figure 14. Expected scales in the FOURIER space.

We expect two different pieces: one for $\boldsymbol{\xi} = O(n)$, and one for $\boldsymbol{\xi} = O(n^2)$. Because you should have the scale n and n^2 in the FOURIER transform space.

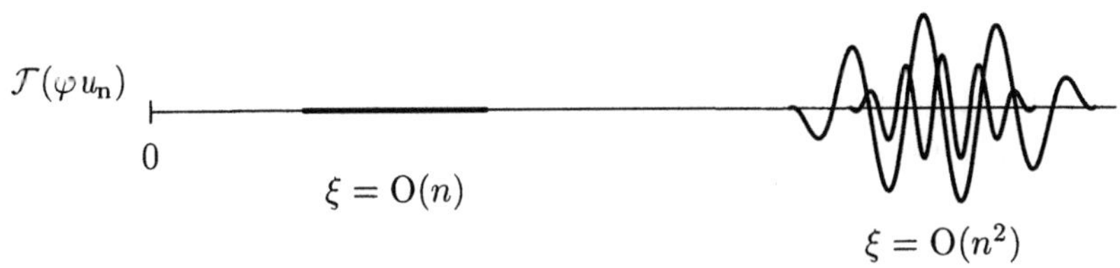

Figure 15. The actual situation in the FOURIER space.

And what I am saying is that there is nothing at the scale $\boldsymbol{\xi} = O(n)$. The scale $O(n)$ is in the FOURIER transform, but it is just not there. What happens there? The scale $\frac{1}{n}$ from the real space is at the scale n^2 in the FOURIER space. What happens is not that you do not see the scale corresponding to $\frac{1}{n}$, you do not see it at $\boldsymbol{\xi} = O(n)$, i.e. at distance $O(n)$ from zero. The scale of order $\frac{1}{n}$ is at distance $O(n^2)$ from zero.

This is a common fact you learn in physics, the simplest example is related to the tuning of pianos. When you tune a piano, you have a diapason (a tuning fork), which vibrates at a given frequency, e.g. 450 Hz for low A in the middle of the piano. You vibrate the fork and the piano wire at the same time, and if the frequency is slightly different, say 451 Hz for the piano wire, you hear some strange humming sound, somewhere at frequency one, but for this to occur one must put the diapason in contact with the piano, for creating a coupling between the two oscillations. Then you have to change the tension of piano wire, until you get it right.

This relies on the fact that

$$\cos ax - \cos bx = -2\sin\frac{a-b}{2}x\sin\frac{a+b}{2}x\,,$$

or

$$\cos ax + \cos bx = 2\cos\frac{a-b}{2}x\cos\frac{a+b}{2}x\,,$$

and you hear the difference of two periods, but multiplied by something.

If you look at the FOURIER transform of the product, you see the sum and the difference. This is what the physicists call *beats*.

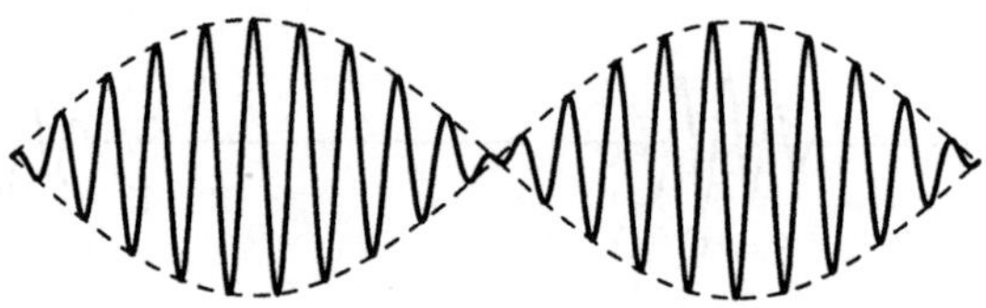

Figure 16. The classical picture of *beats*, as found in Physics books.

What happens in our example above is just that. And why there is a difference? Because the two scales interact. So, if I look at the function

$$f(x/\varepsilon_n) + g(x/\delta_n)$$

and apply the FOURIER transform, I see two things: $\mathcal{F}f$ at the distance $O(\frac{1}{\varepsilon_n})$, and $\mathcal{F}g$ at $O(\frac{1}{\delta_n})$. We assume that $\delta_n \ll \varepsilon_n$.

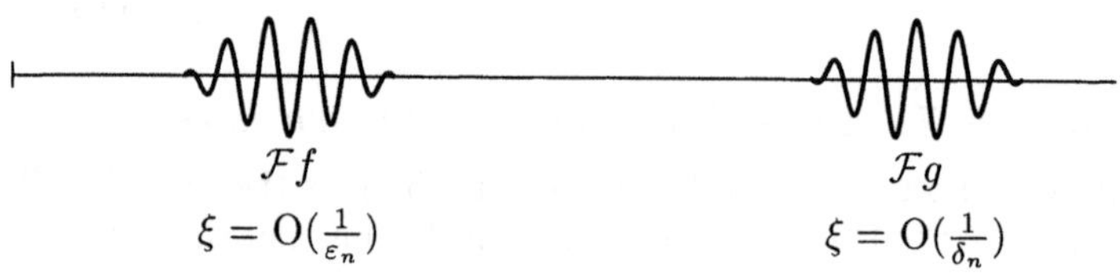

Figure 17. The FOURIER space image of two different scales.

But, if you look at the product, you lose everything at exact scales, but the two scales get mixed. This explains to me something which I had not understood when I learned Physics. When you learn about spectroscopy, you are told that the light rays you see (which were discussed yesterday in Rainer WEHRSE's talk) are due to the interactions with the electrons. But why do you not see an effect created by protons? I expected to see both the frequency of the electrons and the frequency of protons. But you do not see it.

You do not see it just because of this phenomenon. The electrons interact with the protons, and the information gets mixed. Therefore, what you learn here is that the idea to analyse such a situation is not to use two semiclassical measures, like one for the scale ε, and one for the scale $\varepsilon^{\frac{1}{3}}$; you have to use only the smaller scale. This scale interacts with the other, because it is not something which oscillates with size ε. That we see everywhere. You see probably some transition period. Therefore these scales interact. If they interact, then you have this kind of phenomenon.

There is a need for different measures, which will have some hierarchy of scales: the *hierarchical microlocal measures*. They have not yet been introduced, and there are no important theorems about them. But the above example shows that you cannot consider this model with only one characteristic length, to analyse sequences which have more than one characteristic length. You can do it, but you are going to miss something.

We realised that we still need more variants in order to attack this kind of problems.

6.2 Another look at problem modelled on NAVIER-STOKES' system

A few years ago I realised that we needed another type of variants, and I was looking at the problem which I have described earlier in the static case.

Recall (now we add the time derivative, and we should supply some initial conditions as well; for more details cf. [47]):

$$\begin{cases} \dfrac{\partial \mathsf{u}^n}{\partial t} - \nu \triangle \mathsf{u}^n + \mathsf{u}^n \times \mathsf{rot}\,(\mathsf{v}_0 + \mathsf{w}^n) + \nabla p^n = \mathsf{f} \\ \qquad\qquad\qquad\qquad\qquad\qquad\qquad \mathsf{div}\,\mathsf{u}^n = 0\,. \end{cases}$$

It was explained earlier that a term with **Mu** appears in the effective equation for the stationary case. The answer can be expressed in the terms of the H-measure associated to the sequence v_n. I have also done the evolution case, and reached the similar result. I have done it in $\mathbf{R}^3$, as we do need estimates on the pressure.

It is a technical problem what are the right boundary conditions for the pressure, but if you work in the whole space, there are good bounds. I was told later by Wolf VON WAHL that he has written a paper (cf. [60]) where he obtained the estimates on the pressure for every domain Ω. But, when I wrote that in 1984, I needed estimates on the pressure. I found something similar, that there is another term which appears, except I was not able to show that this term is a symmetric matrix (i.e. that $\mathbf{M}^\top = \mathbf{M}$). I wanted to check this with the tool of H-measure, and I did the computation with Chun LIU and Konstantina TRIVISA, who were postdocs at Carnegie Mellon University, and we tried to see whether we shall obtain the same answer. And the answer is no, it requires a different variant of H-measures.

Before we do that, see that here we are putting oscillations in the system (forcing it), and, of course, these are oscillations in $\mathbf{x}$ and t. In order to create oscillations in $\nabla \mathsf{u}$, you need some kind of forcing. But, the equation is parabolic (the principal part is $\partial_t - \triangle$), so it has some regularising effect. If you start from initial data with oscillations, they will disappear in time. However, if you force them all the time, the older will be disappearing, while the newer will still be there. That will force oscillations in the gradient of u. And these are the ones you have to be more careful with.

If you do the FOURIER transform (now in both $\mathbf{x}$ and t, so you have $\boldsymbol{\xi}$ and τ in the dual), what you find is that $|\tau| \sim |\boldsymbol{\xi}|^2$. The scaling after the FOURIER transform is different.

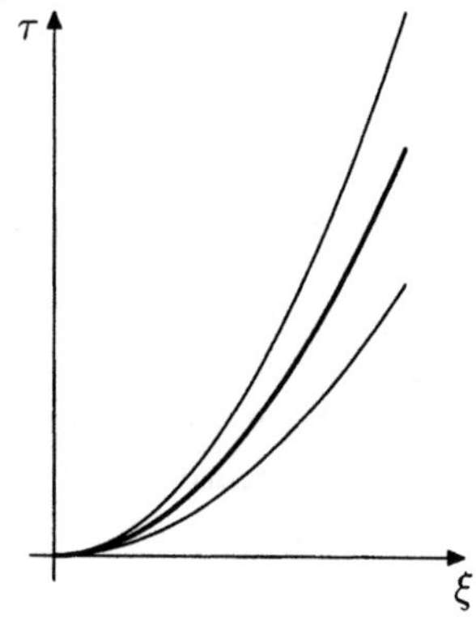

Figure 18. The proper scaling in the Fourier space for the equation with principal part $\partial_t - \Delta$.

What you do is to localise your function u^n by φ and take the Fourier transform $\mathcal{F}(\varphi \mathsf{u}_n)$. You know that the information goes to infinity, but we are going to make a difference between $\boldsymbol{\xi}$ and τ.

It goes to infinity, but I am not interested any more in looking at what happens in the directions on the unit sphere, i.e. in points of the form:

$$\left(\frac{\boldsymbol{\xi}}{\sqrt{|\boldsymbol{\xi}|^2+\tau^2}}, \frac{\tau}{\sqrt{|\boldsymbol{\xi}|^2+\tau^2}}\right),$$

which was fine in the case where $\boldsymbol{\xi}$ and τ were similar as for the wave equation or first order equation. We have to go to infinity preserving the relation $|\tau| \sim |\boldsymbol{\xi}|^2$. Here we have to take the directions on a different surface, i.e. the points of the form

$$\left(\frac{\boldsymbol{\xi}}{\sqrt{|\boldsymbol{\xi}|^2+\tau}}, \frac{\tau}{|\boldsymbol{\xi}|^2+\tau}\right).$$

Consequently, for the test function we shall no longer take

$$\psi\left(\frac{\boldsymbol{\xi}}{\sqrt{|\boldsymbol{\xi}|^2+\tau^2}}, \frac{\tau}{\sqrt{|\boldsymbol{\xi}|^2+\tau^2}}\right),$$

but

$$\psi\left(\frac{\boldsymbol{\xi}}{\sqrt{|\boldsymbol{\xi}|^2+\tau}}, \frac{\tau}{|\boldsymbol{\xi}|^2+\tau}\right).$$

By this we shall achieve that

$$\psi(\lambda\boldsymbol{\xi}, \lambda^2\tau) = \psi(\boldsymbol{\xi}, \tau),$$

which is the right scaling in this situation.

This would be a way to understand how do the oscillations relate, and then you should do the commutation lemma for $[M_b, P_a]$ to be compact, except that the function a is constant on a parabola, and not on rays. Looking at the proof one can convince oneself that it is exactly the same.

Therefore you have a variant here, which is forced on you because the equation is different. And the oscillations in $\mathbf{x}$ and t, when you look what they do, they tend to go to infinity, along some path, which is different, as $\boldsymbol{\xi}$ and τ do not play the same rôle.

This was an important observation.

6.3 Relations between YOUNG measures and H-measures

One important question, which I have already mentioned, but I have not got time to explain yet, is what are the mutual relations between YOUNG measures and H-measures, or their variants (some missing details could be found in [54]).

By now we have seen that, in order to analyse the oscillations, we might have to use something which was not exactly in my definition, or in that of Patrick GÉRARD, but something that is in the same spirit, and adapted to the problem. There are the examples where this occurs, and in 1991 I improved some remark that I had done earlier with François MURAT (see [35]), and which is related to a problem in micromagnetics, the field on which Irene FONSECA has talked about earlier here.

There was some work by Richard D. JAMES and David KINDERLEHRER [24], but they had not really attacked the technical problem, so I repeat quickly the equations

$$\begin{cases} \mathsf{B} = \mathsf{H} + \mathsf{m} \,, \\ \mathsf{H} = \nabla u \,, \end{cases}$$

where $|\mathsf{m}|$ is constant, even though it is supposed to depend on temperature, at least according to physicists, and we take that constant to be 1.

So, the scalar magnetic potential u satisfies

$$\mathsf{div}\,(\nabla u + \chi_\Omega \mathsf{m}) = 0 \,,$$

where χ_Ω appears just for us to remember that m exists only in the material, which occupies the domain Ω. This equation is in $\mathbf{R}^3$, and the energy is an integral, consisting of the following parts:

1. $\int_{\mathbf{R}^3} |\nabla u|^2 d\mathbf{x}$, which is the *magnetic energy*,
2. $\int_\Omega \varphi(\mathbf{x}, \mathsf{m})\, d\mathbf{x}$, the *anisotropy part* (depending on the direction of m),
3. $-\mathsf{H}_e \cdot \mathsf{m}$, *exterior field energy*, and
4. $\varepsilon^2 \int |\nabla \mathsf{m}|^2$, which is the *exchange energy*.

I was interested in understanding what happens when ε tends to zero.

For $\varepsilon > 0$ this is a compactness problem, and the existence is easy, using the compactness in m. And u is just a function of m.

The important thing that I observe is that the mapping which associates ∇u to m is given by a pseudodifferential operator in the class which I have considered (the equation is with constant coefficients).

The sum of the above energies I denote by $J_\varepsilon(\mathsf{m})$, and look for the minimising sequences. Having one such sequence, I should see what happens in each of the terms.

So, we take a sequence m^n which converges to m^∞ in L^∞ weak $*$, but now we know that the limit satisfies only $|\mathsf{m}^\infty| \leq 1$. The weak limits take values in convex hulls; for the sphere the convex hull is the whole ball.

Then $\varphi(\mathbf{x}, \mathsf{m}^n)$ goes to a different limit l_φ, and if you know the Young measure associated to the sequence (m^n) you know that l_φ. But the Young measure knows a lot too much, as it knows the limit for any function of m, and here we only need four functions, three components of m and the function φ. That is in the case where φ does not depend on $\mathbf{x}$, otherwise it is a bit more complicated.

Thus, the energy term with φ can certainly be handled by Young measures, but you do not need all of it. This is something useful to realise, because it is quite a technical task to approximate Young measures. However, it is not clear that you really need the complete Young measure. Maybe something else will be enough. Anyhow, Young measures will take care of that term.

But they will not help us in understanding the limit of $|\nabla u^n|^2$. Of course,

$$\lim \int |\nabla u^n|^2 \geq \int |\nabla u^\infty|^2 ,$$

where u^∞ is obtained by the same relation from m^∞. It is not enough to know that the limit is bigger. You want to know how much bigger it is, in order to do something quantitative.

Therefore you want to understand the extra quantity, and as ∇u is obtained by a pseudodifferential operator from m, you need a product of two pseudodifferential operators, which can be computed by H-measures. So, we have:

$$\int_{\mathbf{R}^3} |\nabla u^n|^2 d\mathbf{x} \longrightarrow \int_{\mathbf{R}^3} |\nabla u^\infty|^2 d\mathbf{x} + \int_{\Omega \times S^2} \boldsymbol{\xi} \otimes \boldsymbol{\xi} \, d\boldsymbol{\mu}(\mathbf{x}, \boldsymbol{\xi}) .$$

The second term on the right hand side is nonnegative, as $\boldsymbol{\mu}$ is Hermitian and nonnegative, and it is zero if $\mathsf{div}\, \mathsf{m}^n$ converges strongly in $\mathrm{H}^{-1}_{\mathrm{loc}}(\Omega)$. Therefore, we have a way to express the limit energy, and can define a relaxed problem.

The control space for this relaxed problem consists of pairs $(\nu, \boldsymbol{\mu})$, where ν is a Young measure, while $\boldsymbol{\mu}$ is an H-measure, both being associated to the sequence satisfying the constraint $|\mathsf{m}^n| = 1$.

On this set of pairs we keep the same equation, but now in the functional we have another term

$$\int_\Omega \int_{\mathbf{R}^3} \varphi(\mathbf{x}, \mathsf{m})\, d\nu_{\mathbf{x}}(\mathsf{m})\, d\mathbf{x} + \int_{\Omega \times S^2} \boldsymbol{\xi} \otimes \boldsymbol{\xi}\, d\boldsymbol{\mu}(\mathbf{x}, \boldsymbol{\xi}) \,.$$

You know that this problem has a solution, because you have just increased the class of solutions in the right way. At this level we are slightly troubled, as we have no idea what are the pairs. Which are the admissible pairs, and which are not? Admissible pairs have not yet been completely characterised.

It is important to note, however, that you do not need to know all the pairs. You need to be able to construct the good ones.

We are going to minimise, and the lemma is the following.

Lemma. *Let $(\nu, \boldsymbol{\mu})$ be an admissible pair. Then there is another admissible pair $(\nu, \boldsymbol{\mu}_0)$ such that*

$$\int_{\Omega \times S^2} \boldsymbol{\xi} \otimes \boldsymbol{\xi}\, d\boldsymbol{\mu}_0(\mathbf{x}, \boldsymbol{\xi}) = 0 \,.$$

So, by keeping the same proportion, while changing the shape of the pieces, you keep the same YOUNG measure, but you change the H-measure.

What is important is that for a given ν you are trying to minimise this quantity. And, of course, it is nonnegative. If you find one for which it is zero, you can be sure that you cannot do any better.

What does that mean? That means that when you have a microstructure, the domain is divided into small pieces, the YOUNG measure seeing only the proportions. But the H-measure sees something about the interfaces, or where they are.

We cannot construct all the H-measures which are allowed, but we can construct enough of them, and in particular the one, $\boldsymbol{\mu}_0$, which minimises the integral in the Lemma.

We say that among all the microstructures which have the same proportion locally, we can change their shape. For example, you might decide to start by a layered material in the direction $\boldsymbol{\xi}$. For such a material, you will have a DIRAC mass for the H-measure in the direction $\boldsymbol{\xi}$ which is perpendicular to the layers. If you turn the layers, you move $\boldsymbol{\xi}$ on the sphere. Therefore, you may change the sequence in such a way that you do not change the local proportion. You cut the small pieces and then you rearrange their directions; the size of the pieces tends to zero, so the YOUNG measure will not be changed. You want to do this in an efficient way, for the H-measure. So, if there is one that satisfies that you cannot do better for your functional, because this term is nonnegative, and the best you can do is to put it down to zero.

This was just the first relaxed problem. Now we are going to describe the second relaxed problem, which is related to ν. This one is easy, as it consists in constructing the convex envelope for the function φ, and describing what is going on.

Remark. The effective energy becomes:

$$\inf_{\nu} \int_{\mathbf{R}^3} |\nabla u^\infty|^2 d\mathbf{x} + \int_\Omega \int_{\mathbf{R}^3} \varphi(\mathbf{x}, \mathsf{m})\, d\nu_{\mathbf{x}}(\mathsf{m})\, d\mathbf{x} - \int_\Omega \mathsf{h}_e \cdot \mathsf{m}$$
$$= \int_{\mathbf{R}^3} |\nabla u^\infty|^2 + \int_\Omega \varphi^{**}(\mathsf{m}) - \int_\Omega \mathsf{h}_e \cdot \mathsf{m} \, .$$

So, either there is a relaxed problem for which you might have some m^∞ inside, and this analysis says that there are some microstructures for which you do not pay here, and this is equivalent to the fact that the divergence of m^n converges strongly in $\mathrm{H}^{-1}_{\mathrm{loc}}(\Omega)$.

And this corresponds to what physicists always want to do, they always want to kill the magnetic field outside, and this can be done explicitly in some cases, but not always.

What JAMES and KINDERLEHRER have done was to more precisely reinterpret what physicists have already done, in terms of oscillations of u^n. They could have only done that for $\mathsf{H}_e = 0$, where, of course, the solution of the relaxed problem is zero, and therefore you want to create it by alternating up and down, so that the limit is zero. That is the way how they have done it.

6.4 Small amplitude Homogenisation (revisited)

That was a slight improvement of something I have noticed with François MURAT long time before, that if you have a sequence of characteristic functions $\chi_n \rightharpoonup \theta$ (weakly $*$ in L^∞, for example), then you look at $u_n := \chi_n - \theta$, and associate to it (or to its subsequence) an H-measure μ. Then the integral in $\boldsymbol{\xi}$;

$$\int_{S^{N-1}} d\mu(\cdot, \boldsymbol{\xi}) = \lim u_n^2 = \theta(1-\theta) \, .$$

This is the only constraint necessary, so that for a given H-measure μ with this property there is a sequence of characteristic functions such that they converge weakly to θ and generate the same H-measure μ.

This is based on the small amplitude Homogenisation, and on classical constructions in Homogenisation of multilayered materials. I think that Sergio SPAGNOLO and Antonio MARINO [43] were the first to use this implicitly, and I was using it with François MURAT in the early seventies [35]. We reused it after some students, Phillipe BRAIDY and Didier POUILLOUX [9], have simplified some argument we had for confocal ellipsoids.

There are connections which are true in some cases, but wrong in some others, that you can construct everything by multiple layering. The important technical things about layering is that there is an explicit formula for layers.

We discussed it in the case where you mix isotropic conductive materials. We said that you had to do average of $\frac{1}{a}$ in the direction perpendicular to the layers. That is what you understand from the Physics.

On the other hand, if you put some anisotropic coefficient in each of the layers, **A** and **B**, with the proportions ϑ and $1-\vartheta$, then you want to say that this is equivalent to some **C**, and the formula is a bit more technical. Let the direction perpendicular to the layers be denoted by e. The formula reads:

$$\mathbf{C} = \vartheta\mathbf{A} + (1-\vartheta)\mathbf{B} - \vartheta(1-\vartheta)(\mathbf{B}-\mathbf{A})\frac{\mathsf{e}\otimes\mathsf{e}}{(1-\vartheta)\mathbf{A}\mathsf{e}\cdot\mathsf{e} + \vartheta\mathbf{B}\mathsf{e}\cdot\mathsf{e}}(\mathbf{B}-\mathbf{A})\,.$$

This shows that the result is tricky if you are mixing anisotropic materials whose eigenvectors are not in the right direction. It is not necessary that **A** and **B** are symmetric for this formula.

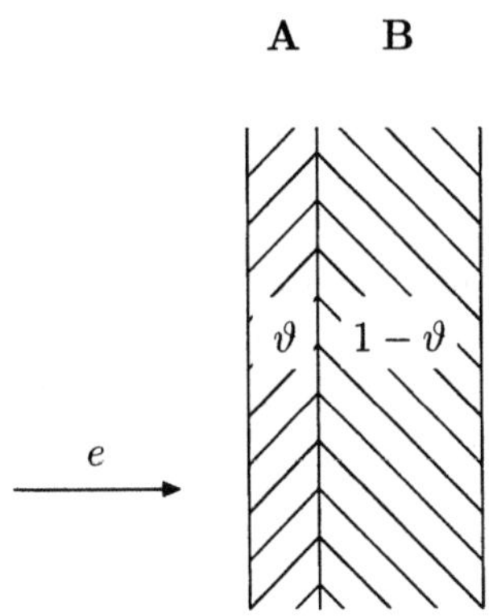

Figure 19. Multiple layers (layers within layers).

How can one use such a formula? It is awful. You can write down a similar formula for mixing n materials in such a way, and it is even more complicated. What is important is that when you do the small amplitude Homogenisation, you would denote **A** and **B** as $\mathbf{M}_i$, which works even for more than two materials, where

$$\mathbf{M}_i := \mathbf{A}_* + \gamma\mathbf{N}_i + \gamma^2\mathbf{P}_i + O(\gamma^3)\,,$$

and for the effective coefficients you would get:

$$\mathbf{A}^{\text{eff}} = \mathbf{A}_* + \gamma\left(\sum_{i=1}^{n}\mathbf{N}_i\right) + \gamma^2[\text{formula}] + O(\gamma^3)\,,$$

where [formula] denotes some known formula similar to the one obtained above in the case of two materials. What you are interested in here is not exactly that formula, but after writing down the TAYLOR expansion of **A** and

B at order two, you want to know what is the TAYLOR expansion of effective coefficients up to the same order.

You see that $\mathbf{B} - \mathbf{A}$ is going to be of order one in γ, and you know the term of order one. For the term of order zero, it is $\mathbf{A}_*$. Thus this terrible mess above disappears, and you can describe all multiple layerings at order two, and that is much easier. Then you realise that when you mix different materials, you are mostly putting the point on the unit sphere in the direction of e (which is the same as $-\mathsf{e}$), and therefore if you put in the layers some material which already corresponds to some H-measures, what you are doing is making a combination of H-measures, and adding eventually a DIRAC mass in the direction e if they do not have the same average. Because if they are of the same order in γ, this term will not be there, of course. So you may just have that, you can put on the sphere the form you want, and that is the first basic idea.

But, in order to do more general things, you have to do that in a slightly more general way, and then you interpret that by saying that what we are doing is identifying the H-measure, or actually identifying the real part of that H-measure, because you have to be aware of the fact that in all these examples that we have constructed, the H-measure always charges the point $\boldsymbol{\xi}$ and the point $-\boldsymbol{\xi}$ in the same way.

Of course, you might say that we should have defined the H-measure differently, but for the propagation, say for the wave equation, you are interested in how much information you send in the direction $\boldsymbol{\xi}$, and not both in $\boldsymbol{\xi}$ and $-\boldsymbol{\xi}$, as you always want to have the same thing in both directions. But you cannot avoid that if you take only real data for the wave equation. So, if you look at the theorems for propagation, and you prepare your solution as real sequences, then what you find at the end is that there is as much energy being sent in the direction $\boldsymbol{\xi}$ as there is in the direction $-\boldsymbol{\xi}$.

And physicists know that, that in order to see the difference, you have to do something special. This might be just a parabolic mirror, and then what you send is in a certain direction after the reflection on your mirror. Like a car headlight, sending the beam of light in front. You cannot actually send the light only in front; you have always to send light in the opposite direction also. All this is true only for linear equations; nonlinear effects, like the mirror, can make a difference between direction $\boldsymbol{\xi}$ and direction $-\boldsymbol{\xi}$.

The above analysis shows that we can construct a lot of H-measures, and you can control what are the materials that you have used.

If you have a sequence u^n of functions taking their values in a finite set $K = \{k_1, \ldots, k_p\}$, the YOUNG measure ν is going to be a sum of DIRAC masses:

$$\nu = \eta_1 \delta_{k_1} + \ldots + \eta_p \delta_{k_p} .$$

By this you can construct all those multilayered homogenised materials, identify the term with γ^2, and we deduce what kind of H-measure you can get. And then we have here to do some approximation; you have to use an argument of metrisability for bounded sequences, and at the end you have this trick, that for the same ν you can construct a lot of different sequences with different H-measures μ.

6.5 Nonlocal effects in Homogenisation

In the little time that is left, I want to mention something that is useful for the applications, even though not directly related to H-measures. But the YOUNG measures do play a rôle there.

These are the nonlocal effects which appear in Homogenisation. That is understood mostly in the linear situation.

The study of such problems started some time ago; the first results I have heard of were from the early seventies by Henri SANCHEZ-PALENCIA (see [37]). He studied different models: acoustics and electromagnetism. In the late seventies, Jacques-Louis LIONS in his course created some academic problems where the limit of a sequence of partial differential operators was not a partial differential operator.

What I realised in the late seventies actually was that there was some qualitative reason for which this was important, but not in the same problem. And the context was to understand Physics, the laws of absorption and emission of particles. This is an important piece of Physics, where one tries to apply the law of probability in nature, in order to explain spectroscopy by absorption, emission, and so on. What I have understood is that this was perhaps only a language to describe some kind of memory effect. If you have an equation, where on the right hand side you get something like

$$\int_0^t K(t-s)u(s)\,ds\ ,$$

then we are talking about memory effects.

For the existence, it is not a problem, as you can use the fixed point argument, and there are other methods as well. But physicists are usually very happy if they know a probabilistic game which creates equations like that. And they mistake the proof that such a solution can be approached in this way with the fact that this is the only way it can appear. My feeling was that this type of problem was just the sign that the effective equation has some memory effect term. Knowing that this is what happens for some first order problems, then there are some natural questions raised in it, which are not completely understood.

What happens if you look at a linear degenerate elliptic problem, like a first order equation,

$$u^n_t + \mathsf{a}^n \cdot \nabla u^n = f\ ,$$

where a^n oscillates. What you obtain is that the limit equation is different:

$$u^\infty_t + \mathsf{a}^\infty \cdot \nabla u^\infty + K * u^\infty = f\ .$$

The correction in this example is nonlocal (the memory effect).

This happens also for partially elliptic problems where the oscillations are not in the right direction, and I constructed an example with François MURAT in 1984; consider

$$-a(t)u_{xx} + b(t)u = f(x,t) \ ,$$

where a and b are positive, bounded from below, so this is an elliptic problem for any t. We put the oscillations in a and b. In the argument it is important that I do not use functions of x as coefficients, so we can use the FOURIER transform $\mathcal{F}$ in x:

$$\left(4\pi^2|\xi|^2 a^n(t) + b^n(t)\right)\mathcal{F}u^n = \mathcal{F}f \ ,$$

so that $\mathcal{F}u^n$ can be expressed explicitly as

$$\mathcal{F}u^n = \frac{\mathcal{F}f}{4\pi^2|\xi|^2 a^n(t) + b^n(t)} \ .$$

As a^n and b^n oscillate, the limit in this expression can be obtained by using the YOUNG measure of the pair, and the limit is usually not of the same form. And when you do the inverse FOURIER transform, you are going to find a different type of problem. What we did find as a limiting problem was:

$$-a^{\text{eff}}u_{xx} + b^{\text{eff}}u + H(x,t) * u = f \ .$$

The above convolution, as was the case with the FOURIER transform, was only in x. The apparition of the diffusion term here has a strange effect. Andrey PIATNITSKI lectured yesterday on something similar with probability; there were different layers, etc. I am not sure whether this is the same, as I unfortunately do not understand the probability.

It is important to realise that we can put different scales in this model. You imagine that if you go very fast, you can look at the equation along the whole line. Fluctuations in t are not rapid enough, so that you know that you get to the diffusion in x term. But then, when you make a and b to oscillate, it creates that extra term here in the direction of the layers. When the functions oscillate rapidly, in the limit you must take an average over the whole line.

Before that I studied even simpler problem, which was an ordinary differential equation

$$\partial_t u + a^n(x)u = f \ ,$$

where x is like a parameter. When you take the weak limit, it means that you mix some averages of things that come more or less independently, because we were looking at the oscillations. So there is no way that you are going to explain what is going on here by an ordinary differential equation, and the problem you get at the end is the inverse of the LAPLACE transform, which is used in solving the equation, which will give you for the effective equation

$$\partial_t u + a^{\text{eff}}(x)u - \int_0^1 K(x, t-s)u(s)\,ds = f \ .$$

We have some type of nonlocal effect, and it is important to control what kind of kernels K can appear, which was related to my idea of looking at spectroscopy in this way.

Some of my students have worked on some generalisations; Nenad ANTONIĆ [4] worked on the problem when you have a term in u_x in the elliptic case, and that forced a change of the method; Luisa MASCARENHAS [30] worked on similar problems as well.

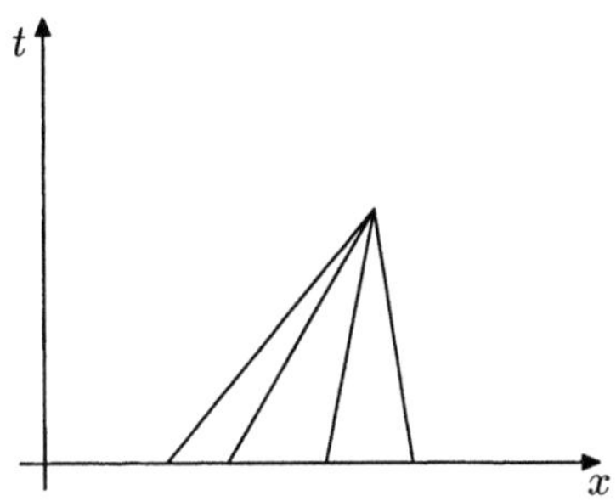

Figure 20. Past lines to point (x, t) with different speeds λ.

The most interesting was the work of Youcef AMIRAT, Kamel HAMDACHE and Abdelhamid ZIANI [3] on the transport equation

$$u_t + a^n(y)u_x = f(x, t) \ ,$$

which is related to slightly more realistic problems. They were supposed to look at the propagation with some speed, which varies in the perpendicular direction y. What you do here is to first apply the FOURIER transform in x, and then the LAPLACE transform in t, and you apply the same analysis as I did, and then the limiting equation is quite interesting:

$$u_t^\infty + a^\infty(y)u_x^\infty - \int_0^t \int_{\mathbf{R}} u_{xx}^\infty(x - \lambda(t-s), y, s)\,d\mu_y(\lambda)ds = f \ .$$

You get a nonlocal effect in x and t, where you integrate over the past lines coming with different speeds λ.

You have a measure μ_y which is a nonlinear function of the YOUNG measure of the sequence a^n, and then you integrate in s. This is very different from the earlier example which was parabolic, with infinite propagation speed. Here the propagation speed is finite. It has the scaling like tu_{xx}, which was used for conservation laws by Constantine DAFERMOS (see [12]).

You get some nonlocal term, and this is more general than partial differential equations.

This motivates applications for spectroscopy, waves in nonhomogeneous materials, seismics, and turbulence effects, which are related to studying questions like

$$\begin{cases} u_t + \mathsf{a}^n \cdot \nabla u = f \\ \operatorname{div} \mathsf{a}^n = 0 \,. \end{cases}$$

After Sergio SPAGNOLO had successfully studied sequences of elliptic problems

$$-\operatorname{div}\left(\mathbf{A}^n(\mathbf{x})\nabla(u^n)\right) = f^n$$

in Ω or parabolic problems

$$\frac{\partial u^n}{\partial t} - \operatorname{div}\left(\mathbf{A}^n(\mathbf{x},t)\nabla(u^n)\right) = f^n$$

in $\Omega \times \langle 0, T\rangle$, with suitable boundary conditions (and initial data in the parabolic case) and general bounded positive definite matrices $\mathbf{A}^n$, he stumbled on the apparently similar case of second order *wave* equations

$$\frac{\partial^2 u^n}{\partial t^2} - \operatorname{div}\left(\mathbf{A}^n(\mathbf{x},t)\nabla(u^n)\right) = f^n\,,$$

with suitable boundary conditions and initial data, for general bounded and symmetric positive definite matrices $\mathbf{A}^n$. The reason was that available theorems giving existence and uniqueness of solutions assume that the coefficients $\mathbf{A}^n$ are regular in t (the case where $\mathbf{A}^n$ is independent of t is easy); he worked then at understanding if one could define solutions without regularity hypotheses on the coefficients, but the results that he obtained, finding solutions in duals of GEVREY classes for example, are not suitable for Homogenisation.

It is actually important to study some cases of time dependent coefficients (but not arbitrary ones), in connection with classical experiments, which have resulted in the development of non classical theories, the experiments of spectroscopy. In an experiment of spectroscopy, one sends light through a gas, and therefore a natural *classical* setting for this problem would be to consider the MAXWELL system, and to study the question of multiple scattering of light on the *atoms* or *molecules* constituting this gas. This is not an easy problem, and one stumbles immediately on a first difficulty due to the fact that the *atoms* or *molecules* constituting the gas, whatever they are, are supposed to be moving, and one should therefore develop a theory for the MAXWELL system (or for the scalar wave equation as a first step), with coefficients that do vary with t, but those variations with respect to t should certainly not be arbitrary.

A first step is to consider the case of a rigid and homogeneous object traveling with fixed velocity v in an infinite and uniform medium; in that case the coefficients are of the form $\mathbf{A}(\mathbf{x} - \mathsf{v}\,t)$, and as $\mathbf{A}$ is discontinuous at the surface of the object, these coefficients do not satisfy the usual hypothesis that their derivative with respect to t is bounded, but one can easily overcome this difficulty by performing a Galilean transformation or a LORENTZ transformation, so that one is back to the case where the coefficients are independent of t. Using as a simplified model the wave equation in all $\mathbf{R}^N$, the corresponding energy estimate would come out by (formally) multiplying the equation by $u_t + \mathsf{v} \cdot \nabla u$, and the coefficients are only partially regular and such that $(\partial_t + \mathsf{v} \cdot \nabla)A$ is bounded. Of course, it is important that the velocity v stays much smaller than the characteristic velocity of the wave equation itself. If there were a large number m of some of these rigid and homogeneous objects, with various shapes, each traveling at its own constant velocity, then the existence and uniqueness of a solution would follow easily from the finite propagation speed property of the wave equation, as long as collisions between these objects would be avoided; in that case, the energy estimate would still be obtained by (formally) multiplying the equation by the same quantity, but v would be variable now.

It is an open question to find a class of coefficients depending on $(\mathbf{x}, t)$, generalising the preceding case of a finite number of moving rigid inclusions avoiding collisions, for which existence, uniqueness and uniform bounds can be obtained. Could one extend then such a theory to the case of elastic bodies, for which one might have to introduce effects of electrostriction and magnetostriction? Could one answer questions of Homogenisation for these classes of equations, and will they explain some of the effects that physicists have observed?

When I started studying memory effects, one reason was that I had guessed that the absorption and spontaneous emission effects that physicists had invented, in order to explain what they had observed in experiments of spectroscopy, were precisely the sign of an effective equation having a memory effect; I could not tell what equations one should consider for describing these phenomena, but as a first step towards studying the propagation of waves in a material with moving objects in it, I had chosen to study first order hyperbolic equations with variable coefficients.

Part of the challenge is that one has no idea about what these moving objects are, as most of what physicists claim about particles, atoms or molecules, result from some dogmatic theory like Quantum Mechanics, and I think that some of these dogmas, which had been thought natural at some time, are just mistakes on the path of discovery, and that they will be replaced by more realistic ones, probably after a more complete mathematical theory of microstructures in solutions of partial differential equations will be developed. Anyway, I hope that for performing the Homogenisation programme that I have sketched, it will not be necessary to know exactly what these objects are. The concept of H-measures permits to compute in a more accurate way some effective coefficients and it also permits to describe the evolution of some microstructures in solutions of hyperbolic systems in a way which explains why these microstructures can behave like *particles*. After I had successfully created this concept, Graeme MILTON had pointed out to me that in order to compute scattering coefficients I needed information on three-point correlations, while H-measures are only related to two-point correlations, and therefore his feeling was that one needs a more precise mathematical tool to describe questions about spectroscopy, and it is necessary to go beyond H-measures.

References

1. Grégoire ALLAIRE: *Homogenization and two-scale convergence*, *SIAM Journal of Mathematical Analysis* **23** (1992) 1482–1518
2. Grégoire ALLAIRE, François MURAT: *Homogenization of the Neumann problem with nonisolated holes*, *Asymptotic Analysis* **7** (1993), no. 2, 81–95.
3. Youcef AMIRAT, Kamel HAMDACHE, Abdelhamid ZIANI: *Étude d'une équation de transport à mémoire*, *Comptes Rendus de l'Academie des sciences Paris Sér. I Math.* **311** (1990), no. 11, 685–688.
4. Nenad ANTONIĆ: *Memory effects in homogenisation: Linear second-order equations*, *Archive for Rational Mechanics and Analysis* **125** (1993) 1–24
5. Nenad ANTONIĆ: *H-measures applied to symmetric systems*, *Proceedings of the Royal Society of Edinburgh* **126A** (1996) 1133–1155
6. Ivo BABUŠKA: *Homogenization and its applications. Mathematical and computational problems*, in *Numerical solutions of partial differential equations — III*, pp. 89–116, Academic Press, 1976
7. Alain BENSOUSSAN, Jacques-Louis LIONS, George PAPANICOLAOU: *Asymptotic analysis for periodic structures*, North-Holland, 1978
8. David J. BERGMAN: *Bulk physical properties of composite media*, in *Les méthodes de l'homogénéisation: théorie et applications en physique*, (*Bréau-sans-Nappe*, 1983), pp. 1–128, *Collect. Dir. Etudes Rech. Elec. France*, 57, *Eyrolles, Paris*, 1985
9. Phillipe BRAIDY, Didier POUILLOUX: unpublished memoir
10. Alberto CALDERÓN: *Commutators of singular integral operators*, *Proceedings of the National Academy of Sciences of USA* **53** (1965) 1092–1099
11. Yves COLIN DE VERDIÈRE: *Ergodicité et fonctions propres du laplacien*, *Communications in Mathematical Physics* **102** (1985), no. 3, 497–502.
12. Constantine M. DAFERMOS: *Hyperbolic conservation laws in continuum physics*, Springer, 2000
13. Ronald J. DIPERNA, Andrew J. MAJDA: *Oscillations and concentration in weak solutions of the incompressible fluid equations*, *Communications in Mathematical Physics* **108** (1987) 667–689
14. Paul Adrien Maurice DIRAC: *The principles of Quantum Mechanics*, Clarendon Press, 1956
15. Gilles FRANCFORT, François MURAT: *Oscillations and energy densities in the wave equation*, *Communications in Partial Differential Equations* **17** (1992) 1785–1865
16. Gilles FRANCFORT, François MURAT, Luc TARTAR: *Fourth order moments of nonnegative measures on S^2 and applications*, *Archive for Rational Mechanics and Analysis* **131** (1995) 305–333
17. Patrick GÉRARD: *Mesures semi-classiques et ondes de Bloch*, in *Séminaire Equations aux Dérivées Partielles*, École Polytechnique, Palaiseau, exp. XVI (1990)
18. Patrick GÉRARD: *Microlocal defect measures*, *Communications in Partial Differential Equations* **16** (1991) 1761–1794
19. Patrick GÉRARD: *Oscillations and concentration effects in semilinear dispersive wave equations*, *Journal of Functional Analysis* **141** (1996) 60–98
20. Patrick GÉRARD, Erik LEICHTNAM: *Ergodic properties of eigenfunctions for the Dirichlet problem*, *Duke Mathematical Journal* **71** (1993) 559–607

21. Zvi HASHIN, Shmuel SHTRIKMAN: *A variational approach to the theory of effective magnetic permeability of multiphase materials, Journal of Applied Physics* **35** (1962) 3125–3131
22. Lars HÖRMANDER: The analysis of linear partial differential operators I–IV, Springer, 1983–1985
23. John K. HUNTER, Joseph B. KELLER: *Weakly nonlinear high frequency waves, Communications in Pure and Applied Mathematics* **36** (1983) 547–569
24. Richard D. JAMES, David KINDERLEHRER: *Frustration in ferromagnetic materials, Continuum Mechanics and Thermodynamics* **2** (1990) 215–239
25. Joseph B. KELLER, George C. PAPANICOLAOU, Leonid V. RYZHIK: *Transport equations for elastic and other waves in random medium, Wave motion* **24** (1996) 327–330
26. Joseph B. KELLER, George C. PAPANICOLAOU, Leonid V. RYZHIK: *Stability of the P-to-S energy ratio in the diffusive regime, Bull. Seismological Soc. Amer.* **86** (1996) 1107–1115
27. Lev LANDAU, Evgenij LIFSCHITZ: *Électrodynamique des milieux continus*, Mir, 1969
28. Pierre-Louis LIONS: *The concentration-compactness principle in the calculus of variations: The locally compact case I & II, Annales de l'Institute Henri Poincaré* **1 C** (1984) 109–145, 223–283; *The limit case I & II, Revista Matemática Iberoamericana* **1** (1985) No. 145–201, No. 45–121
29. Pierre-Louis LIONS, Thierry PAUL: *Sur les mesures de Wigner, Revista Matemática Iberoamericana* **9** (1993) 553–618
30. Luisa MASCARENHAS: *A linear homogenisation problem with time dependent coefficient, Transactions of the American Mathematical Society* **281** (1984) 179–195
31. Graeme W. MILTON: *Theory of composites*, Cambridge, 2001
32. Stefan MÜLLER: *Minimizing sequences for nonconvex functionals, phase transitions and singular perturbations*, Problems involving change of type (Stuttgart, 1988), 31–44, Lecture Notes in Phys., 359, Springer, Berlin, 1990.
33. François MURAT, Luc TARTAR: *Calcul des variations et homogénéisation*, in *Les méthodes de l'homogénéisation: théorie et applications en physique, (Bréau-sans-Nappe,* 1983), pp. 319-369, *Collect. Dir. Etudes Rech. Elec. France,* 57, *Eyrolles, Paris,* 1985
34. François MURAT, Luc TARTAR: *H-convergence*, translation of lecture notes from 1978, in *Topics in the mathematical modeling of composite materials, Cherkaev-Kohn (eds.),* Vol. 31 of *Progress in nonlinear differential equations and their applications,* 21–43, Birkhaüser, Boston, 1997
35. François MURAT, Luc TARTAR: work still in preparation
36. G. NGUETSENG: *A general convergence result for a functional related to the theory of homogenization, SIAM Journal of Mathematical Analysis* **20** (1989) 608–623
37. Henri SANCHEZ-PALENCIA: *Non-homogeneous media and vibration theory,* Springer, 1980
38. Alexander SHNIRELMAN: *Ergodic properties of eigenfunctions, Uspehi matematičeskih nauk* **29** (1974), no. 6(180), 181–182.
39. Laurent SCHWARTZ: *Théorie des distributions,* Hermann, 1957
40. Sergej L. SOBOLEV: *Applications of functional analysis in mathematical physics,* American Mathematical Society, 1963

41. Sergio SPAGNOLO: *Sul limite delle soluzioni di problemi di Cauchy relativi all'equazione del calore*, *Annali della Scuola Normale Superiore Pisa* **21** (1967) 657–699
42. Sergio SPAGNOLO: *Sulla convergenza di soluzioni di equazioni paraboliche ed ellittiche*, *Annali della Scuola Normale Superiore Pisa* **22** (1968) 571–597
43. Sergio SPAGNOLO, Antonio MARINO: *Un tipo di approssimazione dell'operatore* $\sum_{ij} D_i(a_{ij}(x)D_j)$ *con operatori* $\sum_j D_j(\beta(x)D_j)$, *Annali della Scuola Normale Superiore Pisa* **23** (1969) 657–673
44. Luc TARTAR: *Compensated compactness and applications to partial differential equations*, in *Nonlinear analysis and mechanics: Heriot-Watt Symposium, Vol. IV*, pp. 136-212, *Pitman Res. Notes Math. Ser.* 39, Longman, 1979
45. Luc TARTAR: *Estimation fines des coefficients homogénéisés*, in *E. De Giorgi colloquium*, pp. 168–187, Pitman, 1985
46. Luc TARTAR: *Remarks on homogenization*, in *Homogenization and effective moduli of materials and media*, pp. 228–246, *IMA Vol. Math. Appl. 1*, Springer, 1986
47. Luc TARTAR: *Remarks on oscillations and Stokes' equation*, in *Macroscopic modelling of turbulent flows*, pp. 24–31, *Lecture Notes in Physics 230*, Springer, 1985
48. Luc TARTAR: *Nonlocal effects induced by homogenization*, in *Partial differential equations and the calculus of variations, Vol. II*, pp. 925-938, *Progr. Nonlinear Differential Equations Appl. 2*, Birkhaüser, 1989
49. Luc TARTAR: *H-measures and small amplitude homogenization*, in *Random Media and Composites*, (*Leesburg, VA*, 1988) 89–99, *SIAM, Philadelphia, PA* 1989
50. Luc TARTAR: *Propagation of oscillations and concentration effects in partial differential equations*, in *Ordinary and Partial Differential Equations, Vol. III*, (*Dundee*, 1990), pp. 208–219, *Pitman Res. Notes Math. Ser.* 254, Longman, 1991
51. Luc TARTAR: *H-measures, a new approach for studying homogenisation, oscillations and concentration effects in partial differential equations*, *Proceedings of the Royal Society of Edinburgh* **115A** (1990) 193–230
52. Luc TARTAR: *H-measures and applications*, in *Proceedings of the International Congress of Mathematicians*, pp. 1215–1223, Mathematical Society of Japan, Tokyo, 1991
53. Luc TARTAR: *Memory effects and homogenization*, *Archive for Rational Mechanics and Analysis* **111** (1990) 121–133
54. Luc TARTAR: *Beyond* YOUNG *measures*, *Meccanica* **30** (1995) 505–526
55. Luc TARTAR: *Homogenization and Hyperbolicity*, Dedicated to Ennio De Giorgi, *Ann. Scuola Norm. Sup. Pisa Cl. Sci.* (4) **25** (1997) 785–805
56. Luc TARTAR: *How to describe oscillations of solutions of nonlinear partial differential equations*, in *Transactions of the Sixth Army Conference on Applied Mathematics and Computing* (*Boulder, CO.* 1988), 1133-1141, *ARO Rep.* 89-1, *U.S. Army Res. Office, Research Triangle Park, NC*, 1989
57. Luc TARTAR: *Approximation of H-measures*, Conference in honour of Roland GLOWINSKI (Tours, May 1997); text finished too late to be included in the proceedings. Research Report 97-204, Department of Mathematical Sciences, Carnegie Mellon University, Pittsburgh, PA 15213

58. Luc TARTAR: H-measures and applications, in *Nonlinear partial differential equations and their applications, Collège de France Seminar, Vol XI*, pp. 282–290, *Pitman Res. Notes Math. Ser.* 299, Longman, 1994
59. Luc TARTAR: *Homogenization, Compensated Compactness, and H-measures*, lecture notes in preparation
60. Wolf VON WAHL: *The equations of* NAVIER-STOKES *and abstract parabolic equations,* Friedrich Vieweg & Sohn, Braunschweig, 1985
61. Jack WARGA: *Optimal control of differential and functional equations*, Academic press, 1972
62. Vasilij V. ŽIKOV, Sergej M. KOZLOV, Olga A. OLEINIK: *Homogenization of differential operators and integral functionals*, Springer, 1994

Statistical Mechanics of Violent Relaxation in Stellar Systems

Pierre-Henri Chavanis

Laboratoire de Physique Quantique, Université Paul Sabatier, 118 route de Narbonne, 31062 Toulouse, France

Abstract. We discuss the statistical mechanics of violent relaxation in stellar systems following the pioneering work of Lynden-Bell (1967). The solutions of the gravitational Vlasov-Poisson system develop finer and finer filaments so that a statistical description is appropriate to smooth out the small-scales and describe the "coarse-grained" dynamics. In a coarse-grained sense, the system is expected to reach an equilibrium state of a Fermi-Dirac type within a few dynamical times. We describe in detail the equilibrium phase diagram and the nature of phase transitions which occur in self-gravitating systems. Then, we introduce a small-scale parametrization of the Vlasov equation and propose a set of relaxation equations for the coarse-grained dynamics. These relaxation equations, of a generalized Fokker-Planck type, are derived from a Maximum Entropy Production Principle (MEPP). We make a link with the quasilinear theory of the Vlasov-Poisson system and derive a truncated model appropriate to collisionless systems subject to tidal forces. With the aid of this kinetic theory, we qualitatively discuss the concept of "incomplete relaxation" and the limitations of Lynden-Bell's theory.

Contents

1 Introduction

It has long been realized that galaxies, and self-gravitating systems in general, follow a kind of organization despite the diversity of their initial conditions and their environment. This organization is illustrated by morphological classification schemes such as the Hubble sequence and by simple rules which govern the structure of individual self-gravitating systems. For example, elliptical galaxies display a quasi-universal luminosity profile described by de Vaucouleurs' $R^{1/4}$ law and most of globular clusters are well fitted by the Michie-King model [8]. The question that naturally emerges is, what determines the particular configuration to which a self-gravitating system settles. It is possible that their present configuration crucially depends on the conditions that prevail at their birth and on the details of their evolution. However, in view of their apparent regularity, it is tempting to investigate whether their organization can be favoured by some fundamental physical principles like those of thermodynamics and statistical physics. We ask therefore if the actual states of self-gravitating systems are not simply more probable than any other possible configuration, i.e. if they cannot be considered as maximum entropy states. This thermodynamical approach may be particularly relevant for globular clusters and elliptical galaxies which are described by a distribution function that is almost isothermal [8]. In the case of globular clusters, the relaxation proceeds via two-body encounters and this collisional evolution is well-described by kinetic equations of a Fokker-Planck-Landau type for which a H-theorem[1] is available. By contrast, for elliptical galaxies, two-body encounters are completely negligible (the corresponding relaxation time t_{coll} exceeds the age of the universe by many orders of magnitude) and the galaxy dynamics is described by the *Vlasov equation*, i.e. collisionless Boltzmann equation [8]. Since the Vlasov equation rigorously conserves entropy, a relaxation towards an isothermal distribution looks at first sight relatively surprising. Yet, the inner regions of elliptical galaxies appear to be isothermal and this fact stemed as a mystery for a long time.

In a seminal paper, Lynden-Bell [43] argued that the violently changing gravitational field of a newly formed galaxy leads to a redistribution of energies between stars and provides a mechanism analogous to a relaxation in a gas. The importance of this form of relaxation had previously been stressed by a number of authors including Hénon and King but Lynden-Bell showed for the first time the relevance of a statistical description. He argued that the Vlasov-Poisson system develops an intricate *mixing process* in phase space associated with the heavily damped oscillations of a protogalaxy initially far from mechanical equilibrium and collapsing under its own gravity. As a result, the solutions of the Vlasov equation are not smooth but involve intermingled filaments at smaller and smaller scales. In this sense, there is no convergence towards equilibrium but rather the formation of a fractal-like structure in phase space. However, if we introduce a macroscopic level of description and make a local average of the distribution function over the

---

[1] The H-theorem, proved by Boltzmann for an ideal gas, states that entropy is a monotonically increasing function of time. Boltzmann's kinetic theory of gases therefore provides a direct justification of the second principle of thermodynamics. The generalization of this theorem for self-gravitating systems rests on simplifying assumptions which are difficult to rigorously justify [38].

filaments, the resulting "coarse-grained" distribution function is smooth and is expected to reach a maximum entropy state (i.e. most mixed state) on a very short time scale of the order of the dynamical time t_D. This process is called *violent relaxation* and is acknowledged to account for the regularity of elliptical galaxies or other collisionless self-gravitating systems. Lynden-Bell predicted that the equilibrium state should be described by a Fermi-Dirac distribution function or a superposition of Fermi-Dirac distributions. Here, degeneracy is due not to quantum mechanics but to the Liouville theorem that prevents the smooth distribution function from exceeding the maximum of its initial value. In the non degenerate limit, the Fermi-Dirac distribution functions reduce to Maxwellians. The prediction of an isothermal distribution for collisionless stellar systems was considered as a triumph in the 1960's and laid the foundation of a new type of statistical mechanics. Of course, the validity of the theory is conditioned by a hypothesis of ergodicity which may not be completely fulfilled. This is the complicated problem of "incomplete relaxation" which limitates the power of prediction of Lynden-Bell's approach. However, as we shall see, these difficulties should not throw doubts on the importance of this statistical description. A similar relaxation process is at work in two-dimensional turbulence (described by the 2D Euler equation) and can explain the organization and maintenance of coherent vortices, such as the Great Red Spot of Jupiter, which are common features of large-scale geophysical or astrophysical flows [53,46,23,52,26,6,7]. The mathematical relevance of this statistical description has been given by Robert [50] introducing the concept of Young measures. The formal analogy between two-dimensional vortices and stellar systems has been discussed by Chavanis [11,14,17,27].

This paper is organized as follows. In section 2, we introduce the gravitational Vlasov-Poisson system and list its main properties. In section 3, we present the statistical approach of Lynden-Bell [43] to the problem of violent relaxation. In section 4, we show that the Fermi-Dirac equilibrium distribution predicted by Lynden-Bell is not entirely satisfactory since it has an infinite mass. We must therefore invoke *incomplete relaxation* and introduce truncated models. In the artificial situation in which the system is enclosed within a spherical box, we can calculate the Fermi-Dirac spheres explicitly and prove the existence of a global entropy maximum for each value of energy. For low energies, this equilibrium state has a degenerate core surrounded by a dilute atmosphere, as calculated by Chavanis & Sommeria [25]. More generally, we determine the complete equilibrium phase diagram and discuss the nature of phase transitions in self-gravitating systems. In section 5, we describe the coarse-grained relaxation of collisionless stellar systems towards statistical equilibrium in terms of a generalized Fokker-Planck equation. This relaxation equation is derived from a phenomenological Maximum Entropy Production Principle (MEPP) and involves a diffusion in velocity space compensated by a nonlinear friction. In section 6, we present a quasilinear theory of the Vlasov-Poisson system and show that it leads to a kinetic equation of a Landau type. When the system is close to equilibrium (so that a thermal bath approximation can be implemented) this equation reduces to the Fokker-Planck equation of the thermodynamical approach and the diffusion coefficient can be explicitly evaluated. This provides a new, self-consistent, equation for the "coarse-grained" dynamics of stellar systems where small scales have been smoothed-out in an optimal way. In section 7, we use this kinetic model to derive the distribution function of a tidally truncated collisionless stellar system. This truncated model preserves the main features of Lynden-Bell's distribution (including degeneracy) but has a finite mass, avoiding the artifice of a spherical container. Other truncated models attempting to take into account incomplete relaxation are discussed.

2 The gravitational Vlasov-Poisson system

For most stellar systems, the encounters between stars are negligible [8] and the galaxy dynamics is described by the self-consistent Vlasov-Poisson system

$$\frac{\partial f}{\partial t} + \mathbf{v}\frac{\partial f}{\partial \mathbf{r}} + \mathbf{F}\frac{\partial f}{\partial \mathbf{v}} = 0, \tag{1}$$

$$\Delta\Phi = 4\pi G \int f d^3\mathbf{v}. \tag{2}$$

Here, $f(\mathbf{r},\mathbf{v},t)$ denotes the distribution function (defined such that $fd^3\mathbf{r}d^3\mathbf{v}$ gives the total mass of stars with position $\mathbf{r}$ and velocity $\mathbf{v}$ at time t), $\mathbf{F}(\mathbf{r},t) = -\nabla\Phi$ is the gravitational force (by unit of mass) experienced by a star and $\Phi(\mathbf{r},t)$ is the gravitational potential related to the star density $\rho(\mathbf{r},t) = \int fd^3\mathbf{v}$ by the Newton-Poisson equation (2). The Vlasov equation (1) simply states that, in the absence of encounters, the distribution function f is conserved by the flow in phase space. This can be written $Df/Dt = 0$ where $D/Dt = \partial/\partial t + \mathbf{U}_6\nabla_6$ is the material derivative and $\mathbf{U}_6 = (\mathbf{v},\mathbf{F})$ is a generalized velocity field in the 6-dimensional phase space $(\mathbf{r},\mathbf{v})$ [by definition, $\nabla_6 = (\partial/\partial\mathbf{r}, \partial/\partial\mathbf{v})$ is the generalized nabla operator]. Since the flow is incompressible, i.e. $\nabla_6\mathbf{U}_6 = 0$, the hypervolume of a "fluid" particle is conserved. Since, in addition, a fluid particle conserves the distribution function, this implies that the total mass (or hypervolume) of all phase elements with phase density between f and $f+\delta f$ is conserved. This is equivalent to the conservation of the Casimir integrals

$$I_h = \int h(f) d^3\mathbf{r}d^3\mathbf{v}, \tag{3}$$

for any continuous function $h(f)$ (they include in particular the total mass $M = \int fd^3\mathbf{r}d^3\mathbf{v}$). It is also straightforward to check that the Vlasov-Poisson system conserves the total energy (kinetic + potential)

$$E = \int \frac{1}{2} f v^2 d^3\mathbf{r}d^3\mathbf{v} + \frac{1}{2}\int f\Phi d^3\mathbf{r}d^3\mathbf{v} = K + W, \tag{4}$$

the angular momentum

$$\mathbf{L} = \int f(\mathbf{r} \wedge \mathbf{v}) d^3\mathbf{r} d^3\mathbf{v}, \tag{5}$$

and the impulse

$$\mathbf{P} = \int f\mathbf{v} d^3\mathbf{r} d^3\mathbf{v}. \tag{6}$$

In the following, we shall work in the barycentric frame of reference and assume that the system is non rotating so that the conservation of $\mathbf{L}$ and $\mathbf{P}$ can be ignored (see Refs. [43,27] for a generalization).

3 Lynden-Bell's approach of violent relaxation

When the initial condition is far from equilibrium, the Vlasov-Poisson system develops a complicated mixing process in phase space and generates intermingled filaments due to stirring effects. Therefore, a *deterministic* description of the flow in phase space requires a rapidly increasing amount of information as time goes on. For that reason, it is appropriate to undertake a *probabilistic* description in order to smooth out the small-scales (fine-grained) and concentrate on the locally averaged (coarse-grained) quantities. This statistical analysis has been considered a long time ago by Lynden-Bell [43]. The statistical mechanics of continuous systems is not as firmly established as in the usual case of N-body systems but Robert [50] has recently developed a mathematical justification of this procedure in terms of Young measures. We shall describe below the argumentation of Lynden-Bell, which is more intuitive in a first approach.

Starting from some arbitrary initial condition, the distribution function is stirred in phase space but conserves its values η (levels of phase density) and the corresponding hypervolumes $\gamma(\eta)d\eta$ as a property of the Vlasov equation. Let us introduce the probability density $\rho(\mathbf{r}, \mathbf{v}, \eta)$ of finding the level of phase density η in a small neighborhood of the position $\mathbf{r}, \mathbf{v}$ in phase space. This probability density can be viewed as the local area proportion occupied by the phase level η and it must satisfy at each point the normalization condition

$$\int \rho(\mathbf{r}, \mathbf{v}, \eta) d\eta = 1. \tag{7}$$

The locally averaged distribution function is then expressed in terms of the probability density as

$$\overline{f}(\mathbf{r}, \mathbf{v}) = \int \rho(\mathbf{r}, \mathbf{v}, \eta) \eta d\eta, \tag{8}$$

and the associated (macroscopic) gravitational potential satisfies

$$\Delta\overline{\Phi} = 4\pi G \int \overline{f} d^3\mathbf{v}. \tag{9}$$

Since the gravitational potential is expressed by space integrals of the density, it smoothes out the fluctuations of the distribution function, supposed at very fine scale, so Φ has negligible fluctuations. It is then possible to express the conserved quantities of the Vlasov equation as integrals of the macroscopic fields. These conserved quantities are the global probability distributions of phase density $\gamma(\eta)$ (i.e., the total hypervolume occupied by each level η of phase density)

$$\gamma(\eta) = \int \rho(\mathbf{r}, \mathbf{v}, \eta) d^3\mathbf{r} d^3\mathbf{v}, \tag{10}$$

and the total energy

$$E = \int \frac{1}{2} \overline{f} v^2 d^3\mathbf{r} d^3\mathbf{v} + \frac{1}{2} \int \overline{f}\ \overline{\Phi} d^3\mathbf{r} d^3\mathbf{v}. \tag{11}$$

As discussed above the gravitational potential can be considered as smooth, so we can express the energy in terms of the coarse-grained functions $\overline{f}$ and $\overline{\Phi}$ neglecting the internal energy of the fluctuations.

To determine the equilibrium distribution of the system, we need to introduce an entropy functional like in ordinary statistical mechanics. As is customary, Lynden-Bell defines the mixing entropy as the logarithm of the number of microscopic configurations associated with the same macroscopic state (characterized by the probability density $\rho(\mathbf{r}, \mathbf{v}, \eta)$). After introducing a counting "à la Boltzmann", he arrives at the expression [43]:

$$S = -\int \rho(\mathbf{r}, \mathbf{v}, \eta) \ln \rho(\mathbf{r}, \mathbf{v}, \eta) d\eta d^3\mathbf{r} d^3\mathbf{v}, \tag{12}$$

where the integral extends over phase space and over all the levels of phase elements. A mathematical justification of this entropy has been given by Robert [50]. The most likely distribution to be reached at equilibrium is then obtained by maximizing the mixing entropy (12) subject to the constraints (10), (11) and the normalization condition (7). This variational problem is treated by introducing Lagrange multipliers so that the first variations satisfy

$$\delta S - \beta \delta E - \int \alpha(\eta) \delta\gamma(\eta) d\eta - \int \zeta(\mathbf{r}, \mathbf{v}) \delta \left(\int \rho d\eta \right) d^3\mathbf{r} d^3\mathbf{v} = 0, \tag{13}$$

where β is the inverse temperature and $\alpha(\eta)$ the "chemical potential" of species η. The resulting optimal probability density is a Gibbs state which has the form [43,27]:

$$\rho(\mathbf{r}, \mathbf{v}, \eta) = \frac{e^{-\alpha(\eta) - \beta\eta(\frac{v^2}{2} + \overline{\Phi})}}{\int e^{-\alpha(\eta) - \beta\eta(\frac{v^2}{2} + \overline{\Phi})} d\eta}. \tag{14}$$

The previous analysis gives a well defined procedure to compute the statistical equilibrium states. The gravitational field is obtained by solving the

Poisson equation (2) with the distribution function (8) determined by the Gibbs state (14). The solution depends on the Lagrange multipliers β and $\alpha(\eta)$ which must be related to the conserved quantities E and $\gamma(\eta)$ by equations (11), (10). This procedure determines *critical points* of entropy. Whether these critical points are maxima or not is decided by the sign of the second variations of entropy.

Following Lynden-Bell [43], we consider a particular situation that presents interesting features and for which the previous problem can be studied in detail. Keeping only two levels $f = \eta_0$ and $f = 0$ is convenient to simplify the discussion and is probably representative of more general cases. Within this "patch" approximation, the mixing entropy reduces to

$$S = -\int \left\{ \frac{\overline{f}}{\eta_0} \ln \frac{\overline{f}}{\eta_0} + \left(1 - \frac{\overline{f}}{\eta_0}\right) \ln\left(1 - \frac{\overline{f}}{\eta_0}\right) \right\} d^3\mathbf{r} d^3\mathbf{v}, \qquad (15)$$

and the equilibrium coarse-grained distribution $\overline{f} = \rho(\mathbf{r}, \mathbf{v}, \eta_0)\eta_0$ takes explicitly the form

$$\overline{f} = \frac{\eta_0}{1 + \lambda e^{\beta\eta_0\epsilon}}, \qquad (16)$$

where $\epsilon = \frac{v^2}{2} + \overline{\Phi}$ is the energy of a star (per unit of mass) and $\lambda \equiv e^{\alpha(\eta_0) - \alpha(0)} > 0$ an equivalent Lagrange multiplier. Equation (16) is, apart from a reinterpretation of the constants, the distribution function for the self-gravitating Fermi-Dirac gas. Here, the exclusion principle $\overline{f} \leq \eta_0$ is due to the incompressibility constraint (7) not to quantum mechanics. Because of the averaging procedure, the coarse-grained distribution function can only *decrease* by internal mixing, as vaccum is incorporated into the patch, and this results in an "effective" exclusion principle. Lynden-Bell's distribution therefore corresponds to a 4-th type of statistics since the "fluid" particles are distinguishable but subject to an exclusion principle. However, formally, this distribution coincides with the Fermi-Dirac distribution. In the fully degenerate case, this equilibrium has been studied extensively in connection with white dwarf stars [9]. The structure of equilibrium may depend crucially on the degree of degeneracy (see section 4) but Lynden-Bell [43] gives arguments according to which stellar systems would be non degenerate [2]. In that limit, $\overline{f} \ll \eta_0$, and equation (16) reduces to the Maxwell-Boltzmann statistics

$$\overline{f} = A e^{-\beta\eta_0\epsilon}. \qquad (17)$$

This was in fact the initial goal of Lynden-Bell in 1967: his theory of "violent relaxation" was able to justify a Maxwell-Boltzmann equilibrium distribution, without recourse to collisions, on a short time scale $\sim t_D \ll t_{coll}$ consistent with the age of ellipticals. In addition, the individual mass of the stars never appears in his theory based on the Vlasov equation. Therefore,

[2] In fact, his arguments do not apply to galactic nuclei where his type of degeneracy may be important.

the equilibrium state (17) does not lead to a segregation by mass in contrast with a collisional relaxation. This is in agreement with the observed light distributions in elliptical galaxies that would show greater colour differences if a marked segregation by mass was established.

4 Computation of Fermi-Dirac spheres

The first problem to be tackled is obviously the computation of self-gravitating Fermi-Dirac spheres. Let us first consider the non degenerate limit $\overline{f} \ll \eta_0$ corresponding to a dilute system. In that case, the mixing entropy (15) reduces to the ordinary Boltzmann entropy

$$S = -\int f \ln f d^3\mathbf{r} d^3\mathbf{v}, \tag{18}$$

whose maximization at fixed mass and energy leads to the Maxwell-Boltzmann statistics (17). Substituting this optimal distribution in the Poisson equation (2), we find that the gravitational potential is determined at equilibrium by the differential equation

$$\Delta\Phi = 4\pi G A' e^{-\beta\Phi}, \tag{19}$$

where the Lagrange multipliers A' and β have to be related to the total mass and total energy of the system. The Boltzmann-Poisson equation (19) has been studied extensively in the context of isothermal gaseous spheres [9] and in the case of collisional stellar systems such as globular clusters [45]. We can check by direct substitution that the distribution

$$\Phi_s(r) = \frac{1}{\beta}\ln(2\pi G\beta A r^2), \qquad \rho_s(r) = \frac{1}{2\pi G\beta r^2}, \tag{20}$$

is an exact solution of equation (19) known as the *singular isothermal sphere* [8]. Since $\rho \sim r^{-2}$ at large distances, the total mass of the system $M = \int_0^{+\infty} \rho 4\pi r^2 dr$ is infinite! More generally, we can show that any solution of the Boltzmann-Poisson equation (19) behaves like the singular sphere as $r \to +\infty$ and has therefore an infinite mass. This means that no equilibrium can exist in an unbounded domain: the density can spread indefinitely while conserving energy and increasing entropy [8].

In practice, the infinite mass problem does not arise if we realize that the relaxation of the system is necessarily *incomplete* [43]. There are two major reasons for incomplete relaxation: (i) The mean field relaxation process is dependant on the strength of the variations in potential. As these die out, the relaxation ceases and it is likely that the system may find a stable steady state before the relaxation process is completed. Therefore, the relaxation is effective only in a finite region of space (roughly the main body of the galaxy) and during a finite period of time (while the galaxy is dynamically unsteady). Orbits which lie partly outside the relaxing region and have periods longer than the time for which the galaxy is unsteady will not acquire their full quotas of stars. (ii) in practice, the galaxy will not be isolated but will be subject to the tides of other systems. Therefore, high energy stars will escape the system being ultimately captured by the gravity of a nearby object. These two independent effects have similar consequences and will produce a modification of the distribution function at high energies. Some truncated models can be obtained by developing a kinetic theory of encounterless relaxation (sections 5-7). This kinetic approach will provide a precise framework to understand what limits relaxation and why complete equilibrium is not reached in general.

However, for the present, we shall avoid the infinite mass problem by confining artificially the system within a box of radius R. The calculation of finite isothermal spheres has been carried out by Antonov [1], Lynden-Bell & Wood [45], Katz [39], Padmanabhan [47,48], de Vega & Sanchez [30] and Chavanis [18] (see Chavanis [21] for an extension in general relativity). These studies were performed in the framework of gaseous stars or collisional stellar systems (e.g., globular clusters) but they extend in principle to violently relaxed collisionless stellar systems described by the distribution function (17). Therefore, we shall give a brief summary of these classical results before considering the case of the Fermi-Dirac distribution (16). The phase diagram is represented in Fig. 1 where the inverse temperature is plotted as a function of minus the energy. It is possible to prove the following results: (i) there is no global maximum for the Boltzmann entropy. (ii) there are not even critical points for the Boltzmann entropy if $\Lambda = -ER/GM^2 > 0.335$. (iii) local entropy maxima (LEM) exist if $\Lambda = -ER/GM^2 < 0.335$; they have a density contrast $\mathcal{R} = \rho(0)/\rho(R) < 709$ (upper branch). (iv) critical points of entropy with density contrast $\mathcal{R} > 709$ are unstable saddle points SP (spiraling curve). Conclusions (i) and (ii) have been called "gravothermal catastrophe" or "Antonov instability". When $\Lambda = -ER/GM^2 > 0.335$, there is no hydrostatic equilibrium and the system is expected to collapse and overheat. This is a natural evolution (in a thermodynamical sense) because a self-gravitating system can always increase entropy by taking a "core-halo" structure and by making its core denser and denser (and hotter and hotter) [8]. As discussed by Lynden-Bell & Wood [45], this instability is probably related to the negative specific heats of self-gravitating systems: by losing heat, the core grows hotter and evolves away from equilibrium.

The "gravothermal catastrophe" picture has been confirmed by sophisticated numerical simulations of globular clusters that introduce a precise description of heat transfers between the "core" and the "halo" using moment equations [41], orbit averaged Fokker-Planck equation [29] or fluid equations for a thermally conducting gas [44]. In these studies, the collapse proceeds self-similarly (with power law behaviours) and the central density becomes infinite in a finite time. This singularity has been known as "core collapse" and many globular clusters have probably experienced core collapse. This is

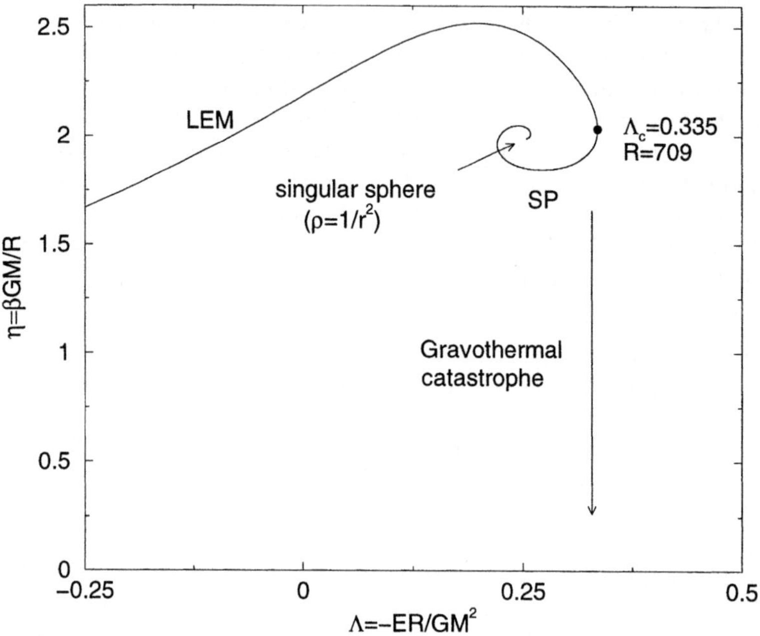

Fig. 1. Equilibrium phase diagram for classical isothermal spheres.

certainly the most exciting theoretical aspect of the collisional evolution of stellar systems. In practice, the formation of hard binaries can release sufficient energy to stop the collapse and even drive a reexpansion of the system [32]. Then, a series of gravothermal oscillations should follow but it takes times much larger than the age of globular clusters [3]. It has to be noted that, although the central density tends to infinity, the mass contained in the core tends to zero so, in this sense, the gravothermal catastrophe is a rather unspectacular process [8]. However, the gravothermal catastrophe may also be at work in dense clusters of compact stars (neutron stars or stellar mass black holes) embedded in evolved galactic nuclei and it can presumably initiate the formation of supermassive black holes via the collapse of such clusters. Indeed, when the central redshift (related to the density contrast) exceeds a critical value $z_c \sim 0.5$, a dynamical instability of general-relativistic origin sets in and the star cluster undergoes a catastrophic collapse to a black hole on a dynamical time [49]. It has been suggested that this process leads naturally to the birth of supermassive black holes of the right size to explain quasars and AGNs [56].

Since collisionless stellar systems undergoing violent relaxation are described by an *isothermal* distribution function of the Fermi-Dirac type (16), it is particularly relevant to ask whether galaxies in their birth stages can undergo a form of gravothermal catastrophe and if so what they do about it.

The collapse of galaxies was described qualitatively by Lynden-Bell & Wood [45]: "...The centers will then begin to separate into a core - a sort of separate body which might even be called a nucleus. This will cease to shrink when it becomes degenerate in Lynden-Bell's sense. The system will then have a core-halo structure which is an equilibrium of an isothermal Fermi-Dirac gas sphere. These will show a variety of different nuclear concentrations depending on the degeneracy parameter. It is evident that theory developed along these lines has the chance of making sense of the variety of different galactic nuclei."

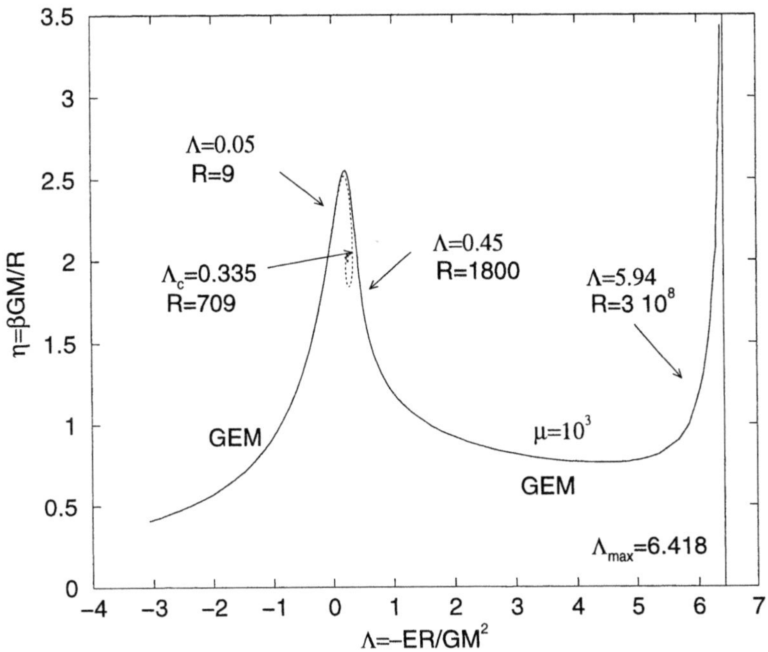

Fig. 2. Equilibrium phase diagram for Fermi-Dirac spheres ($\mu = 10^3$)

The computation of isothermal Fermi-Dirac spheres has been performed only recently by Chavanis & Sommeria [25]. It can be proved that a global entropy maximum exists for all accessible values of energy [25,51]. Therefore, degeneracy has a stabilizing role and is able to stop the "gravothermal catastrophe". The equilibrium diagram is represented in Figs. 2-3 and now depends on the degeneracy parameter $\mu = \eta_0/\langle f \rangle \equiv \eta_0\sqrt{512\pi^4 G^3 M R^3}$, where η_0 is the maximum allowable value of the distribution function and $\langle f \rangle$ its typical average value in the box of radius R. We see that the inclusion of degeneracy has the effect of unwinding the spiral (dashed line). When μ is small (Fig. 2), there is only one critical point of entropy for each value of energy and it

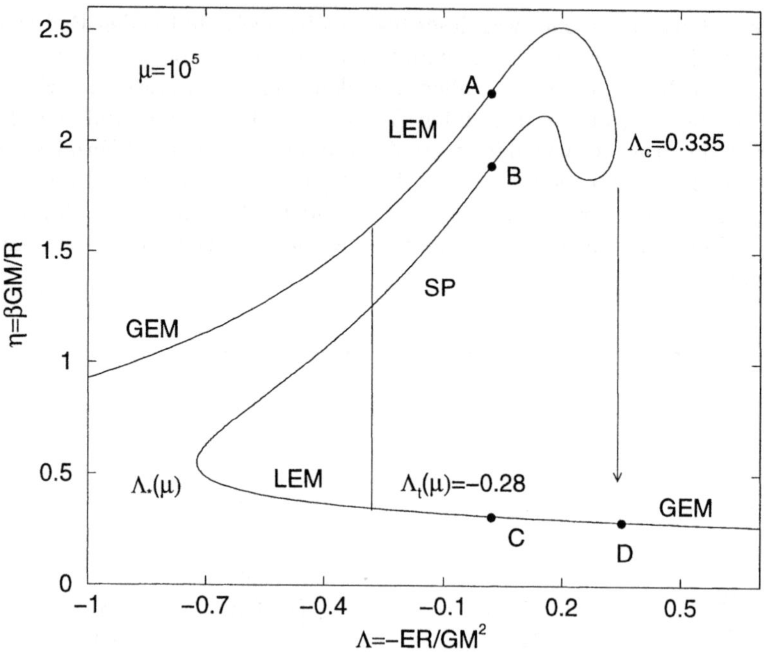

Fig. 3. Equilibrium phase diagram for Fermi-Dirac spheres ($\mu = 10^5$)

is a global entropy maximum (GEM). For small values of Λ (high energies) the solutions almost coincide with the classical, non degenerate, isothermal spheres. When Λ is increased (low energies) the solutions take a "core-halo" structure with a partially degenerate core surrounded by a dilute Maxwellian atmosphere. It is now possible to overcome the critical energy $\Lambda_c = 0.335$ and the critical density contrast $\mathcal{R} \simeq 709$ found by Antonov for a classical isothermal gas. In that region, the specific heat $C = dE/dT$ is negative. As energy decreases further, more and more mass is concentrated into the nucleus (which becomes more and more degenerate) until a minimum accessible energy, corresponding to $\Lambda_{max}(\mu)$, at which the nucleus contains all the mass. In that case, the atmosphere has been "swallowed" and the system has the same structure as a cold white dwarf star [9]. This is a relatively singular limit since the density drops to zero at a finite radius whereas for partially degenerate systems, the density decays like r^{-2} at large distances. When the degeneracy parameter μ is large (Fig. 3), there are now several critical points of entropy for each single value of energy in the range $\Lambda_*(\mu) < \Lambda < \Lambda_c$. The solutions on the upper branch (points A) are non degenerate and have a smooth density profile; they form the "gaseous" phase. The solutions on the lower branch (points C) have a "core-halo" structure with a degenerate nucleus and a dilute atmosphere; they form the "condensed" phase. According to the theorem of Katz [39], they are both entropy maxima while the intermediate solutions, points B, are unstable saddle points (SP). These points are similar to points A, except that they contain a small embryonic nucleus (with small mass and energy) which plays the role of a "germ" in the langage of phase transitions. For small values of Λ, the non degenerate solutions (upper branch) are global entropy maxima (GEM). They become local entropy maxima (metastable states) at a certain $\Lambda_t \geq \Lambda_*$ depending on the degeneracy parameter. At that Λ_t, the degenerate solutions (lower branch) that were only local entropy maxima (LEM) suddenly become global entropy maxima. We expect therefore a phase transition to occur between the "gaseous" phase (upper branch) and the "condensed" phase (lower branch) at Λ_t. For $\Lambda_t(\mu) < \Lambda < \Lambda_c$, the non degenerate solutions (points A) are metastable but we may suspect [25] that they are long-lived so that they *are* physical. It is plausible that these metastable states will be selected by the dynamics even if states with higher entropy exist. This depends on a complicated notion of "bassin of attraction" as studied by Chavanis, Rosier and Sire [22] with the aid of a simple model of gravitational dynamics. Therefore, the true phase transition will occur at $\Lambda_c = 0.335$ at which the gaseous phase completely disappears: at that point, the gravothermal catastrophe sets in but, for collisionless stellar systems (or for fermions), the core ultimately ceases to shrink when it becomes degenerate. In that case, the system falls on to a global entropy maximum which is the true equilibrium state for these systems (point D). This global entropy maximum has a "core-halo" structure with a degenerate core and a dilute atmosphere. A simple analytical model of these phase transitions has been proposed in Ref. [20] and provides a fairly good agreement with the full numerical solution. A particularity of self-gravitating systems, which are by essence non extensive, is that the statistical ensembles (microcanonical and canonical) are not interchangeable. Therefore, the description of the equilibrium diagram is different whether the system evolves at fixed energy of fixed temperature. A discussion of this interesting phenomenon can be found in the review of Padmanabhan [48] and in Chavanis [20].

For astrophysical purposes, it is still a matter of debate to decide whether collisionless stellar systems like elliptical galaxies are degenerate (in the sense of Lynden-Bell) or not. Since degeneracy can stabilize the system without changing its overall structure at large distances, we have suggested that degeneracy could play a role in galactic nuclei [25]. The recent simulations of Leeuwin and Athanassoula [42] and the theoretical model of Stiavelli [57] are consistent with this idea especially if the nucleus of elliptical galaxies contains a primordial massive black hole. Indeed, the effect of degeneracy on the distribution of stars surrounding the black hole can explain the cusps observed at the center of galaxies [57]. This form of degeneracy is also relevant for massive neutrinos in Dark Matter models where it competes with quantum degeneracy [40]. In fact, the Fermi-Dirac distribution of massive neutrinos in Dark Matter models (which form a collisionless self-gravitating system) might be justified more by the process of "violent relaxation" [43] than by quantum mechanics [34,4]. As shown by Chavanis & Sommeria [25], violent relaxation can lead to the formation of a dense degenerate nucleus with a small radius and a large mass. This massive degenerate nucleus could be an alternative to black holes at the center of galaxies [25,5] .

5 The maximum entropy production principle

Basically, a collisionless stellar system is described in a self-consistent mean field approximation by the Vlasov-Poisson system (1), (2). In principle, these coupled equations determine completely the evolution of the distribution function $f(\mathbf{r},\mathbf{v},t)$. However, as discussed in section 3, we are not interested in practice by the finely striated structure of the flow in phase space but only by its macroscopic, i.e. smoothed-out, structure. Indeed, the observations and the numerical simulations are always realized with a finite resolution. Moreover, the "coarse-grained" distribution function $\overline{f}$ is likely to converge towards an equilibrium state contrary to the exact distribution f which develops smaller and smaller scales.

If we decompose the distribution function and the gravitational potential in a mean and fluctuating part ($f=\overline{f}+\tilde{f}$, $\Phi=\overline{\Phi}+\tilde{\Phi}$) and take the local average of the Vlasov equation (1), we readily obtain an equation of the form

$$\frac{\partial\overline{f}}{\partial t}+\mathbf{v}\frac{\partial\overline{f}}{\partial\mathbf{r}}+\overline{\mathbf{F}}\frac{\partial\overline{f}}{\partial\mathbf{v}}=-\frac{\partial\mathbf{J}_f}{\partial\mathbf{v}},\tag{21}$$

for the "coarse-grained" distribution function with a diffusion current $\mathbf{J}_f=\overline{\tilde{f}\tilde{\mathbf{F}}}$ related to the correlations of the "fine-grained" fluctuations.

The problem in hand consists in determining a closed form for the diffusion current $\mathbf{J}_f$. Its detailed expression $\overline{\tilde{f}\tilde{\mathbf{F}}}$ depends on the subdynamics and is therefore extremely difficult to capture from first principles. Indeed, the "violent relaxation" is a very nonlinear and very chaotic process ruling out any attempt to implement *perturbative* methods if we are far from equilibrium. For that reason, there is a strong incentive to explore *variational* methods which are considerably simpler and give a more intuitive understanding of the problem. We propose to describe the relaxation of the coarse-grained distribution function $\overline{f}$ towards the Gibbs state (14) with a Maximum Entropy Production Principle [27]. This thermodynamical principle assumes that: "during its evolution, the system tends to maximize its rate of entropy production $\dot{S}$ while satisfying all the constraints imposed by the dynamics". There is no rigorous justification for this principle and it is important therefore to confront the MEPP with kinetic theories (when they are available) in order to determine its range of validity. In any case, the MEPP can be considered as a convenient tool to build relaxation equations which are mathematically well-behaved and which can serve as numerical algorithms to calculate maximum entropy states. It is also "perfectible" in the sense that any new information about the dynamics of the system can be incorporated. This principle is reminiscent of Jaynes [36] ideas and is a clear extension of the well known principle of statistical mechanics according to which: "at equilibrium, the system is in a maximum entropy state consistent with all the constraints". This is also the most natural extension of Lynden-Bell's theory out of equilibrium.

Like for equilibrium states, the evolution of the flow in phase space is described by a set of local probabilities $\rho(\mathbf{r},\mathbf{v},\eta,t)$ and the locally averaged gravitational potential $\overline{\Phi}(\mathbf{r},t)$ is obtained by the integration of equation (2) where $f(\mathbf{r},\mathbf{v},t)$ is replaced by $\overline{f}(\mathbf{r},\mathbf{v},t) = \int \rho(\mathbf{r},\mathbf{v},\eta,t)\eta d\eta$. The phase elements are thus transported in phase space by the corresponding averaged velocity field $\overline{\mathbf{U}}_6 = (\mathbf{v}, -\nabla\overline{\Phi})$ and we suppose that, in addition, they undergo a diffusion process. This diffusion occurs only in velocity space, due to the fluctuations of the gravitational field Φ. There is no diffusion in position space since the velocity $\mathbf{v}$ is a pure coordinate and does not fluctuate. As a result, the probability densities will satisfy a convection-diffusion equation of the form

$$\frac{\partial\rho}{\partial t} + \mathbf{v}\frac{\partial\rho}{\partial\mathbf{r}} + \overline{\mathbf{F}}\frac{\partial\rho}{\partial\mathbf{v}} = -\frac{\partial\mathbf{J}}{\partial\mathbf{v}}, \tag{22}$$

where $\mathbf{J}(\mathbf{r},\mathbf{v},\eta,t)$ is the diffusion current of the phase element η. This equation conserves the total hypervolume (in phase space) occupied by each phase element. Multiplying equation (22) by η and integrating over the levels of phase density returns equation (21) with $\mathbf{J}_f = \int \mathbf{J}(\mathbf{r},\mathbf{v},\eta,t)\eta d\eta$.

To apply the MEPP, we need first to compute the rate of change of entropy during the convection-diffusion process. Taking the time derivative of equation (12), expressing $\partial_t\rho$ by equation (22) and noting that $\rho\ln\rho$ is conserved by the advective term, we get

$$\dot{S} = -\int \mathbf{J}\frac{\partial\ln\rho}{\partial\mathbf{v}} d^3\mathbf{r}d^3\mathbf{v}d\eta. \tag{23}$$

We now determine $\mathbf{J}$ such that, for a given field ρ at each time t, $\mathbf{J}$ maximizes the entropy production $\dot{S}$ with appropriate dynamical constraints, which are:

- the conservation of the local normalization condition (7) implying

$$\int \mathbf{J}(\mathbf{r},\mathbf{v},\eta,t)d\eta = 0, \tag{24}$$

- the conservation of energy expressed from equations (11) and (21) as

$$\dot{E} = \int \mathbf{J}_f\mathbf{v}d^3\mathbf{r}d^3\mathbf{v} = 0, \tag{25}$$

- a limitation on the eddy flux $|\mathbf{J}|$, characterized by a bound $C(\mathbf{r},\mathbf{v},t)$, which exists but is not specified

$$\int \frac{J^2}{2\rho}d\eta \leq C(\mathbf{r},\mathbf{v},t). \tag{26}$$

This variational problem can be solved by introducing (at each time t) Lagrange multipliers $\boldsymbol{\zeta}$, β, $1/D$ for the three respective constraints. It can be shown by a convexity argument that reaching the bound (26) is always favorable for increasing $\dot{S}$, so that this constraint can be replaced by an equality. The condition

$$\delta\dot{S} - \beta\delta\dot{E} - \int \frac{1}{D}\delta\left(\int \frac{J^2}{2\rho}d\eta\right)d^3\mathbf{r}d^3\mathbf{v} - \int \boldsymbol{\zeta}\delta\left(\int \mathbf{J}d\eta\right)d^3\mathbf{r}d^3\mathbf{v} = 0, \quad (27)$$

yields an optimal current of the form [27]:

$$\mathbf{J} = -D(\mathbf{r},\mathbf{v},t)\left[\frac{\partial\rho}{\partial\mathbf{v}} + \beta(\eta - \overline{f})\rho\mathbf{v}\right]. \quad (28)$$

The Lagrange multiplier $\boldsymbol{\zeta}$ has been eliminated, using the condition (24) of local normalization conservation. At equilibrium, the diffusion currents must vanish and we can check that this yields the Gibbs state (14) with β the corresponding inverse temperature [27]. During the evolution, this quantity varies with time and is determined by the condition of energy conservation (25). Introducing (28) in the constraint (25), we get

$$\beta(t) = -\frac{\int D\frac{\partial\overline{f}}{\partial\mathbf{v}}\mathbf{v}d^3\mathbf{r}d^3\mathbf{v}}{\int D(\overline{f^2} - \overline{f}^2)v^2 d^3\mathbf{r}d^3\mathbf{v}}, \quad (29)$$

where $\overline{f^n} = \int \rho\eta^n d\eta$ are the local moments of the fine-grained distribution function. We have thus obtained a complete set of relaxation equations which exactly conserve the distribution of phase levels and energy. The rate of entropy production can be put in the form [27]:

$$\dot{S} = \int \frac{J^2}{D\rho}d^3\mathbf{r}d^3\mathbf{v}d\eta, \quad (30)$$

so the diffusion coefficient D must be positive for entropy increase. Except for its sign, the diffusion coefficient is not determined by the thermodynamical approach as it is related to the unknown bound $C(\mathbf{r},\mathbf{v},t)$.

These relaxation equations (22), (28) and (29) can be simplified in the case of a single non zero density level η_0 and provide a new non linear equation for the evolution of the coarse-grained distribution function $\overline{f} = \rho\eta_0$:

$$\frac{\partial\overline{f}}{\partial t} + \mathbf{v}\frac{\partial\overline{f}}{\partial\mathbf{r}} + \mathbf{F}\frac{\partial\overline{f}}{\partial\mathbf{v}} = \frac{\partial}{\partial\mathbf{v}}\left[D\left(\frac{\partial\overline{f}}{\partial\mathbf{v}} + \beta\overline{f}(\eta_0 - \overline{f})\mathbf{v}\right)\right]. \quad (31)$$

In the non degenerate limit ($\overline{f} \ll \eta_0$), equation (31) takes the form of a Fokker-Planck equation

$$\frac{\partial\overline{f}}{\partial t} + \mathbf{v}\frac{\partial\overline{f}}{\partial\mathbf{r}} + \mathbf{F}\frac{\partial\overline{f}}{\partial\mathbf{v}} = \frac{\partial}{\partial\mathbf{v}}\left[D\left(\frac{\partial\overline{f}}{\partial\mathbf{v}} + \beta\eta_0\overline{f}\mathbf{v}\right)\right]. \quad (32)$$

This Fokker-Planck equation (sometimes called the Kramers-Chandrasekhar equation) is well-known in the case of collisional stellar systems (without the bar on f!) and is usually derived from a Markov hypothesis and a stochastic Langevin equation [10]. It can also be obtained from the following argument: because of close encounters, the stars undergo brownian motion and diffuse in velocity space (responsible for the first term in the right hand side). However, under the influence of this diffusion, the kinetic energy per star will diverge as $< v^2/2 >\sim 3Dt$ and one is forced to introduce an "ad hoc" dynamical friction (second term in the right hand side) with a friction coefficient $\xi = \beta D\eta_0$ (Einstein relation) in order to compensate this divergence and recover the Maxwellian distribution of velocities at equilibrium. From these two considerations (which express the fluctuation-dissipation theorem) results the ordinary Fokker-Planck equation (32). The Fokker-Planck equation is justified here by an argument of a very wide scope that does not directly refer to the subdynamics: it appears to maximize the rate of entropy production at each time with appropriate constraints. Therefore, it applies equally well to collisional or (coarse-grained) collisionless stellar systems. In this description, the diffusion term and the "dynamical friction" directly result from the variational principle and are associated with the variations of S and E respectively. Furthermore, the inverse temperature β appears as a Lagrange multiplier associated with the conservation of energy and the Einstein relation is automatically satisfied. In addition, our procedure can take into account the degeneracy of collisionless stellar systems, keeping equation (31) instead of the non degenerate limit (32). In the degenerate case, the friction is non linear in $\overline{f}$ so that equation (31) is *not* a Fokker-Planck equation in a strict sense. This nonlinearity is necessary to recover the Fermi-Dirac distribution function at equilibrium (while a linear friction drives the system towards the Maxwell-Boltzmann distribution). This generalization is important because degeneracy is specific to collisionless systems and may be crucial for the existence of an equilibrium state (see section 4).

In equations (31) (32), β is not constant but evolves with time so as to conserve the total energy. In the non degenerate limit, and assuming that D is constant, we get after a part integration

$$\beta(t) = -\frac{\int \frac{\partial \overline{f}}{\partial \mathbf{v}} \mathbf{v} d^3\mathbf{r} d^3\mathbf{v}}{\int \eta_0 \overline{f} v^2 d^3\mathbf{r} d^3\mathbf{v}} = \frac{3M}{2\eta_0 K(t)} \tag{33}$$

This equation relates the formal Lagrange multiplier $\beta(t)$ to the inverse of the *average* kinetic energy. This is of course satisfying on a physical point of view. An alternative Fokker-Planck equation involving a *local* temperature $T(\mathbf{r}, t)$ instead of $\beta(t)^{-1}$ has been proposed by Clemmow & Dougherty [28] in the case of collisional systems. The energy is assumed to be locally conserved by the collisions, which is valid when the mean free path is much smaller than the size of the system. By contrast, this hypothesis does not seem to be justified for the violent relaxation of a collisionless system, which is rather a global process.

Let us briefly review the main properties of our relaxation equations. First of all, they rigorously satisfy the conservation of energy and phase space hypervolumes like the Vlasov-Poisson system (the conservation of impulse and angular momentum can also be satisfied by introducing appropriate Lagrange multipliers [27]). Moreover, they guarantee the increase of entropy at each time ($\dot{S} \geq 0$) with an optimal rate. Of course, this H-theorem is true for the coarse-grained entropy $S = -\int \rho \ln \rho d^3\mathbf{r}d^3\mathbf{v}d\eta$ and not for the fine-grained entropy $S_{f.g} = -\int f \ln f d^3\mathbf{r}d^3\mathbf{v}$ which is constant, as the integral of any function of f. The source of irreversibility is due to the coarse-graining procedure that smoothes out the small-scales and erases the microscopic details of the evolution. Accordingly, the relaxation equations (22), (28) are (29) are likely to drive the system towards an equilibrium state (the Gibbs state (14)) contrary to the Vlasov equation that develops finer and finer scales. In mathematical terms, this means that the distribution function f converges to an equilibrium state $\overline{f}$ *in a weak sense.* If fact, the situation is more complicated since the Gibbs state does not exist in an infinite domain (section 4). We shall see, however, in section 7 that the diffusion coefficient is proportional to the fluctuations of the distribution function (equation (69)). Now, in physical situations, these fluctuations vanish before the systems had time to relax completely. As a result, the relaxation stops and the system remains frozen in a subdomain of phase space. It is only in this subdomain (corresponding to the main body of the galaxy) that the Gibbs state is justified. This provides a physical mechanism for confining galaxies and justifying truncated models. Alternatively, if the galaxy is not isolated but subject to the tides of a neighboring object, a tidally truncated model can be explicitly derived from these relaxation equations (section 7). It has a finite mass while preserving the essential features of Lynden-Bell's distribution function.

The relaxation equations (22), (28) and (29) can be simplified in two particular limits when: (i) the initial condition is approximated by a single level of phase density η_0 surrounded by vaccum (ii) degeneracy, in Lynden-Bell's sense, is neglected. These simplifications lead to the Fokker-Planck equation (32). If the galaxy has sufficiently large energy, this equation will drive the system towards a Maxwell-Boltzmann equilibrium state (17). However, if the energy falls below a critical value, the Fokker-Planck equation does not reach any equilibrium state anymore and the system can achieve ever increasing values of entropy by developing core collapse (Antonov instability). For collisionless stellar systems, this "gravothermal catastrophe" should stop when the center of the galaxy becomes degenerate (see section 4). In that case, the Fokker-Planck equation (32) is not valid anymore and must be replaced by the degenerate relaxation equation (31). This equation converges towards the Fermi-Dirac equilibrium state (16), which exists for all values of energy.

In summary, the MEPP is able to yield relevant relaxation equations for the coarse-grained dynamics of collisionless stellar systems experiencing violent relaxation. This relatively elegant and simple variational principle shows that the global structure of the relaxation equations is determined by purely thermodynamical arguments. All explicit reference to the subdynamics is encapsulated in the diffusion coefficient which cannot be captured by the MEPP (it appears as a Lagrange multiplier related to an unknown bound on the diffusion flux). It must be therefore calculated with a kinetic model such as the quasilinear theory described in the next section.

6 The quasilinear theory

The quasilinear theory of the Vlasov-Poisson system was first considered by Kadomtsev & Pogutse [37] for a homogeneous Coulombian plasma and extended by Severne & Luwel [55] for an inhomogeneous gravitational system. This theory was further discussed by Chavanis [13] in an attempt to make a link with the Maximum Entropy Production Principle. We shall give here a simple account of the quasilinear theory. Further details can be found in Refs. [37,55,13]. Our objective is to obtain an expression for the diffusion current $\mathbf{J}$ by working directly on the Vlasov-Poisson system, i.e. without *assuming* that the entropy increases as is done in the thermodynamical approach.

Since the diffusion current $\mathbf{J}_f = \overline{\tilde{f}\tilde{\mathbf{F}}}$ is related to the fine-grained fluctuations of the distribution function, any systematic calculation starting from the Vlasov equation (1) must necessarily introduce an evolution equation for $\tilde{f}$. This equation is simply obtained by substracting equation (21) from equation (1). This yields

$$\frac{\partial \tilde{f}}{\partial t} + \mathbf{v}\frac{\partial \tilde{f}}{\partial \mathbf{r}} + \overline{\mathbf{F}}\frac{\partial \tilde{f}}{\partial \mathbf{v}} = -\tilde{\mathbf{F}}\frac{\partial \overline{f}}{\partial \mathbf{v}} - \tilde{\mathbf{F}}\frac{\partial \tilde{f}}{\partial \mathbf{v}} + \overline{\tilde{\mathbf{F}}\frac{\partial \tilde{f}}{\partial \mathbf{v}}}. \tag{34}$$

The essence of the quasilinear theory is to assume that the fluctuations are weak and neglect the nonlinear terms in equation (34) altogether. In that case, equations (21) and (34) reduce to the coupled system

$$\frac{\partial \overline{f}}{\partial t} + L\overline{f} = -\frac{\partial}{\partial \mathbf{v}}\overline{\tilde{\mathbf{F}}\tilde{f}}, \tag{35}$$

$$\frac{\partial \tilde{f}}{\partial t} + L\tilde{f} = -\tilde{\mathbf{F}}\frac{\partial \overline{f}}{\partial \mathbf{v}}, \tag{36}$$

where $L = \mathbf{v}\frac{\partial}{\partial \mathbf{r}} + \overline{\mathbf{F}}\frac{\partial}{\partial \mathbf{v}}$ is the advection operator in phase space. Physically, these equations describe the coupling between a subdynamics (here the small scale fluctuations $\tilde{f}$) and a macrodynamics (described by the coarse-grained distribution function $\overline{f}$).

Introducing the Greenian

$$G(t_2,t_1) \equiv \exp\left\{ -\int_{t_1}^{t_2} dt L(t) \right\}, \tag{37}$$

we can immediately write down a formal solution of equation (36), namely

$$\tilde{f}(\mathbf{r},\mathbf{v},t) = G(t,0)\tilde{f}(\mathbf{r},\mathbf{v},0) - \int_0^t ds G(t,t-s)\tilde{\mathbf{F}}(\mathbf{r},t-s)\frac{\partial \overline{f}}{\partial \mathbf{v}}(\mathbf{r},\mathbf{v},t-s). \tag{38}$$

Although very compact, this formal expression is in fact extremely complicated. Indeed, all the difficulty is encapsulated in the Greenian $G(t,t-s)$ which supposes that we can solve the smoothed-out Lagrangian flow

$$\frac{d\mathbf{r}}{dt} = \mathbf{v}, \qquad \frac{d\mathbf{v}}{dt} = \overline{\mathbf{F}}, \tag{39}$$

between t and $t-s$. In practice, this is impossible and we will have to make some approximations.

The objective now is to substitute the formal result (38) back into equation (35) and make a closure approximation in order to obtain a self-consistant equation for $\overline{f}(\mathbf{r},\mathbf{v},t)$. If the fluctuating force $\tilde{\mathbf{F}}$ were external to the system, we would simply obtain a diffusion equation

$$\frac{\partial \overline{f}}{\partial t} + L\overline{f} = \frac{\partial}{\partial v^{\mu}}\left(D^{\mu\nu}\frac{\partial \overline{f}}{\partial v^{\nu}} \right), \tag{40}$$

with a diffusion coefficient given by a Kubo formula

$$D^{\mu\nu} = \int_0^t ds \overline{\tilde{F}^{\mu}(\mathbf{r},t)\tilde{F}^{\nu}(\mathbf{r},t-s)}. \tag{41}$$

However, in the case of the Vlasov-Poisson system, the gravitational force is in fact produced by the distribution of matter itself and this coupling will give rise to a friction term in addition to the pure diffusion. Indeed, we have

$$\tilde{\mathbf{F}}(\mathbf{r},t) = \int \mathbf{F}(\mathbf{r}' \to \mathbf{r})\tilde{f}(\mathbf{r}',\mathbf{v}',t)d^3\mathbf{r}'d^3\mathbf{v}', \tag{42}$$

where

$$\mathbf{F}(\mathbf{r}' \to \mathbf{r}) = G\frac{\mathbf{r}'-\mathbf{r}}{|\mathbf{r}'-\mathbf{r}|^3}, \tag{43}$$

represents the force created by a (field) star in $\mathbf{r}'$ on a (test) star in $\mathbf{r}$ (Newton's law). Therefore, considering equations (38) and (42), we see that the fluctuations of the distribution function $\tilde{f}(\mathbf{r},\mathbf{v},t)$ are given by an iterative process: $\tilde{f}(t)$ depends on $\tilde{\mathbf{F}}(t-s)$ which itself depends on $\tilde{f}(t-s)$ etc. We shall solve this problem perturbatively in an expansion in powers of G. This is the equivalent of the "weak coupling approximation" in plasma physics. To order G^2, we obtain after some rearrangements

$$\frac{\partial \overline{f}}{\partial t} + L\overline{f} = \frac{\partial}{\partial v^{\mu}} \int_0^t ds \int d^3\mathbf{r}' d^3\mathbf{v}' d^3\mathbf{r}'' d^3\mathbf{v}'' F^{\mu}(\mathbf{r}' \to \mathbf{r}) G'(t, t-s) G(t, t-s)$$
$$\times \Big\{ F^{\nu}(\mathbf{r}'' \to \mathbf{r}) \overline{\tilde{f}(\mathbf{r}', \mathbf{v}', t-s) \tilde{f}(\mathbf{r}'', \mathbf{v}'', t-s)} \frac{\partial \overline{f}}{\partial v^{\nu}}(\mathbf{r}, \mathbf{v}, t-s)$$
$$+ F^{\nu}(\mathbf{r}'' \to \mathbf{r}') \overline{\tilde{f}(\mathbf{r}, \mathbf{v}, t-s) \tilde{f}(\mathbf{r}'', \mathbf{v}'', t-s)} \frac{\partial \overline{f}}{\partial v'^{\nu}}(\mathbf{r}', \mathbf{v}', t-s) \Big\}. \tag{44}$$

In this expression, the Greenian G refers to the fluid particle $\mathbf{r}(t), \mathbf{v}(t)$ and the Greenian G' to the fluid particle $\mathbf{r}'(t), \mathbf{v}'(t)$. To close the system, it remains for one to evaluate the correlation function $\overline{\tilde{f}(\mathbf{r}, \mathbf{v}, t) \tilde{f}(\mathbf{r}', \mathbf{v}', t)}$. We shall assume that the mixing in phase space is sufficiently efficient so that the scale of the kinematic correlations is small with respect to the coarse-graining mesh size. In that case,

$$\overline{\tilde{f}(\mathbf{r}, \mathbf{v}, t) \tilde{f}(\mathbf{r}', \mathbf{v}', t)} = \epsilon_r^3 \epsilon_v^3 \delta(\mathbf{r} - \mathbf{r}') \delta(\mathbf{v} - \mathbf{v}') \overline{\tilde{f}^2}(\mathbf{r}, \mathbf{v}, t), \tag{45}$$

where ϵ_r and ϵ_v are the resolution scales in position and velocity respectively. Now,

$$\overline{\tilde{f}^2} = \overline{(f - \overline{f})^2} = \overline{f^2} - \overline{f}^2. \tag{46}$$

We shall assume for simplicity that the initial condition in phase space consists of a patch of uniform distribution function ($f = \eta_0$) surrounded by vaccum ($f = 0$). This is the two-level approximation already considered in sections 3 and 5. In that case $\overline{f^2} = \overline{\eta_0 \times f} = \eta_0 \overline{f}$ and, therefore,

$$\overline{\tilde{f}(\mathbf{r}, \mathbf{v}, t) \tilde{f}(\mathbf{r}', \mathbf{v}', t)} = \epsilon_r^3 \epsilon_v^3 \delta(\mathbf{r} - \mathbf{r}') \delta(\mathbf{v} - \mathbf{v}') \overline{f}(\eta_0 - \overline{f}). \tag{47}$$

Substituting this expression in equation (44) and carrying out the integrations on $\mathbf{r}''$ and $\mathbf{v}''$, we obtain

$$\frac{\partial \overline{f}}{\partial t} + L\overline{f} = \epsilon_r^3 \epsilon_v^3 \frac{\partial}{\partial v^{\mu}} \int_0^t ds \int d^3\mathbf{r}' d^3\mathbf{v}' F^{\mu}(\mathbf{r}' \to \mathbf{r})_t F^{\nu}(\mathbf{r}' \to \mathbf{r})_{t-s}$$
$$\times \left\{ \overline{f}'(\eta_0 - \overline{f}') \frac{\partial \overline{f}}{\partial v^{\nu}} - \overline{f}(\eta_0 - \overline{f}) \frac{\partial \overline{f}'}{\partial v'^{\nu}} \right\}_{t-s}. \tag{48}$$

We have written $\overline{f}'_{t-s} \equiv \overline{f}(\mathbf{r}'(t-s), \mathbf{v}'(t-s), t-s)$, $\overline{f}_{t-s} \equiv \overline{f}(\mathbf{r}(t-s), \mathbf{v}(t-s), t-s)$, $F^{\mu}(\mathbf{r}' \to \mathbf{r})_t \equiv F^{\mu}(\mathbf{r}'(t) \to \mathbf{r}(t))$ and $F^{\nu}(\mathbf{r}' \to \mathbf{r})_{t-s} \equiv F^{\nu}(\mathbf{r}'(t-s) \to \mathbf{r}(t-s))$ where $\mathbf{r}(t-s)$ and $\mathbf{v}(t-s)$ are the position and velocity at time $t-s$ of the fluid particle located in $\mathbf{r} = \mathbf{r}(t)$, $\mathbf{v} = \mathbf{v}(t)$ at time t. It is determined by the characteristics (39) of the smoothed-out Lagrangian flow.

Equation (48) is a non Markovian integrodifferential equation: the value of $\overline{f}$ in $\mathbf{r}$, $\mathbf{v}$ at time t depends on the value of the whole field $\overline{f}(\mathbf{r}', \mathbf{v}', t-s)$ at earlier times. If the decorrelation time τ is short, we can make a *Markov approximation* and replace the bracket at time $t-s$ by its value taken at time t. Noting furthermore that the integral is dominated by the contribution of field stars close to the star under consideration (i.e. when $\mathbf{r}' \to \mathbf{r}$), we shall make a *local approximation* and replace $\overline{f}'(\eta_0 - \overline{f}')$ and $\frac{\partial \overline{f}'}{\partial v'^\nu}$ by their values taken at $\mathbf{r}$. In that case, the foregoing equation simplifies in

$$\frac{\partial \overline{f}}{\partial t} + L\overline{f} = \epsilon_r^3 \epsilon_v^3 \frac{\partial}{\partial v^\mu} \int_0^t ds \int d^3\mathbf{r}' d^3\mathbf{v}' F^\mu(\mathbf{r}' \to \mathbf{r})_t F^\nu(\mathbf{r}' \to \mathbf{r})_{t-s}$$
$$\times \left\{ \overline{f}'(\eta_0 - \overline{f}') \frac{\partial \overline{f}}{\partial v^\nu} - \overline{f}(\eta_0 - \overline{f}) \frac{\partial \overline{f}'}{\partial v'^\nu} \right\}_t, \quad (49)$$

where, now, $\overline{f}' = \overline{f}(\mathbf{r}, \mathbf{v}', t)$. The explicit reference to the past evolution of the system is only retained in the memory function

$$\int_0^t ds \int d^2\mathbf{r}' F^\mu(\mathbf{r}' \to \mathbf{r})_t F^\nu(\mathbf{r}' \to \mathbf{r})_{t-s}.$$

This function can be calculated explicitly if we assume that, between $t-s$ and t, the stars follow linear trajectories, so that $\mathbf{v}(t-s) = \mathbf{v}$ and $\mathbf{r}(t-s) = \mathbf{r} - \mathbf{v}s$ [2]. This leads to the generalized Landau equation

$$\frac{\partial \overline{f}}{\partial t} + L\overline{f} = \frac{\partial}{\partial v^\mu} \int d^3\mathbf{v}' K^{\mu\nu} \left\{ \overline{f}'(\eta_0 - \overline{f}') \frac{\partial \overline{f}}{\partial v^\nu} - \overline{f}(\eta_0 - \overline{f}) \frac{\partial \overline{f}'}{\partial v'^\nu} \right\}, \quad (50)$$

where $K^{\mu\nu}$ is the tensor

$$K^{\mu\nu} = 2\pi G^2 \epsilon_r^3 \epsilon_v^3 \ln \Lambda \frac{1}{u} \left(\delta^{\mu\nu} - \frac{u^\mu u^\nu}{u^2} \right), \quad (51)$$

and $\mathbf{u} = \mathbf{v}' - \mathbf{v}$, $\ln \Lambda = \ln(R/\epsilon_r)$. This equation applies to inhomogeneous systems but, as a result of the local approximation, the effect of inhomogeneity is only retained in the advective term. Equation (50) is very similar to the well-known Landau equation of collisional self-gravitating systems and electric charges [38]. There are nevertheless two important differences: (i) The friction term is non linear in $\overline{f}$, accounting for the degeneracy discovered by Lynden-Bell at equilibrium. (ii) The diffusion coefficient is proportional to the mass $\eta_0 \epsilon_r^3 \epsilon_v^3$ of a macrocell completely filled by the phase fluid instead of the mass m of an individual star in the ordinary Landau equation. In general $\eta_0 \epsilon_r^3 \epsilon_v^3 \gg m$ so that the relaxation by phase mixing is much more rapid than the collisional relaxation. From the above theory, we find that the collisionless relaxation operates on a time scale $\sim t_D$, the dynamical time, whereas the collisional relaxation operates on a time scale $t_{coll} \sim \frac{N}{\ln N} t_D \gg t_D$ where

$N \sim 10^{12}$ is the number of stars in the galaxy. Therefore, the relaxation by phase mixing really corresponds to a "violent relaxation" [43].

It can be shown that the generalized Landau equation (50) conserves mass and energy [2]. In fact, as a result of the local approximation, the energy is conserved locally, as if the system were homogeneous, and we have

$$\int \left(\frac{\partial \overline{f}}{\partial t}\right)_{c.g.} \frac{v^2}{2} d^3\mathbf{v} = \int \mathbf{J}_f \mathbf{v} d^3\mathbf{v} = 0, \qquad \forall \mathbf{r}. \tag{52}$$

It is also easy to show that equation (50) satisfies a H-theorem: $\dot{S} \geq 0$. When a stationary state is reached $\dot{S} = 0$ and the Fermi-Dirac distribution (16) is obtained, in agreement with Lynden-Bell's statistical theory [43]. This provides therefore another way of justifying his results from a dynamical point of view which does not explicitly rely on a maximization of entropy (the *H*-theorem is not assumed but *derived* from the kinetic theory). It can be noted that these properties result from the *symmetry* of the Landau collision term and not from Lagrange multipliers like in the thermodynamical approach. This is more satisfactory on a physical point of view.

It is important to stress, however, that this quasilinear theory cannot describe the early, very non linear and very chaotic, stages of the "violent relaxation". Indeed, the detailed study of Severne & Luwel [55] reveals that the various approximations introduced in the theory make equation (50) applicable only for $t \gg t_D$, where t_D is the dynamical time. Since the relaxation time is precisely of order t_D, the quasilinear theory is only *marginally* applicable and can describe, at most, the late quiescent phases of the relaxation, when the fluctuations have weaken. By contrast, it is plausible that equation (31) is more general and more appropriate to the context of "violent relaxation" since it results from a thermodynamical approach which exploits at best the chaoticity of the system and the complete lack of information that we have to face at small scales. As a clear difference, it should be noted that the MEPP takes into account only the *global* conservation of energy (the Lagrange multiplier $\beta(t)$ is a kind of inverse average temperature determined by an integration over the whole system) whereas the approximations introduced in the kinetic model lead to a *local* conservation of energy. This strong locality cannot account for the rather collective processes which are involved in the violent relaxation and may unveil a failure of the quasilinear theory. However, in section 7, we show that the two approaches are consistant if we are close to equilibrium and we use the quasilinear theory as a model to determine an explicit expression for the diffusion coefficient that appears in equation (31).

7 Truncated models for collisionless stellar systems

The equations provided by the MEPP and by the quasilinear theory have a very different mathematical structure. The relaxation equation (31) is a partial differential equation whereas the generalized Landau equation (50) is an integrodifferential equation: the value of $\overline{f}(\mathbf{v})$ at time $t+dt$ depends explicitly on the *whole* distribution of velocities $\overline{f}(\mathbf{v}')$ at time t through an integration over $\mathbf{v}'$. The usual way to transform an integrodifferential equation into a partial differential equation is to make a guess for the function appearing in the integral and refine the initial guess by successive iterations. In practice, we simply make *one* sensible guess. Therefore, if we are close to equilibrium (and this is in fact dictated by the conditions of validity of the quasilinear theory), it seems natural to replace the distribution function $\overline{f}'$ by the Fermi-Dirac distribution

$$\overline{f}' = \frac{\eta_0}{1+\lambda e^{\beta\eta_0\frac{v'^2}{2}}}. \tag{53}$$

In more physical terms, this amounts to a "thermal bath approximation": the stars have not yet relaxed completely, but when we focus on the relaxation of a given stellar fluid particle (described by $\overline{f}$) we can consider, in a first approximation, that the rest of the system (described by $\overline{f}'$) is at equilibrium. With this thermal bath approximation, the generalized Landau equation (50) reduces to the nonlinear partial differential equation [13]:

$$\frac{\partial\overline{f}}{\partial t}+\mathbf{v}\frac{\partial\overline{f}}{\partial\mathbf{r}}+\overline{\mathbf{F}}\frac{\partial\overline{f}}{\partial\mathbf{v}}=\frac{\partial}{\partial\mathbf{v}}\left[D\left(\frac{\partial\overline{f}}{\partial\mathbf{v}}+\beta\overline{f}(\eta_0-\overline{f})\mathbf{v}\right)\right], \tag{54}$$

with a diffusion coefficient

$$D=\frac{16\sqrt{2}\pi^2G^2\epsilon_r^3\epsilon_v^3\ln\Lambda}{\eta_0^{1/2}\beta^{5/2}v^3}\int_0^{\beta\eta_0\frac{v^2}{2}}\frac{\sqrt{x}}{1+\lambda e^x}dx. \tag{55}$$

Equation (54) is precisely the equation derived from the MEPP. Together with the explicit expression of the diffusion coefficient (55) it provides a self-consistent model for the "coarse-grained" dynamics of collisionless stellar systems experiencing violent relaxation. More general equations including anisotropic effects can also be obtained from this formalism [13].

We shall now assume that the galaxy is not isolated but subject to the tides of other systems. In that case, high energy stars that have elongated orbits are removed by the gravity of these objects. We seek therefore a stationary solution of equation (54) satisfying the boundary condition $\overline{f}(\epsilon_m)=0$, where ϵ_m is the escape energy above which $\overline{f}=0$. This solution will provide a truncated distribution function with a finite mass [13]. In fact, this problem was already tackled by Lynden-Bell in his seminal paper [43]. He proposed to describe the evolution of the coarse-grained distribution function $\overline{f}$ by the ordinary Fokker-Planck equation (32) with the heuristic argument that the fluctuations of the gravitational potential during violent relaxation play the same role as collisions. With the additional (heuristic) argument that $D\sim\frac{1}{v^3}$

for large velocities, he could obtain a stationary solution of a Michie-King type [8]:

$$\overline{f} = \begin{cases} A(e^{-\beta\eta_0\epsilon} - e^{-\beta\eta_0\epsilon_m}), & \epsilon \leq \epsilon_m, \\ 0, & \epsilon \geq \epsilon_m. \end{cases} \tag{56}$$

Since stars with $\epsilon \geq \epsilon_m$ are removed by the tidal field, this distribution function provides a depletion of high energy states and solves the infinite mass problem.

Our present approach justifies the two phenomenological arguments of Lynden-Bell since equation (54) reduces to the Fokker-Planck equation if we assume $\overline{f} \ll \eta_0$ (no degeneracy) and equation (55) gives a diffusion coefficient $D \sim \frac{1}{v^3}$ for large velocities. There is however a kind of "gap" in Lynden-Bell's approach since equation (56) does not reduce to the Fermi-Dirac statistics (16) in the limit of low energies. Using the more general equation (54) which properly accounts for degeneracy effects, we now try to build up a "one piece" distribution function which makes the bridge between Lynden-Bell's statistics (16) for $\epsilon \ll \epsilon_m$ and the Michie-King model (56) for $\epsilon \sim \epsilon_m$.

During the stage of violent relaxation, the stars extract their energy from the rapid fluctuations of the gravitational field. By this process, some stars may acquire very high energies and escape from the system (being ultimately captured by the gravity of other systems). For these stars, $D = \frac{K}{v^3}$ is a good approximation and equation (54) writes

$$\frac{\partial \overline{f}}{\partial t} + \mathbf{v}\frac{\partial \overline{f}}{\partial \mathbf{r}} + \overline{\mathbf{F}}\frac{\partial \overline{f}}{\partial \mathbf{v}} = \frac{\partial}{\partial \mathbf{v}}\left[\frac{K}{v^3}\left(\frac{\partial \overline{f}}{\partial \mathbf{v}} + \beta\overline{f}(\eta_0 - \overline{f})\mathbf{v}\right)\right]. \tag{57}$$

We seek a stationary solution of the form $\overline{f} = \overline{f}(\epsilon)$. Using the identity $\frac{\partial}{\partial \mathbf{v}}(\frac{\mathbf{v}}{v^3}) = 0$ (valid for sufficiently large $|\mathbf{v}|$), we obtain

$$0 = \frac{d}{d\epsilon}\left[K\left(\frac{d\overline{f}}{d\epsilon} + \beta\overline{f}(\eta_0 - \overline{f})\right)\right], \tag{58}$$

or, equivalently,

$$\frac{d\overline{f}}{d\epsilon} + \beta\overline{f}(\eta_0 - \overline{f}) = -J, \tag{59}$$

where J is a constant of integration. If $J = 0$, we recover Lynden-Bell's distribution function (16). If $J \neq 0$, equation (59) accounts physically for an escape of stars at a constant rate J. The system is therefore not truly static since it looses gradually stars but we may consider that the galaxy passes by a succession of quasi-stationary states which are solution of equation (59). As stated previously, this equation is valid only for high energy stars. For lower energies, the system has settled down in a pure equilibrium state and $J = 0$, leading to Lynden-Bell's distribution.

Our goal, now, is to solve equation (59). Put under the form

$$\frac{d\overline{f}}{d\epsilon} + \beta\eta_0\overline{f} - \beta\overline{f}^2 + J = 0, \tag{60}$$

we recognize a Riccati equation [33]. With the change of variables $\overline{f} = -\frac{1}{\beta}\frac{u'}{u}$, it can be converted into a linear partial differential equation of second order

$$\frac{d^2u}{d\epsilon^2} + \eta_0\beta\frac{du}{d\epsilon} - J\beta u = 0. \tag{61}$$

The associated characteristic polynomial $x^2 + \eta_0\beta x - J\beta = 0$ has a strictly positive discriminant $\Delta = \eta_0^2\beta^2 + 4J\beta > 0$. Therefore, the general solution of equation (61) is

$$u = e^{-\frac{\eta_0\beta}{2}\epsilon}(Ae^{\frac{\delta}{2}\epsilon} + Be^{-\frac{\delta}{2}\epsilon}), \tag{62}$$

where $\delta = \sqrt{\Delta}$ and A, B are constants of integration. The solution of equation (60) is therefore

$$\overline{f} = \frac{1}{2\beta}\frac{A(\eta_0\beta - \delta)e^{\frac{\delta}{2}\epsilon} + B(\eta_0\beta + \delta)e^{-\frac{\delta}{2}\epsilon}}{Ae^{\frac{\delta}{2}\epsilon} + Be^{-\frac{\delta}{2}\epsilon}}. \tag{63}$$

Setting $\lambda = A/B$ ($B \neq 0$ otherwise $\overline{f}$ would be constant), it can be rewritten

$$\overline{f} = \frac{1}{2\beta}\frac{\lambda(\eta_0\beta - \delta) + (\eta_0\beta + \delta)e^{-\delta\epsilon}}{\lambda + e^{-\delta\epsilon}}. \tag{64}$$

This distribution function vanishes at the escape energy $\epsilon = \epsilon_m$ defined by

$$\lambda(\eta_0\beta - \delta) + (\eta_0\beta + \delta)e^{-\delta\epsilon_m} = 0. \tag{65}$$

With this new variable, we obtain the result

$$\overline{f} = \lambda\eta_0\frac{e^{-\delta\epsilon} - e^{-\delta\epsilon_m}}{(\lambda - e^{-\delta\epsilon_m})(\lambda + e^{-\delta\epsilon})}, \tag{66}$$

where δ is a solution of

$$\delta = \eta_0\beta\frac{\lambda + e^{-\delta\epsilon_m}}{\lambda - e^{-\delta\epsilon_m}}. \tag{67}$$

Now, for the cases of physical interest $\lambda \gg e^{-\beta\eta_0\epsilon_m}$, which means that degeneracy is negligible for energies close to the escape energy ϵ_m. In that case, $\delta \simeq \eta_0\beta$ and we obtain the final result [13]:

$$\overline{f} = \eta_0\frac{e^{-\beta\eta_0\epsilon} - e^{-\beta\eta_0\epsilon_m}}{\lambda + e^{-\beta\eta_0\epsilon}}. \tag{68}$$

The previous calculation is justified as long as $D \neq 0$, corresponding to relatively strong fluctuations. When the fluctuations die down at the end of the relaxation, the diffusivity D and therefore the diffusion current $|\mathbf{J}|$ go to zero. There is no more evaporation but the distribution function (68) is maintained as a stationary solution of the Vlasov equation. When $\epsilon \sim \epsilon_m$, we recover the Michie-King model (56) and when $\epsilon \ll \epsilon_m$ equation (68) reduces to the Fermi-Dirac distribution function (16). Therefore, the distribution function (68) connects *continuously* the two limits considered by Lynden-Bell [43] in his seminal paper and can serve as a relevant model for (possibly degenerate) collisionless stellar systems. In particular, this distribution function could describe galactic halos limited in extension as a consequence of tidal interactions with other systems [34]. It could also be used to calculate realistic equilibrium states of collisionless stellar systems without the artifice of a material box. However, the main results of the box model [25] should not be dramatically altered.

The previous model describes self-gravitating systems subject to tidal forces. This form of confinement was first introduced in the case of globular clusters tidally trucated by a nearby galaxy [8]. This model can also describe a fraction of elliptical galaxies living in a rich environment. However, for the majority of elliptical galaxies, tidal forces are weak and the galaxy can be considered as isolated. Now, for isolated systems, other processes can account for "incomplete relaxation" and lead to different truncated models. Starting from equation (41), we can show that the diffusion coefficient is proportional to the fluctuations of the distribution function integrated over the velocity:

$$D^{\mu\nu} = 2\pi G^2 \epsilon_r^3 \epsilon_v^3 \ln \Lambda \int \frac{1}{u}\left(\delta^{\mu\nu} - \frac{u^\mu u^\nu}{u^2}\right)(\overline{f'^2} - \overline{f}'^2) d^3\mathbf{v}'. \tag{69}$$

Since these fluctuations rapidly decay as we depart from the relatively well-mixed central region of the galaxy, the diffusion current decreases also and this results in a confinement of the distribution function. It is expected therefore that the relaxation will be effective only in a limited region of space where $\overline{f^2} - \overline{f}^2$ is sufficiently large. On the other hand, as the system develops finer and finer filaments during the mixing process, the diffusion coefficient is expected to decrease in time. The diffusion current takes therefore increasingly small values and the relaxation is slowed down. This decay may be quite rapid and a quantitative treatment of this effect would require a better understanding of the correlation function $\overline{\tilde{f}(\mathbf{r},\mathbf{v},t)\tilde{f}(\mathbf{r}',\mathbf{v}',t)}$ whose expression was simply postulated in section 6. The development of these ideas will lead to other truncated models probably closer to those introduced phenomenologically by, e.g., Stiavelli & Bertin [58], Tremaine [59] and Hjorth & Madsen [31]. In particular, the above discussion is quite consistent with the scenario of incomplete violent relaxation developed by Hjorth & Madsen. These authors introduce a two-step process: (i) in a first step, they assume that violent relaxation proceeds to completion in a *finite* spatial region, of radius r_{max}, which represents roughly the core of the galaxy (where the fluctuations are important). At this stage, the escape energy represents no special threshold so that negative as well as positive energy states are populated in that region. (ii) After the relaxation process is over, positive energy particles leave the system and particles with $\Phi(r_{max}) < \epsilon < 0$ move in orbits beyond r_{max}, thereby changing the distribution function to something significantly 'thinner' than a Boltzmann distribution. The crucial point to realize is that the differential energy distribution $N(\epsilon)$, where $N(\epsilon)d\epsilon$ is the number of stars with energy between ϵ and $\epsilon + d\epsilon$, will be discontinuous at the escape energy $\epsilon = 0$ since there is a finite number of particles within r_{max} after the relaxation process. It can be shown that this discontinuity implies necessarily that $f(\epsilon) \sim (-\epsilon)^{5/2}$ for $\epsilon \to 0^-$ [35]. Therefore, the truncated model consistent with this scenario is [31]:

$$\overline{f} = \begin{cases} Ae^{-\beta\eta_0\epsilon} & \epsilon < \epsilon'; \\ B\ (-\epsilon)^{5/2} & \epsilon' \geq \epsilon < 0; \\ 0 & \epsilon \geq 0 \end{cases} \tag{70}$$

This truncated model corresponds formally to a compositite configuration made of an isothermal core and a polytropic envelope with index $n = 4$. Of course, if the core is degenerate, the Boltzmann distribution for $\epsilon < \epsilon'$ must be replaced by the Fermi-Dirac one. This truncated model gives very good fit with elliptical galaxies and can reproduce the $R^{1/4}$ law [31]. The box model of section 4 can be considered as a simple approximation of this more realistic model, the "box" playing the role of the envelope. In fact, the scenario developed by Hjorth & Madsen is almost equivalent to complete violent relaxation in a finite container with the container removed after the relaxation. This scenario solves the infinite mass problem and rehabilitates the use of statistical mechanics to understand the structure of elliptical galaxies.

8 Conclusion

We have described in this paper the process of violent relaxation in stellar systems from the viewpoint of statistical mechanics, as originally introduced by Lynden-Bell [43]. Lynden-Bell proposed that the structure of galaxies could result from a law of chaos: there is a total lack of information at small scales since the stars have complicated orbits, but the exciting phenomenon is that *microscopic* disorder leads to *macroscopic* order. The same ideas of statistical mechanics have been introduced in two-dimensional turbulence described by the Euler equation [53,46,23] to explain the formation and maintenance of large scale coherent vortices like the Great Red Spot of Jupiter or the cyclones and anticyclones that populate the earth atmosphere [26,6,7]. The formation of self-organized vortices in two-dimensional turbulence can also have applications in the context of planet formation where large-scale vortices present in the solar nebula could efficiently trap dust particles to form the planetesimals and the planets (see a complete list of references in Chavanis [15]). In fact, the statistical mechanics of the 2D Euler equation is equivalent to the theory of Lynden-Bell applying to the Vlasov equation. In a sense, the Vlasov equation is just the Euler equation for a "fluid" evolving in a six-dimensional phase space. In this analogy [11,14,17], the vorticity and the stream function in 2D turbulence play the same role as the distribution function and the gravitational field in galaxies. This analogy concerns not only the equilibrium states (the formation of large-scale structures) but also the relaxation towards equilibrium [27] and the statistics of fluctuations [24]. A kinetic theory of two-dimensional vortices can be developed in analogy with stellar dynamics [19]. In this kinetic theory, a point vortex experiences a diffusion process and a "systematic drift". This "systematic drift" [12] is the counterpart of the "dynamical friction" [10] experienced by a star as a result of close encounters. Relaxation equations equivalent to equations (31) and (50) have been derived in the context of vortex dynamics [54,12,16,19]. In addition, the problem of "incomplete relaxation" is also encountered in 2D turbulence to explain the confinement of a vortex (e.g., a dipole or a tripole) that forms after a rapid merging [27,23]. It has been demonstrated explicitly in two-dimensional turbulence (where the numerical simulations are easier to implement) that the relaxation equations derived from the MEPP and including a space dependant diffusion coefficient related to the fluctuations of the vorticity (similar to Eq. (69)) can account for this kinetic confinement [52]. We believe that the relaxation equations proposed in the stellar context should work as well. The statistical mechanics of continuous vorticity fields also predicts a Fermi-Dirac distribution at equilibrium with the same interpretation of the degeneracy as in Lynden-Bell's theory. Although this degeneracy is hard to evidence for galaxies (and remains controversal), it has been vindicated by various numerical simulations and laboratory experiments of two-dimensional fluids. This suggests that the degenerate version of the theory should also be used in the stellar context. If all these effects are taken into account correctly, it is plausible that the statistical mechanics of 2D vortices and self-gravitating systems has a chance to account for the fascinating process of self-organization in nature.

References

1. Antonov, V.A. (1962) Vest. Leningr. Gos. Univ. **7**, 135
2. Balescu, R. (1963) Statistical Mechanics of Charged Particles, Interscience, New York
3. Bettwieser, E., Sugimoto, D. (1984) Post-collapse evolution and gravothermal oscillation of globular clusters. Mon. Not. R. astr. Soc. **208**, 493
4. Billic, N., Viollier, R.D. (1997) Gravitational phase transition of heavy neutrino matter. Phys. Lett. B **408**, 75
5. Billic, N., Lindebaum, R.J., Tupper, G.B., Viollier, R.D. On the formation of degenerate heavy neutrino stars, to appear in Phys. Lett. B [astro-ph/0106209]
6. Bouchet, F., Sommeria, J. Emergence of intense jets and Jupiter Great Red Spot as maximum entropy structures, submitted to J. Fluid. Mech. [physics/0003079]

7. Bouchet, F., Chavanis, P.H., Sommeria, J. Statistical mechanics of Jupiter's Great Red Spot in the shallow water model, in preparation
8. Binney, J., Tremaine, S. (1987) Galactic Dynamics. Princeton Series in Astrophysics
9. Chandrasekhar, S. (1985) An Introduction to the Theory of Stellar Structure. Dover
10. Chandrasekhar, S. (1943) Stochastic problems in physics and astronomy. Rev. Mod. Phys. **15**, 1
11. Chavanis, P.H. (1996) Contribution à la mécanique statistique des tourbillons bidimensionnels. Analogie avec la relaxation violente des systèmes stellaires, Thèse de doctorat, Ecole Normale Supérieure de Lyon
12. Chavanis, P.H. (1998) Systematic drift experienced by a point vortex in two-dimensional turbulence. Phys. Rev. E **58**, R1199
13. Chavanis, P.H. (1998) On the coarse-grained evolution of collisionless stellar systems. Mon. Not. R. astr. Soc. **300**, 981
14. Chavanis, P.H. (1998) From Jupiter's Great Red Spot to the structure of galaxies: statistical mechanics of two-dimensional vortices and stellar systems. Annals of the New York Academy of Sciences **867**, 120
15. Chavanis, P.H. (2000) Trapping of dust by coherent vortices in the solar nebula. Astron. Astrophys. **356**, 1089
16. Chavanis, P.H. (2000) Quasilinear theory of the 2D Euler equation. Phys. Rev. Lett. **84**, 5512
17. Chavanis P.H. (2001) On the analogy between two-dimensional vortices and stellar systems. In: Kambe, T., Nakano, T., Miyauchi, T. (Eds.) Proceedings of the IUTAM Symposium on Geometry and Statistics of Turbulence, Kluwer Academic Publishers
18. Chavanis, P.H. Gravitational instability of finite isothermal spheres. To appear in Astron. Astrophys. [astro-ph/0103159]
19. Chavanis, P.H. (2001) Kinetic theory of point vortices: diffusion coefficient and systematic drift. Phys. Rev. E **64**, 026309
20. Chavanis, P.H. Phase transitions in self-gravitating systems. Self-gravitating fermions and hard spheres models. Submitted to Phys. Rev. E [cond-mat/0109294]
21. Chavanis, P.H. Gravitational instability of finite isothermal spheres in general relativity. Analogy with neutron stars. To appear in Astron. Astrophys. [astro-ph/0108230]
22. Chavanis, P.H., Rosier, C., Sire, C. Thermodynamics of self-gravitating systems. Submitted to Phys. Rev. E. [cond-mat/0107345]
23. Chavanis, P.H., Sommeria, J. (1998) Classification of robust isolated vortices in two-dimensional hydrodynamics. J. Fluid Mech. **356**, 259
24. Chavanis, P.H., Sire, C. (2000) The statistics of velocity fluctuations arising from a random distribution of point vortices: the speed of fluctuations and the diffusion coefficient. Phys. Rev. E **62**, 490
25. Chavanis, P.H., Sommeria, J. (1998) Degenerate equilibrium states of collisionless stellar systems. Mon. Not. R. astr. Soc. **296**, 569
26. Chavanis, P.H., Sommeria, J. Statistical mechanics of the shallow water system. To appear in Phys. Rev. E [physics/0004056]
27. Chavanis, P.H., Sommeria, J., Robert, R. (1996) Statistical mechanics of two-dimensional vortices and collisionless stellar systems, Astrophys. J. **471**, 385

28. Clemmow, P., Dougherty, J. Electrodynamics of Particles and Plasmas (New-York: Addison-Wesley)
29. Cohn, H. (1980) Late core collapse in star clusters and the gravothermal instability. Astrophys. J. **242**, 765
30. de Vega, H.J., Sanchez, N. Statistical mechanics of the self-gravitating gas: I. Thermodynamical limit and phase diagrams. [astro-ph/0101568]
31. Hjorth, J., Madsen, J. (1993) Statistical mechanics of galaxies. Mon. Not. R. astr. Soc. **265**, 237
32. Inagaki,S., Lynden-Bell, D. (1983) Self-similar solutions for post-collapse evolution of globular clusters. Mon. Not. R. astr. Soc. **205**, 913
33. Ince, E.L. (1956) Ordinary Differential Equations, Dover, New-York
34. Ingrosso, G., Merafina, M., Ruffini, R., Strafella, F. (1992) System of self-gravitating semidegenerate fermions with a cutoff of energy and angular momentum in their distribution function. A&A **258**, 223
35. Jaffe, W. (1987) In: de Zeeuw T. (Ed.) Proc. IAU Symp. 127, Structure and Dynamics of Elliptical Galaxies, Reidel, Dordrecht, 511
36. Jaynes, E.T. In: Rosenkrantz, R. (Ed.) The Minimum Entropy Production Principle. Kluwer, Dordrecht
37. Kadomtsev, B.B., Pogutse, O.P. (1970) Collisionless relaxation in systems with Coulomb interactions. Phys. Rev. Lett. **25**, 1155
38. Kandrup, H.E. (1981) A generalized Landau equation for a system with a self-consistent mean field: derivation from an N-particle Liouville equation. Astrophys. J. **244**, 316
39. Katz, J. (1978) On the number of unstable modes of an equilibrium. Mon. Not. R. astr. Soc. **183**, 765
40. Kull, A., Treumann, R.A., Böringer, H. (1996) Violent relaxation of indistinguishable objects and neutrino hot dark matter in clusters of galaxies. Astrophys. J. Lett. **466**, L1
41. Larson, R.B. (1970) A method for computing the evolution of star clusters. Mon. Not. R. astr. Soc. **147**, 323
42. Leeuwin, F., Athanassoula, E. (2000) Central cusp caused by a supermassive black hole in axisymmetric models of elliptical galaxies. Mon. Not. R. astr. Soc. **417**, 79
43. Lynden-Bell, D. (1967) Statistical mechanics of violent relaxation in stellar systems. Mon. Not. R. astr. Soc. **136**, 101
44. Lynden-Bell, D., Eggleton, P.P., (1980) On the consequences of the gravothermal catastrophe. Mon. Not. R. astr. Soc. **191**, 483
45. Lynden-Bell, D., Wood, R. (1968) The gravothermal catastrophe in isothermal spheres and the onset of red-giants structure for stellar systems. Mon. Not. R. astr. Soc. **138**, 495
46. Miller, J. (1990) Statistical mechanics of the Euler equation in two dimensions. Phys. Rev. Lett. **65**, 2137
47. Padmanabhan, T. (1989) Antonov instability and the gravothermal catastrophe-revisited. Astrophys. J. Supp. **71**, 651
48. Padmanabhan, T. (1990) Statistical mechanics of gravitating systems. Phys. Rep. **188**, 285
49. Rasio, F., Shapiro, S.L., Teukolsky, S.A. (1989) Solving the Vlasov equation in general relativity. Astrophys. J. **344**, 146
50. Robert, R. (1991) A maximum entropy principle for two-dimensional Euler equations. J. Stat. Phys. **65**, 531

51. Robert, R. (1998) On the gravitational collapse of stellar systems. Class. Quantum Grav. **15**, 3827
52. Robert, R., Rosier, C. (1997) The modelling of small scales in 2D turbulent flows: A statistical mechanical approach. J. Stat. Phys. **86**, 481
53. Robert, R., Sommeria, J. (1991) Statistical equilibrium states for two-dimensional flows. J. Fluid Mech. **229**, 291
54. Robert, R., Sommeria, J. (1992) Relaxation towards a statistical equilibrium state in two-dimensional perfect fluid dynamics. Phys. Rev. Lett. **69**, 2776
55. Severne, G., Luwel, M. (1980) Dynamical theory of collisionless relaxation. Astrophys. & Space Sci. **72**, 293
56. Shapiro, S.L., Teukolsky, S.A. (1995) The collapse of dense star clusters to supermassive black holes: the origin of quasars and AGNs. Astrophys. J. **292**, L41
57. Stiavelli, M. (1998) Violent relaxation around a massive black hole. Astrophys. J. Lett. **495**, L91
58. Stiavelli, M., Bertin, G. (1987) Statistical mechanics and equilibrium sequences of ellipticals. Mon. Not. R. astr. Soc. **229**, 61
59. Tremaine, S. (1987) In: de Zeeuw T. (Ed.) Proc. IAU Symp. 127, Structure and Dynamics of Elliptical Galaxies, Reidel, Dordrecht, 367

On the Role of Interface Energies in the Description of Material Behavior

Gianpietro Del Piero

Dipartimento di Ingegneria, Università di Ferrara, 44100 Ferrara, Italia

Abstract. A possibility for a unified theory for phenomena such as plasticity, rupture, and damage is examined here. It is shown that a variety of responses can be obtained starting from a purely elastic scheme and allowing for the creation of discontinuities at the price of an interface energy. Moreover it is conceivable to relate, at least qualitatively, the expression chosen for the interface energy to the fine properties of the structure of matter. The paper is restricted to one-dimensional models; their two and three dimensional counterparts still exhibit challenging problems of mechanical interpretation and mathematical formulation.

Contents

1 Introduction

At the level of constitutive equations relating macroscopic stress and strain measures, *real* materials exhibit a variety of behaviors. In the years, there was a tendency to focus on some prominent properties of material response and to study them individually; the result was the development of independent, sometimes unrelated, branches of Continuum Mechanics, such as Fracture Mechanics, Damage Mechanics, and many theories sharing the name of Plasticity.

Some hope for recovering a unified view is provided by a simple model, based on the assumption that the total energy of a body is the sum of two parts, a *bulk part* representing the elastic strain energy and a *surface part* associated with defects occurring both at the macroscopic and at the microscopic level. In the model, defects are represented by discontinuities in the displacement field. Such discontinuities may be large or small, concentrated at a single surface or distributed over surface-like regions diffused throughout the body. The first case is typical of fracture, and the second of plastic deformation.

At the present stage, the model is restricted to the one-dimensional case. Two or three dimensional generalizations are far from trivial, due to mathematical difficulties. Within this limit, fracture was studied in [3], plasticity in [7], and damage and other forms of progressive failure were considered in [6], [8]. A first comprehensive account was given in [5]. The present communication can be viewed as an updated review of the state of the research; mainly, it is based on the contents of a paper in progress [9], in which emphasis is given to the mathematical structure of the model.

2 Basic assumptions

Consider a bar of length l, and with any axial displacement u of the bar associate the energy

$$E(u) = \int_0^l w(u'(x))dx + \sum_{y \in \Gamma(u)} \theta([u](y)). \tag{1}$$

Here $[u](y)$ is the jump discontinuity of u at the point y, and $\Gamma(u)$ is the set of all jump points for u in $(0,l)$. If we assume that the functional E is defined over the set of all functions with bounded total variation over $(0,l)$, then $\Gamma(u)$ is a finite or countable set.

To avoid interpenetration, all jumps are supposed to be positive

$$[u](y) > 0 \qquad \forall y \in \Gamma(u). \tag{2}$$

The functions w and θ are the *bulk energy density* and the *interface energy*, respectively. The density w is assumed to be convex, to satisfy $w(0) = w'(0) = 0$ at the reference configuration $u' = 0$, and to grow to infinity at the extreme deformations $u' = -1$ and $u' = +\infty$. The interface energy is supposed to be non-negative and lower semicontinuous, and to satisfy $\theta(0) = 0$. Accordingly, the right limit $\theta(0+)$ may be either positive or zero. Both possibilities will be considered in the following.

We suppose that the bar is subject to a prescribed elongation βl

$$\int_0^l u'(x)dx + \sum_{y \in \Gamma(u)} [u](y) = \beta l, \tag{3}$$

and we wish to study the stable or metastable equilibrium configurations of the bar. *Equilibrium configurations* correspond to displacement fields u which satisfy the inequality

$$\lim_{\lambda \to 0+} \frac{1}{\lambda}(E(u + \lambda\eta) - E(u)) \geq 0 \tag{4}$$

for all perturbations η such that $u+\eta$ satisfies the conditions (2), (3) [9]. They are *stable* or *metastable* if they are global or local minima, respectively, for the energy E among all displacement fields satisfying (2) and (3).

Equilibrium configurations are characterized by the conditions

$$u'(x) = \varepsilon \qquad \forall x \in (0,l)\backslash\Gamma(u), \tag{5}$$

$$\theta'([u](y)) = w'(u'(x)) = \sigma \qquad \forall y \in \Gamma(u),\ \forall x \in (0,l)\backslash\Gamma(u), \tag{6}$$

which require that the strain ε be the same at all continuity points for u and that the stress σ be the same at all points of the bar. Moreover, if the given θ satisfies $\theta(0+) = 0$, a necessary condition for equilibrium is

$$w'(\varepsilon) \le \theta'(0+), \tag{7}$$

while a set of sufficient conditions for equilibrium is formed by equations (5), (6) and by inequality (7) with strict inequality sign [9].

3 Energy minima

If we restrict our attention to equilibrium configurations, the expression (1) of the energy and the condition (3) take the simpler form

$$E(u) = lw(\varepsilon) + \sum_{y\in\Gamma(u)} \theta([u](y)), \tag{8}$$

$$\varepsilon + l^{-1} \sum_{y\in\Gamma(u)} [u](x) = \beta. \tag{9}$$

By eliminating ε, we get an expression of the energy depending only upon the jumps. Let us call the jumps $q_1, q_2, \dots$. Because their total number is not specified a priori, it is convenient to represent the jumps by a sequence q whose elements q_k may take the zero value. Then the energy can be rewritten in the form

$$F^{l,\beta}(q) = lw(\beta - l^{-1}\sum_{k=1}^{\infty} q_k) + \sum_{k=1}^{\infty} \theta(q_k), \tag{10}$$

subject to the sole condition

$$q_k \ge 0. \tag{11}$$

Let us look to global minima for $F^{l,\beta}$. It is quite convenient to decompose the problem in the following way. First we look for the minimum of $F^{l,\beta}$ in the set of all sequences which have a prescribed sum a

$$\sum_{k=1}^{\infty} q_k = a, \tag{12}$$

and then we minimize with respect to a. In the first step, the first term in the expression (10) of the energy is known, and therefore it remains to look for the minimum of the functional

$$\Phi(q) := \sum_{k=1}^{\infty} \theta(q_k) \tag{13}$$

under the conditions (11) and (12). Denote by $\theta_s(a)$ the infimum of Φ for the given a. Then the minimum problem for $F^{l,\beta}$ reduces to the minimum problem for the function

$$f^{l,\beta}(a) = lw(\beta - l^{-1}a) + \theta_s(a) \tag{14}$$

among all non-negative a. By the assumptions made on w and θ, this function is coercive and lower semicontinuous; then, a minimizer does always exist. The same is not true for Φ, which may or may not be coercive, and therefore may or may not have minimizers, depending on θ. However, minimizers for Φ do always exist if we restrict the total number of non-null components of q. Moreover, if q^N is a minimizer in the set of all sequences with at most N non null components, it can be proved that $N \to q^N$ is a minimizing sequence for Φ.

Let $a_{l,\beta}$ be a minimizer for $f^{l,\beta}$. It has been proved in [9] that

(i) there is a ϱ in $[0,1]$ such that

$$0 \le a_{l,\beta} \le (\beta + \varrho)l, \tag{15}$$

(ii) if for a given l there is a $\beta_l \ge 0$ such that $a_{l,\beta_l} > 0$, then $a_{l,\beta} > 0$ for all $\beta > \beta_l$,

(iii) β_l is non-increasing with l,

(iv) $\beta_l > 0$ if either $\theta(0+) > 0$ or $\theta(0+) = 0$ and $\theta'(0+) > 0$, and $\beta_l = 0$ if $\theta(0+) = \theta'(0+) = 0$.

Inequality (15) provides an upper bound for the size of a jump in a stable equilibrium configuration of a bar, and (ii) shows that for every given l there is a *fracture threshold* β_l, such that all energy minimizers are unfractured configurations if $\beta < \beta_l$ and fractured configurations if $\beta > \beta_l$. By (iii), the fracture threshold decreases with increasing size of the bar, in agreement with the *size effect* well known in Fracture Mechanics. Finally, (iv) makes a difference between *coherent* and *incoherent* materials, the latter being those incapable of sustaining any elongation without breaking.

4 Subadditivity

The function θ_s introduced in the preceding section is the *subadditive envelope* of θ [2], [9]. Recall that a function θ is subadditive if

$$\theta(a+b) \le \theta(a) + \theta(b) \qquad \forall a, b \ge 0, \tag{16}$$

and that the subadditive envelope θ_s of θ is the greatest subadditive function such that $\theta_s(a) \leq \theta(a)$ for all $a \geq 0$.

The subadditivity properties of the interface energy are fundamental in determining the features of macroscopic material response. Let us denote by subθ the set of all *subadditivity points* of θ:

$$\operatorname{sub}\theta := \{a \geq 0 \mid \theta_s(a) = \theta(a)\}. \tag{17}$$

Denote, further, by $q^{l,\beta}$ the minimizers for $F^{l,\beta}$, if any, and with $N_{l,\beta}$ the minimum number of jumps among all minimizers $q^{l,\beta}$ for the given l and β. It may be proved that [9]

(i) if $a_{l,\beta} \in$ subθ, then $N_{l,\beta} \leq 1$,
(ii) all non-null components of $q^{l,\beta}$ belong to subθ,
(iii) if subθ contains a closed right interval $[0, d]$ of the origin, then there are minimizers $q^{l,\beta}$ for every $l \geq 0$ and $\beta \geq 0$, and $N_{l,\beta} < +\infty$,
(iv) if subθ does not contain any point in an open right interval $(0, d)$ of the origin, then there is a range of values of β and l for which there are no minimizers for $F^{l,\beta}$.

These results will be illustrated by examples in the following sections. Here we observe that (ii) is a restriction on the amplitude of the fracture openings which may occur in a stable equilibrium configuration, and that (iii) and (iv) provide existence and nonexistence conditions for energy minimizers, respectively.

5 Griffith's model

In his celebrated model for brittle fracture [10], Griffith assumes the interface energy

$$\theta(a) = \begin{cases} 0 & \text{if} \quad a = 0, \\ \gamma & \text{if} \quad a > 0, \end{cases} \tag{18}$$

with γ a positive constant, the *fracture toughness* of the material.

The function θ in (18) is concave, and concave functions are subadditive. Then $\theta = \theta_s$, and from the results (i), (iii) enounced in the preceding section we deduce that in Griffith's model there are stable equilibrium configurations with at most one discontinuity point, and there are minimizers $q^{l,\beta}$ for $F^{l,\beta}$ for all $l \geq 0$ and for all $\beta \geq 0$. If we assume for simplicity that the bulk energy density is quadratic

$$w(\varepsilon) = \frac{1}{2}c\varepsilon^2 \tag{19}$$

(to within a correction in a neighborhood of $\varepsilon = -1$ to satisfy the assumption $w((-1)+) = +\infty$), the expression (10) of the total energy becomes

$$F^{l,\beta}(a) = \begin{cases} \frac{1}{2}cl\beta^2 & \text{if} \quad a = 0, \\ \frac{1}{2}cl(\beta - l^{-1}a)^2 + \gamma & \text{if} \quad a > 0. \end{cases} \tag{20}$$

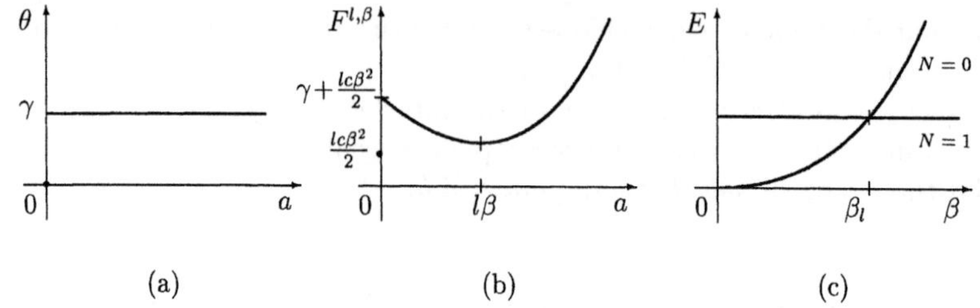

Fig. 1. Interface energy (a), total energy (b), and overall energy-elongation curves (c) in Griffith's model

The graphs of θ and $F^{l,\beta}$ are shown in Figures 1.a and 1.b, respectively. From the latter it is evident that there are two local minima, one at $a = 0$ corresponding to the unfractured equilibrium configuration and one at $a = \beta l$ corresponding to the fractured equilibrium configuration. The energies of the two configurations are plotted in Fig.1.3 as functions of the total elongation β. From the picture we see that the absolute minimum is attained at the unfractured configuration for small β and at the fractured configuration for large β. The fracture threshold

$$\beta_l = (\frac{2\gamma}{cl})^{\frac{1}{2}} \tag{21}$$

can be calculated by equating the expressions of the energy (20) at $a = 0$ and at $a = l\beta$.

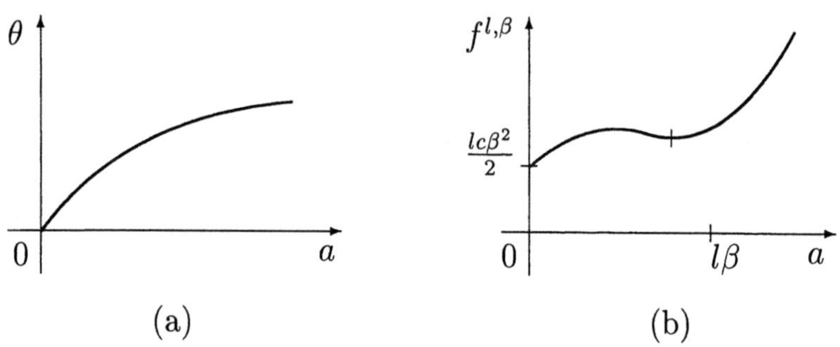

Fig. 2. Interface energy (a) and total energy (b) in Barenblatt's model

6 Barenblatt's model

In Barenblatt's model [1] the interface energy θ is continuous, concave and increasing, as shown in Fig.2.a. Consequently, it is subadditive, the number of the jump points in not greater than one, and there are minimizers $q^{l,\beta}$ for the energy for all $l \geq 0$ and for all $\beta \geq 0$. If we keep the expression (19) for the bulk energy density, we get the following expression of the total energy (14):

$$f^{l,\beta}(a) = \frac{1}{2}cl(\beta - l^{-1}a)^2 + \theta(a). \tag{22}$$

Its graph is shown in Fig.2.b. We see that there are still two local minima, corresponding to unfractured and fractured configurations, respectively. However, the second minimum is not anymore taken at $a = l\beta$ as in Griffith's model, and the first one disappears for sufficiently large β. By equating to zero the derivative of $f^{l,\beta}$ at $a = 0$, we see that this occurs at

$$\beta = c^{-1}\theta'(0+). \tag{23}$$

Beyond this limit, there are no metastable unfractured configurations. Using the equilibrium condition (6), which in the present circumstance becomes

$$\theta'(a) = c(\beta - l^{-1}a) = \sigma, \tag{24}$$

the total energy (22) can be expressed as a function of β. Its derivative is

$$\begin{aligned} \frac{d}{d\beta}f^{l,\beta}(a(\beta)) &= cl(\beta - l^{-1}a(\beta))(1 - l^{-1}\frac{d}{d\beta}a(\beta)) + \theta'(a(\beta))\frac{d}{d\beta}a(\beta) \\ &= cl(\beta - l^{-1}a(\beta)). \end{aligned} \tag{25}$$

By (24), it is equal to the stress multiplied by l:

$$\frac{d}{d\beta}f^{l,\beta}(a(\beta)) = l\sigma(\beta). \tag{26}$$

The overall energy-elongation and stress-elongation curves are shown in Fig.3. They are of two types, depending on the length l of the bar. For small l, Fig.3.a, the energy is increasing with β, there are no local minima other than global minima, and the transition from unfractured ($N = 0$) to fractured ($N = 1$) configurations occurs at the value of β given by (23). For large l, Fig.3.b, the situation is more complicated. There is still a transition from ($N = 0$) to ($N = 1$) at the β given by (23), but the curve ($N = 1$) now begins with an unstable branch which decreases with decreasing β. A metastable branch follows, which becomes stable after the intersection with the curve ($N=0$). The intersection occurs at the fracture threshold β_l. The unfractured configurations with $\beta > \beta_l$ are metastable.

The derivatives of the two functions give the stress-elongation curves shown in Figs.3.c, 3.d. For small l there is a continuous transition between the ascending branch ($N = 0$) and the descending branch ($N = 1$). Thus, beyond the fracture threshold there is a gradual decrease of the stress with increasing elongation of the bar. For large l, the decrease is abrupt at $\beta = \beta_l$. The two behaviors correspond to *ductile* and to *brittle* fracture, respectively. The fact that the governing parameter is the length of the bar confirms the relevance of the *size effect* in fracture.

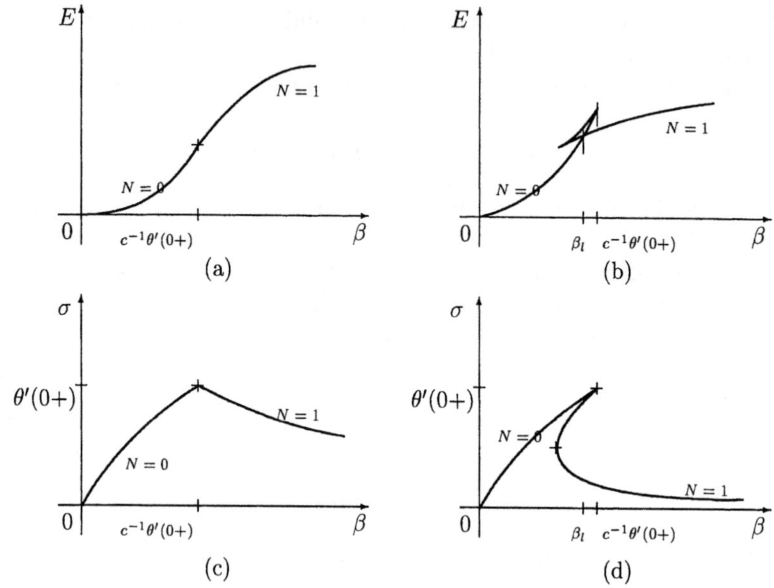

Fig. 3. Overall energy-elongation (a), (b) and stress-elongation curves (c), (d) in Barenblatt's model

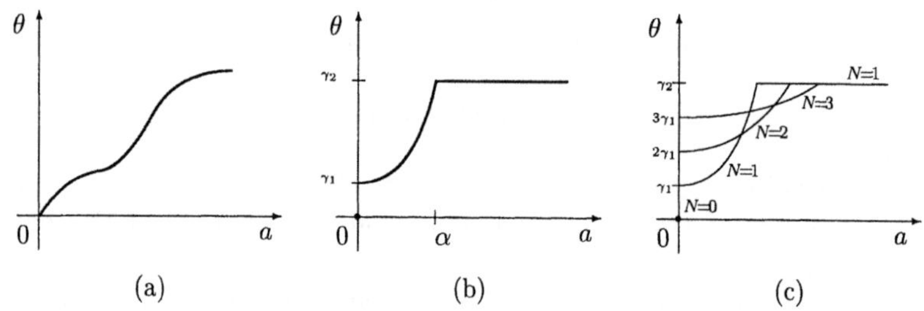

Fig. 4. Bi-modal interface energy (a), the energy (27) (b), and its subadditive envelope (c)

7 Bi-modal energies

A *bi-modal* interface energy is a function which has a convex domain located between two concave domains, as shown in Fig.4.a. This type of energy has two significant energy levels, one corresponding to nucleation of microdefects and one to opening of macrocracks [8]. For simplicity, let us consider the expression

$$\theta(a) = \begin{cases} 0 & \text{if} \quad a = 0, \\ c_1 a^2 + \gamma_1 & \text{if} \quad 0 < a < \alpha, \\ \gamma_2 & \text{if} \quad a > \alpha, \end{cases} \tag{27}$$

represented in Fig.4.b, with

$$c_1 := \frac{\gamma_2 - \gamma_1}{\alpha^2}. \tag{28}$$

In this case, the two significant energy levels are γ_1 and γ_2, and θ may or may not be subadditive, depending on the ratio γ_2/γ_1. Indeed, a simple computation shows that

$$\inf\left\{\sum_{i=1}^{N}\theta(\lambda_i a) \mid \lambda_i > 0, \sum_{i=1}^{N}\lambda_i = 1\right\} = N\theta(\frac{a}{N}), \tag{29}$$

so that

$$\theta_s(a) = \inf_{N\in\mathbb{N}}(N\theta(\frac{a}{N})). \tag{30}$$

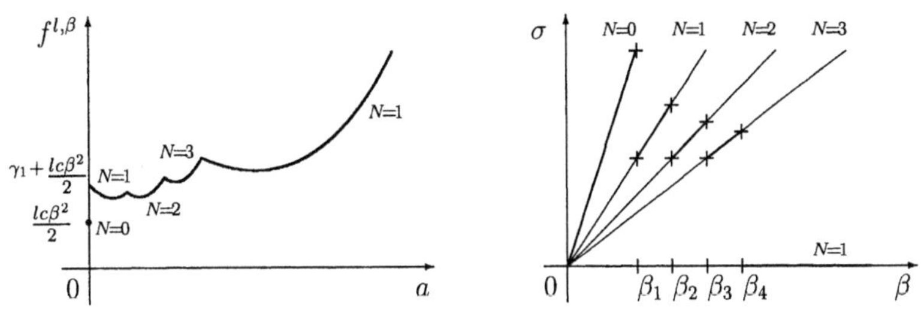

(a) (b)

Fig. 5. Energy-crack opening (a) and stress-elongation curves (b) for the bi-modal energy (27)

In our case, the curves $a \mapsto N\,\theta(a/N)$ are the parabolae

$$N\theta(\frac{a}{N}) = N(\gamma_1 + c_1\frac{a^2}{N^2}) \tag{31}$$

represented in Fig.4.c. At the intersection a_N between the Nth and the $(N+1)$th parabola we have

$$a_N{}^2 = \frac{\gamma_1}{c_1}N(N+1), \qquad N\theta(\frac{a_N}{N}) = \gamma_1(2N+1). \tag{32}$$

For $N = 1$, we have $\theta(a_1) > \gamma_2$ if the ratio γ_2/γ_1 is less than 3. If this is the case, θ is subadditive and the infimum (30) is attained at $N = 1$ for all non-null a. If $3 < \gamma_2/\gamma_1 < 5$, then the same infimum is attained at $N = 1$ for $a < a_1$, at $N = 2$ for $a_1 < a < \overline{a}_2$, with $\overline{a}_2$ such that $2\theta(\overline{a}_2/2) = \gamma_2$, and again at $N = 1$ for $a > \overline{a}_2$. For higher values of γ_2/γ_1, the infimum may be attained at greater values of N; in the example of Fig.4.c, N ranges from 1 to 3.

The total energy (14) is shown in Fig.5.a. We see that there may be several local minima for a given β: one at $a = 0$ with $N = 0$, one at each of the branches $N = 1, 2, 3, ..$, and one at large a with $N = 1$, corresponding to a macroscopically fractured configuration.

Keeping again the expression (19) for the bulk energy and using the equilibrium condition (6), it is possible to express a as a function of β. In the present case, for $0 < a < \alpha$ we have

$$2c_1 a = c(\beta - Nl^{-1}a) = \sigma, \tag{33}$$

and, therefore,

$$\sigma = \frac{2c_1 c}{2c_1 + Ncl^{-1}}\beta. \tag{34}$$

As shown in Fig.5b, the stress-elongation relation is defined by a family of straight lines from the origin, one for each of the N which enter in the definition of θ_s. Each line represents stable equilibrium configurations in the range $\beta_N < \beta < \beta_{N+1}$, with

$$\beta_N = (2c_1 c^{-1} + Nl^{-1})a_N. \tag{35}$$

The decreasing slope of these lines reveals the increasing level of damage in the material.

8 Convex-concave energies

As a final example let us consider an interface energy θ which is convex in a right interval of the origin and concave outside. As shown in Fig.6.a, the subadditive envelope θ_s of θ coincides with the curve θ for $a > d$, and with its tangent at the origin for $a < d$, d being the value of a at which curve and tangent intersect:

$$\theta(d) = \theta'(0+)d. \tag{36}$$

For $a < d$, the energy is

$$f^{l,\beta}(a) = \frac{1}{2}cl(\beta - l^{-1}a)^2 + \theta'(0+)a, \tag{37}$$

and the minimum is attained at

$$a_{l,\beta} = (\beta - c^{-1}\theta'(0+))l. \tag{38}$$

This is indeed the case if $0 < a_{l,\beta} < d$, and this occurs if

$$c^{-1}\theta'(0+) < \beta < c^{-1}\theta'(0+) + l^{-1}d. \tag{39}$$

For $\beta \leq c^{-1}\theta'(0+)$ we have $a_{l,\beta} = 0$, and the stable equilibrium configuration is unfractured. For $\beta \geq c^{-1}\theta'(0+) + l^{-1}d$ we have $a_{l,\beta} \geq d$ and therefore

$\theta(a_{l,\beta}) = \theta_s(a_{l,\beta})$. In this case $a_{l,\beta}$ belongs to subθ and by the result (i) in Section 4 the stable equilibrium configuration has a single fracture.

Let us calculate the number of fractures when β satisfies (39). Let q be any sequence of non-negative numbers q_h whose sum is $a_{l,\beta} < d$. Then $\theta(q_h) < \theta'(0+)\, q_h$ for all h, and therefore

$$\sum_{h=1}^{\infty} \theta(q_h) < \theta'(0+) \sum_{h=1}^{\infty} q_h = \theta'(0+) a_{l,\beta} = \theta_s(a_{l,\beta}). \tag{40}$$

Thus, the function Φ in (13) has no minimizers. This is an example of the situation of nonexistence of minimizers reported in item (iv) of Section 4. However, if we consider the sequence $N \to q^N$ with components

$$q^N{}_h = \begin{cases} N^{-1} a_{l,\beta} & \text{if} \quad h \leq N, \\ 0 & \text{if} \quad h > N, \end{cases} \tag{41}$$

we see that this is a minimizing sequence for Φ. Indeed,

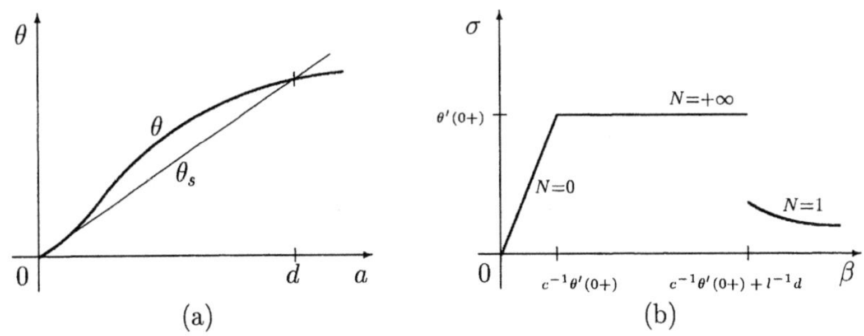

Fig. 6. Convex-concave interface energy and its subadditive envelope (a) and overall stress-elongation curve)(b)

$$\sum_{h=1}^{\infty} \theta(q^N{}_h) = N\theta(N^{-1} a_{l,\beta}) = N(N^{-1}\theta'(0+) a_{l,\beta} + o(N^{-1})), \tag{42}$$

so that in the limit for $N \to \infty$ we recover $\theta'(0+) a_{l,\beta} = \theta_s(a_{l,\beta})$. In [7] it is explained how minimizing sequences can be used to introduce a generalized concept of convergence, and how each minimizing sequence can be associated with an object called a *structured deformation* [4].

The limit situation of an infinite number of infinitely small discontinuities can be interpreted as a plastic deformation. To support this claim, let us evaluate the stress response for β in the range (39). By (37), (38) the energy minimum is

$$f^{l,\beta}(a_{l,\beta}) = \beta l \theta'(0+) - \frac{1}{2} c^{-1} l \theta'(0+)^2, \tag{43}$$

and because by (26) the stress is the derivative of the energy divided by l, we have

$$\sigma = \theta'(0+). \tag{44}$$

The overall stress-elongation curve is shown in Fig.6.b. For $\beta < c^{-1}\theta'(0+)$ we have a linear elastic response, for β in the range (39) we have a plateau of constant stress just like in perfect plasticity, and eventually at $\beta = c^{-1}\theta'(0+) + l^{-1}d$ we have a drop in the stress, to be interpreted as the occurring of macroscopic fracture.

Acknowledgements. This communication is based in part on results from the paper in progress [9], written in collaboration with L. Truskinovsky. The support of the Mathematisches Forschungsinstitut Oberwolfach, who hosted the two of us during the Summer 1999 within the Program *Research in Pairs* is gratefully acknowledged. This research has been supported by the Programma Nazionale Cofinanziato 2000 *Modelli Matematici per la Scienza dei Materiali.*

References

1. Barenblatt G.I. (1962) The mathematical theory of equilibrium cracks in brittle fracture, Adv. Appl. Mech. **7**,55-129.
2. Braides A., Coscia A. (1993) A singular perturbation approach to variational problems in fracture mechanics, Math. Mod. Meth. Appl. Sci **3**, 303-340.
3. Choksi R., Del Piero G., Fonseca I., Owen D.R. (1999) Structured deformations as energy minimizers in models of fracture and hysteresis, Math. and Mech. of Solids **4**, 321-356.
4. Del Piero G., Owen D.R. (1993) Structured deformations of continua, Arch. Rational Mech. Analysis **124**, 99-155.
5. Del Piero G. (1998) Towards a unified approach to fracture, yielding, and damage. In: Inan E., Markov K.Z. (Eds.), Continuum Models and Discrete Systems, World Scientific, Singapore, 67-692.
6. Del Piero G., Truskinovsky L. (1998) A one-dimensional model for localized and distributed failure, J. Phys. IV France **8**, Pr8, 95-102.
7. Del Piero G. (1999) One-dimensional ductile-brittle transition, yielding, and structured deformations. In: Argoul P., Frémond M., Nguyen Q.S. (Eds.) Variations of domains and free boundary problems in solid mechanics, Kluwer, Dordrecht, 203-210.
8. Del Piero G., Truskinovsky L. (2001) Macro and micro cracking in one-dimensional elasticity, Int. J. of Solids and Structures **38**, 1135-1148.
9. Del Piero G., Truskinovsky L., Elastic bars with cohesive energy, in preparation.
10. Griffith A. A. (1920) The phenomenon of rupture and flow in solids, Phil. Trans. Roy. Soc. London, A, **221**, 163-198.

Projection Generated Homogenization

Björn Engquist[1] and Olof Runborg[2]

[1] University of California, Los Angeles, Department of Mathematics, Los Angeles, CA, 90095-1555 and Royal Institute of Technology, Department of Numerical Analysis and Computer Science, 100 44, Stockholm

[2] Princeton University, PACM, Department of Mathematics, Princeton, NJ, 08544-1000

Abstract. Homogenization is a class of analytical techniques for approximating multiscale differential equations. The number of scales are reduced and the resulting equations are easier to analyze or numerically approximate. The classical homogenization technique is based on asymptotic expansions. We shall introduce a homogenization procedure based on projections of differential or discrete approximation operators onto coarser subspaces. The projection procedure is quite general and we give a presentation of a framework in Hilbert spaces. We apply the framework to a simple differential equation with oscillatory coefficients and to two discrete problems. One is wavelet based numerical homogenization and the other a multiscale finite element method.[3]

Contents

1 Introduction

The existence of subgrid scale phenomena in the numerical simulation of differential equations poses considerable difficulties. With subgrid scale phenomena we mean those processes which could influence the solution on the computational grid, but which have length scales shorter than the grid size. One way to deal with these difficulties is to replace the multiscale differential equation by homogenized or effective equations, see e.g. [3]. The homogenized equations are easier to approximate numerically since they do not explicitly contain the smallest scales. Another possibility is to derive numerical methods that mainly operate on the coarser scales. These methods typically originate

[3] The first author was partially supported by the NSF grant 0012151000. The second author was partially supported by the NSF KDI grant DMS 9872890.

from a fine grid representation of the original problem. We shall consider two such methods, wavelet based numerical homogenization [1] and a multiscale finite element method [14,15]. There are also other similar methods based on coarsening techniques from algebraic multigrid, see for example [18,20,21].

In this paper we present a framework for the derivation of both analytical and numerical homogenization. The objective is to find simplified operators that accurately reproduce the effect of subgrid scales. Let L_ε be a linear differential operator for which ε indicates small scales in the coefficients. The solution u_ε of the differential equation

$$L_\varepsilon u_\varepsilon = f_\varepsilon, \tag{1}$$

will typically inherit the small scales from the operator L_ε or the data f_ε. Consider for example the simple model problem

$$L_\varepsilon u_\varepsilon = -\frac{d}{dx}\left(g^\varepsilon(x)\frac{d}{dx}\right)u_\varepsilon(x) = f(x), \qquad 0 < x < 1, \tag{2}$$

$$u_\varepsilon(0) = u_\varepsilon(1) = 0. \tag{3}$$

The coefficient $g^\varepsilon(x)$ can for instance be oscillatory,

$$g^\varepsilon(x) = g(x/\varepsilon), \qquad g(y) \text{ 1-periodic}. \tag{4}$$

Our goal is to find an accurate and computationally efficient finite dimensional approximation of (1),

$$\bar{L}_{\varepsilon\Delta}\bar{u}_{\varepsilon\Delta} = \bar{f}_{\varepsilon\Delta}. \tag{5}$$

Difficulties originate from the small scales or high frequencies in (1). If ε denotes a typical wave length in u_ε and h a typical element size or step size in (5), then h must be substantially smaller than ε in order to resolve the ε-scale in the numerical approximation. This can be costly if ε is small compared to the overall size of the computational domain. Ideally the discretized operator in (5) should be designed such that $\bar{u}_{\varepsilon\Delta}$ is a good approximation of u_ε even if h is not small compared to ε. This goal resembles that of classical analytical homogenization.

Homogenization based on asymptotic derivation is a well established analytical technique. Let us here summarize a few simple results. Consider a set of operators L_ε in (1), indexed by the small parameter ε. The homogenization problem is to find the limit solution $\bar{u}$ and the *homogenized operator* $\bar{L}$ defined by

$$L_\varepsilon u_\varepsilon = f, \qquad \lim_{\varepsilon\to 0} u_\varepsilon = \bar{u}, \qquad \bar{L}\bar{u} = f. \tag{6}$$

In certain cases the convergence above and existence of the homogenized operator can be proved, [3].

For simple model problems, with coefficients that are periodic on the fine scale, exact closed form solutions can be obtained. For instance, with

$g(y)$ positive, 1-periodic and bounded away from zero, we have for the one-dimensional example (2),

$$L_\varepsilon = -\frac{d}{dx}\left(g(x/\varepsilon)\frac{d}{dx}\right), \qquad \bar{L} = -\tilde{g}\frac{d^2}{dx^2}, \qquad \tilde{g} = \left(\int_0^1 \frac{dy}{g(y)}\right)^{-1}. \tag{7}$$

With the same $\tilde{g}$ we get for the hyperbolic operators,

$$L_\varepsilon = \frac{\partial}{\partial t} + g(x/\varepsilon)\frac{\partial}{\partial x}, \qquad \bar{L} = \frac{\partial}{\partial t} + \tilde{g}\frac{\partial}{\partial x}. \tag{8}$$

In higher dimensions, the solution to (6) is a little more complicated, although for some model problems the structure of the homogenized operator can still be written down, as long as the coefficients are periodic or stochastic. In the d-dimensional elliptic case, let $G(y) : \mathbb{R}^d \mapsto \mathbb{R}^{d\times d}$ be uniformly elliptic and 1-periodic in each of its arguments. Let I_d denote the d-dimensional unit square. It can then be shown, [3], that

$$L_\varepsilon = -\nabla\cdot\left(G\left(\frac{x}{\varepsilon}\right)\nabla\right), \quad \bar{L} = -\nabla\cdot(\tilde{g}\nabla), \quad \tilde{g} = \int_{I_d} G(y) - G(y)\frac{d\chi(y)}{dy}dy, \tag{9}$$

where $d\chi/dy$ is the Jacobian of the function $\chi(y) : \mathbb{R}^d \mapsto \mathbb{R}^d$, given by solving the so called cell problem,

$$\nabla\cdot G(y)\frac{d\chi(y)}{dy} = \nabla\cdot G(y), \qquad y \in I_d, \tag{10}$$

with periodic boundary conditions for χ. Note that $\tilde{g}$ is a constant matrix.

The solution u_ε will be of the form

$$u_\varepsilon(x) = \bar{u}(x) + \varepsilon u_1(x, x/\varepsilon) + \varepsilon^2 u_2(x, x/\varepsilon) + \mathcal{O}(\varepsilon^3). \tag{11}$$

This is also the asymptotic ansatz, which is often used in the derivation of $\bar{L}$.

In the following section we introduce the projection technique for homogenzation in a Hilbert space setting. We show, in Sect. 3, that the zero ε-limit of the projection generated operator is the same as the effective operator $\bar{L}$ in classical homogenization for a model differential equation. In Sect. 4 and Sect. 5 we apply the technique to numerical finite dimensional operators.

2 Projection framework

In this section we describe the following approach to homogenization. Consider an equation $Lu = f$ where L is a linear operator, f a right hand side and u a solution that contains fine scales. Let P be a projection operator onto a subspace where the fine scales in the original solution do not exist. Our objective is to find the (projection generated) *homogenized* operator $\bar{L}$

such that $\bar{L}Pu = f$ for all f such that $Pf = f$. (When $Pf \neq f$ we also need to find the homogenized right hand side $\bar{f}$.) We confine ourselves to the case of Hilbert spaces.

Let X be a Hilbert space of functions, typically a Sobolev space. Let $X_0 \subset X$ be a closed subspace representing the coarse part of the functions, and denote by P_x the orthogonal (and symmetric) projection operator in X onto X_0. Let the spaces X_0 and $X_0^\perp$ inherit the innerproduct and norm of X, so that $\|u\|_X = \|u\|_{X_0}$ and $(u,v)_X = (u,v)_{X_0}$ when $u, v \in X_0$, and similar for $X_0^\perp$. In addition, set $Q_x = I_x - P_x$ where I_x is the identity operator in X, and introduce the unitary operator $\mathcal{W}_x$ on X defined by

$$\mathcal{W}_x : X \mapsto X_0^\perp \times X_0, \qquad \mathcal{W}_x u = \begin{pmatrix} Q_x u \\ P_x u \end{pmatrix}. \tag{12}$$

In the same way, define the corresponding operators P_y, Q_y and $\mathcal{W}_y$ for another Hilbert space Y with subspace Y_0. Let $\mathcal{L}(X,Y)$ be the set of bounded linear maps from X to Y. For an operator $L \in \mathcal{L}(X,Y)$, we have

$$\begin{aligned} \mathcal{W}_y L \mathcal{W}_x^* \begin{pmatrix} u \\ v \end{pmatrix} &= \mathcal{W}_y L(P_x + Q_x)(u+v) = \begin{pmatrix} Q_y L(P_x + Q_x)(u+v) \\ P_y L(P_x + Q_x)(u+v) \end{pmatrix} \\ &= \begin{pmatrix} Q_y L(P_x v + Q_x u) \\ P_y L(P_x v + Q_x u) \end{pmatrix} \equiv \begin{pmatrix} A & B \\ C & D \end{pmatrix} \begin{pmatrix} u \\ v \end{pmatrix} \end{aligned} \tag{13}$$

where

$$\begin{aligned} A &= Q_y L Q_x \in \mathcal{L}(X_0^\perp, Y_0^\perp), \qquad & B &= Q_y L P_x \in \mathcal{L}(X_0, Y_0^\perp), \\ C &= P_y L Q_x \in \mathcal{L}(X_0^\perp, Y_0), \qquad & D &= P_y L P_x \in \mathcal{L}(X_0, Y_0). \end{aligned} \tag{14}$$

When A is invertible the following definition can be stated:

Definition 1. Suppose $L \in \mathcal{L}(X,Y)$ and $f \in Y$. When A in (13, 14) is invertible (one-to-one and onto), we define the *homogenized* operator $\bar{L} : X_0 \mapsto Y_0$ as the Schur complement with respect to the decomposition in (13),

$$\bar{L} = D - CA^{-1}B, \tag{15}$$

and the homogenized right hand side as

$$\bar{f} = P_y f - CA^{-1}Q_y f. \tag{16}$$

We will write $\bar{L}_{X,X_0}$ and $\bar{f}_{X,X_0}$ when there is a need to show explicitly between which spaces the homogenization step is made. From Definition 1 we immediately have

Lemma 1. *Suppose $Lu = f$, where $L \in \mathcal{L}(X,Y)$, $u \in X$ and $f \in Y$. If A^{-1} exists,*

$$\bar{L}P_x u = \bar{f}. \tag{17}$$

Proof. Since $Lu = f$ we get

$$\mathcal{W}_y L \mathcal{W}_x^* \mathcal{W}_x u = \mathcal{W}_y f \qquad \Rightarrow \qquad \begin{pmatrix} A & B \\ C & D \end{pmatrix} \begin{pmatrix} Q_x u \\ P_x u \end{pmatrix} = \begin{pmatrix} Q_y f \\ P_y f \end{pmatrix}. \tag{18}$$

Moreover, since A is invertible, this system can be reduced with Gaussian elimination. It yields (17).

Note that $\bar{L}$ and $\bar{f}$ do not exist for all elements of $\mathcal{L}(X, Y)$ since A^{-1} may not exist. In particular, $X_0^\perp$ and $Y_0^\perp$ must have the same cardinality. For some classes of operators A is however indeed invertible. Before we show one such case, let us comment on how to define projections on the dual of X. We denote the duality product between X^* and X by $\langle \cdot, \cdot \rangle$, so that e.g. $\langle u, \cdot \rangle$ represents the bounded linear functional $u \in X^*$. It is clear that a bounded linear functional $\langle f, \cdot \rangle$ on $X_0 \subset X$, can be extended to a bounded linear functional on all of X, simply by adding the projection operator, $\langle f, P_x \cdot \rangle$. In fact, we shall identify X_0^* with a subspace of X^*, such that the projection P_{x^*} in X^* on X_0^* satisfies

$$\langle P_{x^*} f, v \rangle \equiv \langle f, P_x v \rangle, \qquad \forall v \in X. \tag{19}$$

Another way of putting this, is to say that $X_0^* = \mathcal{A}^{-1} X_0$ and $P_{x^*} = \mathcal{A}^{-1} P_x \mathcal{A}$, where $\mathcal{A} : X^* \mapsto X$ is the isomorphic Riesz mapping defined by $\langle u, v \rangle = (\mathcal{A}u, v)_X$ for all $v \in X$.

Proposition 1. *Suppose the operator $L \in \mathcal{L}(X, X^*)$ is given by the weak form*

$$Lu = f \qquad \Leftrightarrow \qquad a[u, v] = \langle f, v \rangle, \qquad \forall v \in X, \tag{20}$$

where $a[\cdot, \cdot] : X \times X \mapsto \mathbb{R}$ is a bilinear form. Let P_{x^*} be defined as in (19) and suppose that for some positive constants c_1 and c_2,

$$\begin{aligned} a[u, v] &\le c_1 \|u\|_X \|v\|_X , && \forall u, v \in X, \\ a[Q_x u, Q_x u] &\ge c_2 \|Q_x u\|_X^2 , && \forall u \in X, \end{aligned} \tag{21}$$

Then L is well-defined and $A : X_0^\perp \mapsto (X_0^\perp)^*$ is one-to-one and onto.

Proof. By the first inequality in (21), the bilinear form $a[u, v]$ is a bounded linear functional on X for each fixed u, so L is well-defined. To show that A is invertible, we need to show that for every $f \in (X_0^\perp)^*$ there is a unique $u \in X_0^\perp$ such that

$$a[u, v] = \langle f, v \rangle, \qquad \forall v \in X_0^\perp. \tag{22}$$

This follows from the Lax–Milgram lemma applied to $X_0^\perp$ if (21) holds.

Some properties kept by the homogenized operator are surjectivity, symmetry and ellipticity, as seen by this proposition. We show it in the case when $X = Y$ and $X_0 = Y_0$.

Proposition 2. *Let $L \in \mathcal{L}(X, X)$ be such that $\bar{L}$ exists. If L is onto, $\bar{L}$ is onto. If L is self-adjoint, $\bar{L}$ is self-adjoint. If L is self-adjoint and also*

$$c_1||u||_X^2 \leq (Lu, u) \leq c_2||u||_X^2, \qquad \forall u \in X. \tag{23}$$

for some positive constants c_1, c_2. Then, for the same constants,

$$c_1||v||_{X_0}^2 \leq \left(\bar{L}v, v\right) \leq (Dv, v) \leq c_2||v||_{X_0}^2, \qquad \forall v \in X_0. \tag{24}$$

Proof. Suppose L is onto. For any $f \in X_0 \subset X$ there exists a $u \in X$ such that $Lu = f$. By Lemma 1 then $\bar{L}Pu = \bar{f}$. From (16) we see that $\bar{f} = f$, so $\bar{L}$ is also onto. Let L be self-adjoint. Then $\mathcal{W}_x L \mathcal{W}_x^*$ is self-adjoint and therefore $A = A^*$, $D = D^*$ and $C = B^*$. When A is invertible this implies that $(A^{-1})^* = A^{-1}$. Hence, $\bar{L}^* = D^* - B^*(A^{-1})^* C^* = D - CA^{-1}B = \bar{L}$.

Next, assume (23). Take $v \in X_0$ and set $u = -A^{-1}Bv \in X_0^{\perp}$. Since $\mathcal{W}_x$ is unitary,

$$\begin{aligned} c_1||v||_{X_0}^2 &\leq c_1||v+u||_X^2 \leq (L(u+v), (u+v)) = \left(\begin{pmatrix} A & B \\ C & D \end{pmatrix}\begin{pmatrix} u \\ v \end{pmatrix}, \begin{pmatrix} u \\ v \end{pmatrix}\right) \\ &= \left(\begin{pmatrix} 0 \\ (-CA^{-1}B + D)v \end{pmatrix}, \begin{pmatrix} -A^{-1}Bv \\ v \end{pmatrix}\right) = \left(\bar{L}v, v\right). \end{aligned} \tag{25}$$

This proves the left inequality in (24). Since $C = B^*$ and A^{-1} is positive by the positivity of L, we have for the right inequality

$$\begin{aligned} \left(\bar{L}v, v\right) &= \left((D - CA^{-1}B)v, v\right) = (Dv, v) - \left(A^{-1}Bv, Bv\right) \\ &\leq (Dv, v) = (Lv, v) \leq c_2||v||_X^2 = c_2||v||_{X_0}^2. \end{aligned} \tag{26}$$

We conclude this section by showing that the homogenization is "independent of the path," again for $X = Y$ and $X_0 = Y_0$.

Proposition 3. *Let X_0 and X_1 be two closed subspaces of X, nested such that $X_0 \subset X_1 \subset X$. Suppose $L \in \mathcal{L}(X, X)$ is onto and that $\bar{L}_{X,X_0}$, $\bar{L}_{X,X_1}$ and $\overline{\bar{L}_{X,X_1}}_{X_1,X_0}$ all exist. If $\bar{L}_{X,X_0}$ is one-to-one, then*

$$\bar{L}_{X,X_0} = \overline{\bar{L}_{X,X_1}}_{X_1,X_0}, \tag{27}$$

$$\bar{f}_{X,X_0} = \overline{\bar{f}_{X,X_1}}_{X_1,X_0}. \tag{28}$$

Proof. Let P_j be the projection onto X_j. Take $u \in X_0$ and let $f = \bar{L}_{X,X_0}u \in X_0$. Since L is onto, there is a $w \in X$ such that $Lw = f$. By Lemma 1, we have

$$\bar{L}_{X,X_0}P_0 w = \bar{f}_{X,X_0}, \qquad \bar{L}_{X,X_1}P_1 w = \bar{f}_{X,X_1}. \tag{29}$$

Since $f \in X_0$ we must have $\bar{f}_{X,X_0} = \bar{f}_{X,X_1} = f$, by (16). Another application of Lemma 1, then gives

$$\overline{\bar{L}_{X,X_1}}_{X_1,X_0} P_0 P_1 w = \bar{f}_{X_1,X_0}. \tag{30}$$

Again, $\bar{f}_{X_1,X_0} = f$ and $P_0 P_1 w = P_0 w$ since $X_0 \subset X_1$. Hence,

$$\bar{L}_{X,X_0} P_0 w = \overline{\bar{L}_{X,X_1}}_{X_1,X_0} P_0 w = f = \bar{L}_{X,X_0} u. \tag{31}$$

But, $\bar{L}_{X,X_0}$ is one-to-one so $P_0 w = u$ and since u was arbitrary, (27) follows. Finally, letting f be an arbitrary element of X, there exists a u such that $Lu = f$ since L is onto and by Lemma 1

$$\bar{f}_{X,X_0} = \bar{L}_{X,X_0} P u = \overline{\bar{L}_{X,X_1}}_{X_1,X_0} P u = \overline{\bar{f}_{X,X_1}}_{X_1,X_0}. \tag{32}$$

This shows (28).

3 Relationship between the homogenization approaches

In this section we analyze how the projection generated homogenized operator relates to the classical homogenized operator. In the elliptic case, there is a striking similarity between the Schur complement in Definition 1 and the classical homogenized operator in (9), repeated here for convenience,

$$PLP - PLQ(QLQ)^{-1}QLP, \tag{33}$$

$$\nabla \left(\int_{I_d} G(y) dy \right) \nabla - \nabla \left(\int_{I_d} G(y) \frac{d\chi(y)}{dy} dy \right) \nabla. \tag{34}$$

Both are written as the average of the original operator minus a correction term, which is computed in a similar way for both operators. For the analytical case, a local elliptic cell problem is solved to get $G\partial_y \chi$, while in the projection case, a positive operator $A = QLQ$ defined on a subspace is inverted to obtain $LQA^{-1}B$. The average over the terms is obtained by integration in the analytical case, and by applying P in the projection case.

The relationship can be made more precise, and we will illustrate it by considering the case set in the one-dimensional torus $\mathbb{T}$, which means we are considering 1-periodic functions. The coarse space will be given by the lowest Fourier modes. Similar results in one dimension have been shown for wavelet bases, see [13,22].

Consider (2) with $g^\varepsilon = g(x/\varepsilon)$ where $g \in L^\infty(\mathbb{T})$ is 1-periodic. Let $X = H^1(\mathbb{T})$ with the dual $X^* = H^{-1}(\mathbb{T})$. The duality product between H^{-1} and H^1 will be written $\langle \cdot, \cdot \rangle$ and the L^2 innerproduct will be denoted by $(\cdot, \cdot)$. Define the operator $L_\varepsilon \in \mathcal{L}(X, X^*)$ by the weak form of the equations, such that $L_\varepsilon u_\varepsilon = f$ is equivalent to

$$a[u, v] := (g(x/\varepsilon)\partial_x u_\varepsilon, \partial_x v) = \langle f, v \rangle, \qquad \forall v \in H^1(\mathbb{T}). \tag{35}$$

We want to compare this with the classical homogenized operator,

$$\bar{L}\bar{u} = f \qquad \Leftrightarrow \qquad \tilde{g}\,(\partial_x \bar{u}, \partial_x v) = \langle f, v \rangle, \qquad \forall v \in H^1(\mathbb{T}), \tag{36}$$

where $\tilde{g}$ was given in (7) and (9),

$$\tilde{g} = \left(\int_0^1 \frac{dy}{g(y)}\right)^{-1} = \int_0^1 g(y) - g(y)\chi_y(y)dy \tag{37}$$

with χ being the solution to (10) in one dimension.

We start by introducing a family of ε-dependent spaces $\{X_0^\varepsilon\}$, which are assumed to be closed subspaces of $H^1(\mathbb{T})$. We denote the orthogonal projection in $H^1(\mathbb{T})$ onto X_0^ε by P_ε. On the dual side we use the subspaces $(X_0^\varepsilon)^*$ as discussed in (19). Their corresponding projections will be denoted by P_{ε^*}. We index the remaining operators by ε in the obvious way, setting $Q_\varepsilon = I - P_\varepsilon$ and

$$\bar{L}_\varepsilon = D_\varepsilon - C_\varepsilon A_\varepsilon^{-1} B_\varepsilon. \tag{38}$$

Our strategy is now to explicitly construct (38) corresponding to (35) for fixed ε and to prove that

$$\lim_{\varepsilon \to 0} \bar{L}_\varepsilon P_\varepsilon u = \bar{L} u \qquad \forall u \in H^1(\mathbb{T}), \tag{39}$$

in H^{-1}. If the spaces $\{X_0^\varepsilon\}$ have some additional structural properties, (40), and if $\bar{L}_\varepsilon$ is elliptic uniformly in ε, (43), the following theorem shows that the solution we get from solving $\bar{L}_\varepsilon \bar{u}_\varepsilon = f$ tends to the homogenized solution $\bar{u}$ strongly in H^1.

Theorem 1. *Let $\{X_0^\varepsilon\}$ be any ε-dependent family of closed subspaces of $H^1(\mathbb{T})$ such that, for all $u \in H_0^1(0,1)$, the corresponding projection operators $\{P_\varepsilon\}$ satisfy*

$$\lim_{\varepsilon \to 0} ||P_\varepsilon u - u||_{H^1} = 0, \qquad P_\varepsilon \mathbf{1} = \mathbf{1}, \tag{40}$$

where $\mathbf{1}$ is the constant function. Let $\bar{L}_\varepsilon$ be the operator in (38) corresponding to (35) and $\{X_0^\varepsilon\}$. Finally, let $\bar{L}$ be the classical homogenized operator (36). If

$$\bar{L}\bar{u} = f = \bar{L}_\varepsilon \bar{u}_\varepsilon, \qquad \bar{u} \in H_0^1(0,1), \quad \bar{u}_\varepsilon \in H_0^1(0,1) \cap X_0^\varepsilon, \tag{41}$$

$$\lim_{\varepsilon \to 0} \left\|(\bar{L}_\varepsilon P_\varepsilon - \bar{L})\bar{u}\right\|_{H^{-1}} = 0, \tag{42}$$

$$\inf_\varepsilon \langle \bar{L}_\varepsilon u, u\rangle \geq \delta \, ||u||_{H^1}^2, \qquad \forall u \in H_0^1 \cap X_0^\varepsilon, \tag{43}$$

then

$$\lim_{\varepsilon \to 0} ||\bar{u} - \bar{u}_\varepsilon||_{H^1} = 0. \tag{44}$$

Proof. By (43), for fixed $\varepsilon > 0$,

$$||u||_{H^1}^2 \leq \frac{1}{\delta} \langle \bar{L}_\varepsilon u, u\rangle \leq \frac{1}{\delta} ||\bar{L}_\varepsilon u||_{H^{-1}} ||u||_{H^1}, \qquad \forall u \in H_0^1 \cap X_0^\varepsilon. \tag{45}$$

So,

$$\|u\|_{H^1} \leq \frac{1}{\delta}\|\bar{L}_\varepsilon u\|_{H^{-1}}, \qquad \forall u \in H_0^1 \cap X_0^\varepsilon. \tag{46}$$

Let P'_ε be the orthogonal projection onto $H_0^1(0,1) \cap X_0^\varepsilon$, defined by $P'_\varepsilon u = P_\varepsilon u - (P_\varepsilon u)(0)$. Then for all $u \in H^1(\mathbb{T})$, by (40),

$$\bar{L}_\varepsilon(P_\varepsilon - P'_\varepsilon)u = (P_\varepsilon u)(0)\bar{L}_\varepsilon \mathbf{1} = (P_\varepsilon u)(0)(P_\varepsilon L_\varepsilon - C_\varepsilon A_\varepsilon^{-1} P_\varepsilon L_\varepsilon)\mathbf{1} = 0, \tag{47}$$

since $L_\varepsilon \mathbf{1} = 0$. Using (46, 47, 41) and the fact that $\bar{u}_\varepsilon \in H_0^1 \cap X_0^\varepsilon$, we obtain

$$\begin{aligned} \|\bar{u} - \bar{u}_\varepsilon\|_{H^1} &\leq \|P'_\varepsilon \bar{u} - \bar{u}\|_{H^1} + \|P'_\varepsilon(\bar{u} - \bar{u}_\varepsilon)\|_{H^1} \\ &\leq \|P'_\varepsilon \bar{u} - \bar{u}\|_{H^1} + \frac{1}{\delta}\|\bar{L}_\varepsilon P'_\varepsilon(\bar{u} - \bar{u}_\varepsilon)\|_{H^{-1}} \\ &= \|P'_\varepsilon \bar{u} - \bar{u}\|_{H^1} + \frac{1}{\delta}\|(\bar{L}_\varepsilon P_\varepsilon - \bar{L})\bar{u}\|_{H^{-1}} \\ &\leq \|P_\varepsilon \bar{u} - \bar{u}\|_{H^1} + |(P_\varepsilon \bar{u})(0)| + \frac{1}{\delta}\|(\bar{L}_\varepsilon P_\varepsilon - \bar{L})\bar{u}\|_{H^{-1}}. \end{aligned} \tag{48}$$

By (40, 42) and the Sobolev inequality $|u|_\infty \leq \|u\|_{H^1}$, the remaining terms vanish when $\varepsilon \to 0$.

In the remainder of the section we let $\{X_0^\varepsilon\}$ be the spaces of band limited functions, defined by

$$X_0^\varepsilon = \left\{ u = \sum_{|k|<1/2\varepsilon} a_k e^{2\pi i k x}, \; a_k \in \mathbb{C} \right\}. \tag{49}$$

Those finite-dimensional spaces are closed subspaces of both $L^2(\mathbb{T})$ and $H^1(\mathbb{T})$. The corresponding orthogonal projections are

$$P_\varepsilon u = \sum_{|k|<1/2\varepsilon} \hat{u}_k e^{2\pi i k x}, \qquad \hat{u}_k = \int_0^1 u(x) e^{-2\pi i k x} dx, \tag{50}$$

where $\hat{u}_k$ is the kth Fourier coefficient of $u(x)$. For this choice of spaces we will actually have that $\bar{L}_\varepsilon = P_{\varepsilon *} \bar{L} P_\varepsilon$.

Before continuing, we introduce some additional notation. We let $y = x/\varepsilon$. A y-dependent function seen as a function of x we will indicate by a superscripted epsilon, e.g. $g(x/\varepsilon) = g^\varepsilon(x)$ as before. A bar above the function denotes its mean value,

$$\bar{g} = \int_0^1 g(x)dx = \hat{g}_0, \tag{51}$$

where $\hat{g}_0$ is the zeroth Fourier coefficient of g.

We now begin by showing some properties of the projections.

Lemma 2. *The projections P_ε and Q_ε commutes with differentiation,*

$$\partial_x P_\varepsilon u = P_\varepsilon \partial_x u, \qquad \partial_x Q_\varepsilon u = Q_\varepsilon \partial_x u, \qquad \forall u \in H^1(\mathbb{T}). \tag{52}$$

For $g \in L^2(\mathbb{T})$ and $\varepsilon = 1/n$ with $n \in \mathbb{Z}^+$,

$$P_\varepsilon g^\varepsilon P_\varepsilon u = \bar{g} P_\varepsilon u, \qquad \forall u \in L^2(\mathbb{T}). \tag{53}$$

For $u \in L^2(\mathbb{T})$,

$$\overline{P_\varepsilon u} = \overline{u}. \tag{54}$$

Finally,

$$||\partial_x Q_\varepsilon u||_{L^2} \geq \frac{\pi}{\varepsilon} ||Q_\varepsilon u||_{L^2}. \tag{55}$$

Proof. Take $u \in H^1(\mathbb{T})$ and let $\{\hat{u}_k\}$ be its Fourier coefficients. Then

$$\begin{aligned}\partial_x P_\varepsilon u &= \partial_x \sum_{|k|<1/2\varepsilon} \hat{u}_k e^{2\pi i k x} = \sum_{|k|<1/2\varepsilon} 2\pi i k \hat{u}_k e^{2\pi i k x} \\ &= P_\varepsilon \sum_k 2\pi i k \hat{u}_k e^{2\pi i k x} = P_\varepsilon \partial_x u. \end{aligned} \tag{56}$$

Since $Q_\varepsilon = I - P_\varepsilon$, we have shown (52). Moreover,

$$P_\varepsilon g^\varepsilon P_\varepsilon u = P_\varepsilon \sum_\ell \sum_{|k|<n/2} \hat{g}_\ell \hat{u}_k e^{2\pi i (k+n\ell)x} = \hat{g}_0 \sum_{|k|<n/2} \hat{u}_k e^{2\pi i k x} = \bar{g} P_\varepsilon u, \tag{57}$$

since $|k + n\ell| \geq n/2$ when $\ell \neq 0$ and $|k| < n/2$. This proves (53). Next, for $u \in L^2$,

$$\overline{P_\varepsilon u} = (P_\varepsilon u, 1)_{L^2} = \sum_{|k|<1/2\varepsilon} \hat{u}_k \left(e^{2\pi i k x}, 1\right) = \hat{u}_0 = \overline{u}, \tag{58}$$

showing (54). Finally, (55) follows from

$$||\partial_x Q_\varepsilon u||_{L^2}^2 = \sum_{|k|\geq 1/2\varepsilon} 4\pi^2 k^2 \hat{u}_k^2 \geq \frac{\pi^2}{\varepsilon^2} \sum_{|k|\geq 1/2\varepsilon} \hat{u}_k^2 = \frac{\pi^2}{\varepsilon^2} ||Q_\varepsilon u||_{L^2}^2 . \tag{59}$$

By (55), we now have

$$a[u,v] \leq ||g||_{L^\infty} ||u||_{H^1} ||v||_{H^1} , \tag{60}$$

$$a[Q_\varepsilon u, Q_\varepsilon u] \geq \operatorname{ess\,inf}_{x\in\mathbb{T}} |g(x)| \ ||\partial_x Q_\varepsilon u||_{L^2}^2 \geq C \operatorname{ess\,inf}_{x\in\mathbb{T}} |g(x)| \, ||Q_\varepsilon u||_{H^1}^2 , \tag{61}$$

so Proposition 1 shows that A_ε is invertible and $\bar{L}_\varepsilon$ exists for all ε. We can then show

Theorem 2. *The projection generated homogenized operator $\bar{L}_\varepsilon$ corresponding to (35) and $\{X_0^\varepsilon\}$ as in (49), satisfies*

$$\bar{L}_\varepsilon = D_\varepsilon - C_\varepsilon A_\varepsilon^{-1} B_\varepsilon = P_{\varepsilon^*} \bar{L} P_\varepsilon, \tag{62}$$

if $\varepsilon = 1/n$ and $n \in \mathbb{Z}^+$. Then we also have the convergence,

$$\lim_{\varepsilon \to 0} \left\| (\bar{L}_\varepsilon P_\varepsilon - \bar{L}) u \right\|_{H^{-1}} = 0, \qquad \forall u \in H^1(\mathbb{T}). \tag{63}$$

Proof. By (52, 53) and (19) we have for D_ε,

$$\begin{aligned}
\langle D_\varepsilon u, v \rangle &= \langle P_{\varepsilon^*} L_\varepsilon P_\varepsilon u, v \rangle = \langle L_\varepsilon P_\varepsilon u, P_\varepsilon v \rangle = (g^\varepsilon \partial_x P_\varepsilon u, \partial_x P_\varepsilon v) \\
&= (g^\varepsilon P_\varepsilon \partial_x u, P_\varepsilon \partial_x v) = (P_\varepsilon g^\varepsilon P_\varepsilon \partial_x u, \partial_x P_\varepsilon v) \\
&= \bar{g} \left(\partial_x P_\varepsilon u, \partial_x P_\varepsilon v \right),
\end{aligned} \tag{64}$$

when $u, v \in H^1(\mathbb{T})$. For the second part of the operator, the equation $A_\varepsilon w = B_\varepsilon u$ needs to be solved. Then $C_\varepsilon A_\varepsilon^{-1} B_\varepsilon u$ is given by $C_\varepsilon w$. In this case we can directly write down the expression for w, namely

$$w(x) = \int_0^x \chi_y \left(\frac{x'}{\varepsilon} \right) \partial_x P_\varepsilon u(x') dx'. \tag{65}$$

where χ is the solution to the cell problem (10). It is well-known that $\chi \in H^1(\mathbb{T})$. By also using (52, 53, 54) and the fact that $P_\varepsilon u \in L^\infty(\mathbb{T})$, we have

$$\begin{aligned}
\|w\|_{L^2} &\le |w|_\infty \le \left(|\chi_y^\varepsilon|, |\partial_x P_\varepsilon u| \right) \le C \, \|\chi\|_{H^1} \, \|P_\varepsilon u\|_{H^1}, & (66) \\
\|\partial_x w\|_{L^2} &\le |\partial_x P_\varepsilon u|_\infty \left\| \chi_y^\varepsilon \right\|_{L^2} \le C |\partial_x P_\varepsilon u|_\infty \, \|\chi\|_{H^1}, & (67) \\
w(1) &= \overline{\chi_y^\varepsilon \partial_x P_\varepsilon u} = \overline{P_\varepsilon \chi_y^\varepsilon P_\varepsilon \partial_x u} = 0 = w(0). & (68)
\end{aligned}$$

Therefore $w \in H^1(\mathbb{T})$. Moreover, by (52) and (53)

$$\partial_x Q_\varepsilon w = \partial_x w - P_\varepsilon \partial_x w = \partial_x w - P_\varepsilon \chi_y^\varepsilon P_\varepsilon \partial_x u = \partial_x w. \tag{69}$$

Now, from (10) it follows that $g^\varepsilon \chi_y^\varepsilon = g^\varepsilon + c_0$ a.e., where c_0 is a constant. Together with (69) this shows that w indeed satisfies $A_\varepsilon w = B_\varepsilon u$,

$$\begin{aligned}
\langle A_\varepsilon w, v \rangle &= \langle L_\varepsilon Q_\varepsilon w, Q_\varepsilon v \rangle = (g^\varepsilon \partial_x Q_\varepsilon w, \partial_x Q_\varepsilon v) = \left(g^\varepsilon \chi_y^\varepsilon \partial_x P_\varepsilon u, \partial_x Q_\varepsilon v \right) \\
&= (g^\varepsilon \partial_x P_\varepsilon u, \partial_x Q_\varepsilon v) + c_0 \left(\partial_x P_\varepsilon u, \partial_x Q_\varepsilon v \right) \\
&= \langle L_\varepsilon P_\varepsilon u, Q_\varepsilon v \rangle + c_0 \left(P_\varepsilon \partial_x u, Q_\varepsilon \partial_x v \right) = \langle B_\varepsilon u, v \rangle .
\end{aligned} \tag{70}$$

Then, finally,

$$\begin{aligned}
\left\langle C_\varepsilon A_\varepsilon^{-1} B_\varepsilon u, v \right\rangle &= \langle C_\varepsilon w, v \rangle = \langle L_\varepsilon Q_\varepsilon w, P_\varepsilon v \rangle = (g^\varepsilon \partial_x Q_\varepsilon w, \partial_x P_\varepsilon v) \\
&= \left(g^\varepsilon \chi_y^\varepsilon \partial_x P_\varepsilon u, \partial_x P_\varepsilon v \right) = \left(P_\varepsilon g^\varepsilon \chi_y^\varepsilon P_\varepsilon \partial_x u, P_\varepsilon \partial_x v \right) \\
&= \overline{g \chi_y} \left(\partial_x P_\varepsilon u, \partial_x P_\varepsilon v \right),
\end{aligned} \tag{71}$$

which gives

$$\left\langle D_\varepsilon - C_\varepsilon A_\varepsilon^{-1} B_\varepsilon u, v\right\rangle = (\bar{g} - \overline{g\chi_y})\,(\partial_x P_\varepsilon u, \partial_x P_\varepsilon v) = \left\langle \bar{L} P_\varepsilon u, P_\varepsilon v\right\rangle$$
$$= \left\langle P_{\varepsilon^*} \bar{L} P_\varepsilon u, v\right\rangle . \tag{72}$$

The convergence (63) follows easily. For $u \in H^1(\mathbb{T})$,

$$\left\|(\bar{L}_\varepsilon P_\varepsilon - \bar{L})u\right\|_{H^{-1}} = \left\|(P_{\varepsilon^*}\bar{L}P_\varepsilon - \bar{L})u\right\|_{H^{-1}} = \sup_{\|v\|_{H^1}=1} \left|\left\langle (P_{\varepsilon^*}\bar{L}P_\varepsilon - \bar{L})u, v\right\rangle\right|$$
$$= \tilde{g} \sup_{\|v\|_{H^1}=1} \left|(P_\varepsilon \partial_x u, P_\varepsilon \partial_x v) - (\partial_x u, \partial_x v)\right|$$
$$\leq \tilde{g}\, \|P_\varepsilon u - u\|_{H^1} , \tag{73}$$

which vanishes with ε.

The result in Theorem 2 together with Theorem 1 now gives the strong convergence in H^1 of $\bar{u}_\varepsilon$ to $\bar{u}$.

Corollary 1. *Let $\bar{L}_\varepsilon$ be the operator in (38) corresponding to (35) and $\{X_0^\varepsilon\}$ as in (49). Let $\bar{L}$ be the classical homogenized operator (36). If*

$$\bar{L}\bar{u} = f = \bar{L}_\varepsilon \bar{u}_\varepsilon, \qquad \bar{u} \in H_0^1(0,1), \quad \bar{u}_\varepsilon \in H_0^1(0,1) \cap X_0^\varepsilon, \tag{74}$$

then

$$\lim_{\varepsilon \to 0} \|\bar{u} - \bar{u}_\varepsilon\|_{H^1} = 0. \tag{75}$$

Proof. Since (42) follows from Theorem 2 we only need to show (40, 43) to obtain (75) from Theorem 1. Standard results in harmonic analysis give (40). By Poincaré's inequality we have $\left\langle \bar{L}u, u\right\rangle \geq \delta \|u\|_{H^1}^2$ for all $u \in H_0^1(0,1)$. Together with (62) in Theorem 2 we then get, for all $u \in H_0^1 \cap X_0^\varepsilon$,

$$\left\langle \bar{L}_\varepsilon u, u\right\rangle = \left\langle \bar{L} P_\varepsilon u, P_\varepsilon u\right\rangle = \left\langle \bar{L}u, u\right\rangle \geq \delta \|u\|_{H^1}^2 . \tag{76}$$

4 Wavelet based homogenization

A wavelet representation lends itself naturally to analyzing the fine and coarse scales of a function. In this section we show how the projection generated homogenization works when we choose wavelet spaces as our coarse and fine space decomposition. We assume a basic knowledge of multiresolution analysis and its related wavelet spaces. For a detailed description we refer the reader to e.g. the book by Daubechies, [10]. In this setting, we let $Lu = f$ be a finite-dimensional approximation of a partial differential equation and apply the machinery in Sect. 2. We also demonstrate how the resulting $\bar{L}$ can be approximated by a sparse matrix and used for computing an approximate solution to the differential equation on a coarse grid.

Let us consider the equation

$$L_{j+1}U = F, \qquad U, F \in V_{j+1}, \qquad L_{j+1} \in \mathcal{L}(V_{j+1}, V_{j+1}), \tag{77}$$

set in the wavelet space V_{j+1}. This equation may originate from a finite difference, finite element or finite volume discretization of a given differential equation. In the simple case of Haar wavlets, U can be identified as a piecewise constant approximation of $u(x)$, the solution to the continuous problem.

For a function in V_{j+1}, the coarse scale is represented by V_j. The fine scale part lies in the wavelet space W_j and, assuming the wavelet system is orthogonal, $V_j \perp W_j$. In the first step we are therefore interested in the homogenized operator given by $\bar{L}_{V_{j+1},V_j}$. Let $\mathcal{W}_j : V_{j+1} \mapsto V_j \times W_j$ be the unitary operator corresponding to (12). As in the infinite dimensional case (18), we apply the transformation $\mathcal{W}_j$ to L_{j+1} from the left and get

$$\mathcal{W}_j L_{j+1} \mathcal{W}_j^* (\mathcal{W}_j U) = \mathcal{W}_j F, \tag{78}$$

or if we decompose $\mathcal{W}_j L_{j+1} \mathcal{W}_j^*$ according to (13),

$$\begin{pmatrix} A_j & B_j \\ C_j & L_j \end{pmatrix} \begin{pmatrix} U_\mathrm{f} \\ U_\mathrm{c} \end{pmatrix} = \begin{pmatrix} F_\mathrm{f} \\ F_\mathrm{c} \end{pmatrix}, \qquad U_\mathrm{f}, F_\mathrm{f} \in W_j, \qquad U_\mathrm{c}, F_\mathrm{c} \in V_j. \tag{79}$$

Here the subindex "f" means projection onto the fine scale subspace, W_j in this case, and subindex "c" stands for projection onto the coarse scale subspace, V_j here. This is a convention we will use frequently. Also note that we changed the notation of the block previously called D_j to L_j, to indicate that $L_j = PL_{j+1}P$ actually corresponds to one type of direct discretization on the coarse scale.

The homogenized "coarse grid operator" is now given by block Gaussian elimination of (79),

$$(L_j - C_j A_j^{-1} B_j) U_\mathrm{c} = F_\mathrm{c} - C_j A_j^{-1} F_\mathrm{f}. \tag{80}$$

For simplicity we will use the notation $\bar{L}_j \equiv \overline{L_{j+1}}_{V_{j+1},V_j}$ and $\bar{F}_j \equiv \bar{F}_{V_{j+1},V_j}$. Hence,

$$\bar{L}_j = L_j - C_j A_j^{-1} B_j, \qquad \bar{F}_j = F_\mathrm{c} - C_j A_j^{-1} F_\mathrm{f}. \tag{81}$$

Since L_j corresponds to a direct coarse discretization, we can interpret the operator $C_j A_j^{-1} B_j$ as a correction term, which includes subgrid phenomena in $\bar{L}_j$.

In the finite dimensional case, we represent the operators using their matrix representations with respect to the wavelet bases. For instance in the one dimensional Haar basis, $\mathcal{W}_j$ is represented by the matrix

$$\mathcal{W}_j = \begin{pmatrix} \tilde{Q}_j \\ \tilde{P}_j \end{pmatrix} = \frac{1}{\sqrt{2}} \begin{pmatrix} 1 & -1 & 0 & \cdots & & \\ 0 & 0 & 1 & -1 & 0 & \cdots \\ \vdots & \vdots & & \ddots & \ddots & \\ 0 & 0 & \cdots & 0 & 1 & -1 \\ 1 & 1 & 0 & \cdots & & \\ 0 & 0 & 1 & 1 & 0 & \cdots \\ \vdots & \vdots & & \ddots & \ddots & \\ 0 & 0 & \cdots & 0 & 1 & 1 \end{pmatrix}. \tag{82}$$

Note, in this representation $\tilde{P}_j^*$ and $\tilde{Q}_j^*$ are bases in V_{j+1} for V_j and W_j, respectively. The corresponding projections in V_{j+1} are $\tilde{P}_j^*\tilde{P}_j$ and $\tilde{Q}_j^*\tilde{Q}_j$. The matrix representation of (79) is a block matrix decomposition of the same form as the operator decomposition.

Before we go on we should note that in general $\bar{L}_j$ will not be represented by a sparse matrix even if L_{j+1} is, because A_j^{-1} will typically be dense. For the homogenization method to be efficient we must be able to approximate $\bar{L}_j$ with a sparse matrix. If this sparse matrix is on banded form, it can be seen as a discretization of a local differential operator acting on the coarse space. In many applications this is indeed true and we will get back to the issue below.

The procedure to obtain $\bar{L}_j$ described above can be applied recursively on $\bar{L}_j$ itself to get $\bar{L}_{j-1}$ and so on. This can easily be verified when L_{j+1} is symmetric positive definite. Then A_j is always invertible by Proposition 1 (with $a[u,v] = (Lu,v)$) and Proposition 2 shows that also $\bar{L}_j$ is symmetric positive definite. By induction $\bar{L}_k$ exists for $k \leq j$. Also note that Proposition 3 gives $\bar{L}_k = \overline{L_{j+1}}_{V_{j+1},V_k}$ and $\bar{F}_k = \bar{F}_{V_{j+1},V_k}$ for $k \leq j$. Moreover, an improvement in the condition number can often be estimated. Typically for standard discretizations

$$\max_{||v||=1} (L_k v, v) < \max_{||u||=1} (L_{j+1}u, u) , \tag{83}$$

when $k \leq j$. Then, from Proposition 2,

$$\begin{aligned}\kappa(\bar{L}_k) &= \frac{\max_{||v||=1} (\bar{L}_k v, v)}{\min_{||v||=1} (\bar{L}_k v, v)} \leq \frac{\max_{||v||=1} (L_k v, v)}{\min_{||v||=1} (L_{j+1} v, v)} \\ &< \frac{\max_{||v||=1} (L_{j+1} v, v)}{\min_{||v||=1} (L_{j+1} v, v)} = \kappa(L_{j+1}).\end{aligned} \tag{84}$$

The two-dimensional wavelet transform

$$\boldsymbol{\mathcal{W}}_j : \boldsymbol{V}_{j+1} \mapsto \boldsymbol{W}_j \times \boldsymbol{V}_j \tag{85}$$

can be written as a tensor product of one-dimensional transforms,

$$\boldsymbol{\mathcal{W}}_j = \mathcal{W}_j \otimes \mathcal{W}_j. \tag{86}$$

A linear operator L_{j+1} that acts on the space $\boldsymbol{V}_{j+1}$ can be decomposed in a way similar to the one-dimensional case. The equation

$$L_{j+1}U = F, \qquad U, F \in \boldsymbol{V}_{j+1}, \qquad L_{j+1} \in \mathcal{L}(\boldsymbol{V}_{j+1}, \boldsymbol{V}_{j+1}) \tag{87}$$

can be transformed to

$$\begin{pmatrix} A_j & B_j \\ C_j & L_j \end{pmatrix} \begin{pmatrix} U_\mathrm{f} \\ U_\mathrm{c} \end{pmatrix} = \begin{pmatrix} F_\mathrm{f} \\ F_\mathrm{c} \end{pmatrix}, \qquad U_\mathrm{f}, F_\mathrm{f} \in \boldsymbol{W}_j, \qquad U_\mathrm{c}, F_\mathrm{c} \in \boldsymbol{V}_j, \tag{88}$$

and the coarse grid operator is again the Schur complement,

$$\overline{L}_{\boldsymbol{V}_{j+1},\boldsymbol{V}_j} \equiv \bar{L}_j = L_j - C_j A_j^{-1} B_j. \tag{89}$$

To get the matrix representation of (88), we take the one-dimensional matrix representations of $\mathcal{W}_j$ and, following (86), compute its Kronecker tensor product with itself. We must thereafter also apply a suitable permutation matrix to the product in order to get the same block structure of the matrix as in (88).

Note that the fine scale part of U can be decomposed as

$$U_{\mathrm{f}} = \begin{pmatrix} U_{\mathrm{ff}} \\ U_{\mathrm{cf}} \\ U_{\mathrm{fc}} \end{pmatrix}, \quad U_{\mathrm{ff}} \in W_j \otimes W_j, \quad U_{\mathrm{cf}} \in V_j \otimes W_j, \quad U_{\mathrm{fc}} \in W_j \otimes V_j. \tag{90}$$

In some cases, the homogenized operator keeps important properties of the original operator. Let the forward and backward undivided differences be defined as

$$\Delta_+ u_i = u_{i+1} - u_i, \qquad \Delta_- u_i = u_i - u_{i-1}. \tag{91}$$

In [11] it was shown that the one-dimensional elliptic model equation

$$-(gu')' = f \tag{92}$$

discretized with

$$-\Delta_+ \mathrm{diag}(g) \Delta_- U = h^2 F \tag{93}$$

will preserve its divergence form during homogenization. That is, we will get

$$\bar{L}_j = \frac{1}{h^2} \Delta_+ H_j \Delta_-, \tag{94}$$

where H_j is a strongly diagonal dominant matrix which can be interpreted as the effective material coefficient related to g. Analogously, for the first order differential operator $g(x)\frac{\partial}{\partial x}$ the discretized form $\mathrm{diag}(g)\Delta_-/h$ is preserved during homogenization,

$$\bar{L}_j = \frac{1}{h} H_j \Delta_-. \tag{95}$$

In two dimensions, the elliptic model equation $-\nabla(g(x,y)\nabla u) = f$ can be discretized as

$$L_{j+1} = -\frac{1}{h^2} \Delta_+^x G \Delta_-^x - \frac{1}{h^2} \Delta_+^y G \Delta_-^y, \qquad L_{j+1} U = F. \tag{96}$$

Then $\bar{L}_j$ is not any longer of exactly the same form as L_{j+1}. The cross-derivatives must also be included. We get

$$\bar{L}_j = -\frac{1}{h^2} \Delta_+^x H^{xx} \Delta_-^x - \frac{1}{h^2} \Delta_+^y H^{yx} \Delta_-^x - \frac{1}{h^2} \Delta_+^x H^{xy} \Delta_-^y - \frac{1}{h^2} \Delta_+^y H^{yy} \Delta_-^y. \tag{97}$$

When the operator L_{j+1} is derived from a finite difference, finite element or finite volume discretization, it is sparse and of a certain structure. In one dimension it might, for instance, be tridiagonal. However, as remarked above the matrix $\bar{L}_j$ is in general not sparse since A_j^{-1} is usually dense. Computing all components of $\bar{L}_j$ would be inefficient. Fortunately, $\bar{L}_j$ will be diagonal dominant in many important cases and we should then be able to find a sparse matrix that is a close approximation of $\bar{L}_j$.

In some simple cases the matrix A_j is in fact diagonal. Examples include operators L of the form

$$Lu = \int_0^x a(t)u(t)dt + b(x)u(x). \tag{98}$$

Let P_j be the orthogonal projection on V_j. Suppose L is discretized in V_{j+1} as $L_{j+1} = P_{j+1}LP_{j+1}$ with a in (98) replaced by $P_{j+1}aP_{j+1}$, and similar for b. When the Haar system is used, A_j is diagonal and $\bar{L}_j$ is of the same form as L_{j+1}. Let $\bar{A}_k$ be related to $\bar{L}_{k+1}$ via (78, 79) in the same way as A_j relates to L_{j+1}. By induction, $\bar{A}_k$ is also diagonal for $k < j$. The operators in (98) turn up for instance in problems with systems of ordinary differential equations and one-dimensional elliptic equations, see [4,13]. In these cases, an explicit recurrence relation between scale levels can be established, which permits the computation of $\bar{L}_k$ on any fixed level k as the starting level, $j+1$, tends to infinity.

For more general problems one must instead rely on the rapid decay of elements in A_j and $\bar{L}_j$ off the diagonal, which is a consequence of the wavelet spaces' good approximation properties. The decay rate of A_j for Calderon–Zygmund and pseudo-differential operators was given by Beylkin, Coifman and Rokhlin in [5]. Letting $L_{j+1} = P_{j+1}LP_{j+1}$, $A_j = \{a^j_{k\ell}\}$, $B_j = \{b^j_{k\ell}\}$ and $C_j = \{c^j_{k\ell}\}$, they show that

$$|a^j_{k\ell}| + |b^j_{k\ell}| + |c^j_{k\ell}| \le \frac{2^{-\lambda j} C_M}{1 + |k-\ell|^{M+1}}, \qquad |k-\ell| \ge \nu, \tag{99}$$

when the wavelet system has M vanishing moments. For Calderon–Zygmund operators $\lambda = 0$ and $\nu = 2M$. For a pseudo-differential operator $\nu = 0$ and if its symbol is $\sigma(x,\xi)$, the value of λ should be taken such that the estimates

$$|\partial_\xi^\alpha \partial_x^\beta \sigma(x,\xi)| \le C_{\alpha,\beta}(1+|\xi|)^{\lambda-\alpha+\beta}, \tag{100}$$

$$|\partial_\xi^\alpha \partial_x^\beta \sigma^*(x,\xi)| \le C_{\alpha,\beta}(1+|\xi|)^{\lambda-\alpha+\beta}, \tag{101}$$

are satisfied for some constants $C_{\alpha,\beta}$. For instance, in the second order elliptic case $\lambda = 2$. Moreover, Beylkin and Coult, [6], showed that if (99) holds with $\lambda = 0$ for A_j, B_j and C_j given by L_{j+1} in (78, 79), then the same estimate also holds for $\bar{A}_{j'}$, $\bar{B}_{j'}$ and $\bar{C}_{j'}$, here given by $P_{j'}\bar{L}_{k+1}P_{j'}$ for $j' \le k < j$ with $j+1$ being the starting homogenization level. Hence, the decay rate is preserved after homogenization.

The decay estimate in [6] for $\bar{A}_{j'}$ is uniform in k and may not be sharp for a fixed k. There is, for example, a general result by Concus, Golub and Meurant, [9], for diagonal dominant, symmetric and tridiagonal matrices. For those cases, which include A_j corresponding to the discretization in (93) of the one-dimensional elliptic operator, the inverse has exponential decay,

$$\left|\left(A_j^{-1}\right)_{k\ell}\right| \leq C\varrho^{|k-\ell|}, \qquad 0 < \varrho < 1. \tag{102}$$

This holds also when the elliptic operator has a lower order term of type $b(x)\partial_x$ discretized with upwinding, [19].

One simple way to approximate $\bar{L}_j$ is to set all components outside a prescribed bandwidth ν equal to zero. Let us define

$$\mathrm{trunc}(M,\nu)_{ij} = \begin{cases} M_{ij}, & \text{if } 2\,|i-j| \leq \nu - 1 \\ 0, & \text{otherwise.} \end{cases} \tag{103}$$

For $\nu = 1$ the matrix is diagonal. For $\nu = 3$ it is a tridiagonal and so on.

Another possible strategy is the basic truncation method used in [5],

$$\mathrm{trunc}_{\mathrm{BCR}}(M,\varepsilon)_{ij} = \begin{cases} M_{ij}, & \text{if } |M_{ij}| > \varepsilon, \\ 0, & \text{otherwise} \end{cases} \tag{104}$$

but it is not practical here since the location of the non-zero elements cannot be controlled.

The matrix $\bar{L}_j$ can also be projected onto banded form in a more effective manner. Let $\{\boldsymbol{v}_j\}_{j=1}^{\nu}$ be a set of linearly independent vectors in $\mathbb{R}^N$. We define the *band projection*, $\mathrm{band}(M,\nu)$, of a matrix $M \in \mathbb{R}^{N\times N}$ as the projection of M onto the subspace of matrices with bandwidth ν such that

$$M\boldsymbol{x} = \mathrm{band}(M,\nu)\boldsymbol{x}, \qquad \forall \boldsymbol{x} \in \mathrm{span}\{\boldsymbol{v}_1, \boldsymbol{v}_2, \ldots, \boldsymbol{v}_\nu\}. \tag{105}$$

In our setting M will usually operate on vectors representing smooth functions, for instance solutions to elliptic equations, and a natural choice for $\boldsymbol{v}_j$ vectors are thus the first ν polynomials,

$$\boldsymbol{v}_j = \{1^{j-1}, 2^{j-1}, \ldots, N^{j-1}\}^T, \qquad j = 1, \ldots, \nu. \tag{106}$$

Smooth solutions to the homogenized problem should be well approximated by these vectors. For the case $\nu = 1$ we get the standard "masslumping" of a matrix, often used in the context of finite element methods.

This technique is similar to the probing technique used by Chan et al., [8]. In that case the vectors $\boldsymbol{v}_j$ are sums of unit vectors. Other probing techniques have been suggested by Axelsson, Pohlman and Wittum, see Chapter 8 in [2]. The choice of $\boldsymbol{v}_j$ vectors could be optimized if there is some a priori knowledge of the homogenized solution. See [1] for further discussion.

It would be expensive to compute $\mathrm{trunc}(\bar{L}_j,\nu)$ using a naive method, such as calculating the full inverse of A_j. By capitalizing on the nearly sparse structure of the matrices involved, it was, however, shown in [6] that $\mathrm{trunc}(\bar{L}_j,\nu)$

can be computed with a cost of only $\mathcal{O}(N)$ operations for N unknowns and fixed accuracy. This method uses a multiresolution based LU decomposition procedure described in [7]. The computation of band($\bar{L}_j, \nu$) can be based on trunc($\bar{L}_j, \mu$), $\mu > \nu$. The additional computational cost is proportional to $(\nu^3 + \mu\nu)N$. The $\nu^3 N$ term corresponds to solving N $\nu \times \nu$ systems and $\mu\nu N$ to computing the right hand sides, see also [2]. Finally, note that the effective cost of using trunc($\bar{L}_j, \nu$) is typically smaller, since the same homogenized operator will often be reused multiple times, for instance with different right hand sides, or in different places of the geometry as a subgrid model.

The truncation methods described above are even more efficient when applied to H_j instead of directly to the homogenized operator. The following proposition shows that when the solution to the homogenized problem belongs to the Sobolev space H^1, the accuracy is one order higher (see [1] for a proof).

Proposition 4. *Suppose $L = \Delta_+ H \Delta_-$ and $LU = h^2 f$. Let $\kappa = ||L^{-1}|| \cdot ||H||$ and*

$$r = \frac{||\delta L||}{||L||} = \frac{||\delta H||}{||H||} < \frac{1}{4\kappa}, \tag{107}$$

where $||\cdot||$ denotes the l_2-norm. Furthermore, let u be any H^1-function such that $U_j = u(jh)$. Then, if

$$(L + \delta L)(U + \delta U_L) = h^2 f, \tag{108}$$

$$\Delta_+(H + \delta H)\Delta_-(U + \delta U_H) = h^2 f \tag{109}$$

the following bounds on the perturbed solutions hold $\forall f$

$$||\delta U_L|| \le \frac{4r\kappa}{1 - 4r\kappa}||U||, \tag{110}$$

$$||\delta U_H|| \le \frac{1}{2} \cdot \frac{4r\kappa}{1 - 4r\kappa}||\Delta_- U|| \le \frac{h}{2} \cdot \frac{4r\kappa}{1 - 4r\kappa}||u||_{H^1}. \tag{111}$$

The bound (110) is sharp.

In two dimensions truncation to simple banded form is in general not adequate, since the full operator will typically be block banded. However, both the crude truncation and the band projection generalizes easily to treat block banded form instead of just banded. Let M be the tensor product of two $N \times N$ matrices. Then we define truncation as

$$\underset{2}{\text{trunc}}(M, \nu)_{ij} = \begin{cases} M_{ij}, & \text{if } 2\,|i - j - rN| \le \nu - 1 - |2r|, \\ 0, & \text{otherwise}, \end{cases} \qquad |2r| + 1 \le \nu. \tag{112}$$

This mimics the typical block structure of a discretized differential operator. For the band projection, the subspace in the one-dimensional definition, is simply replaced by the subspace of matrices with the block banded sparsity pattern defined in (112).

In two dimensions, untangling the various H components of (97) from $\bar{L}$ is more complicated than finding the H in (94) and (95) for one-dimensional problems. There is however no unsurmountable difficulty, and truncating H instead of $\bar{L}$ can be done also for two-dimensional problems.

Applicatons of the wavelet based homogenization technique to subgrid scale moddeling of wave guides and to multigrid methods can be found in [12].

5 Multiscale finite element methods

Assume that a standard Galerkin finite element method, with lower order basis functions, is approxmating the model problem

$$L_\varepsilon u_\varepsilon = -\frac{d}{dx}\left(g(x/\varepsilon)\frac{d}{dx}\right)u_\varepsilon(x) = f(x), \qquad 0 < x < 1, \tag{113}$$

with Dirichlet boundary conditions,

$$u_\varepsilon(0) = u_\varepsilon(1) = 0. \tag{114}$$

Let the element size $h \gg \varepsilon$. The integrals in the stiffness matrix will then approximate integrals where $g(x/\varepsilon)$ is replaced by $\bar{g}$ in (51) (and not $\tilde{g}$ in (37)) resulting in the wrong solution as $h, \varepsilon \to 0$. It is essential that the coarse grid finite element method contains some important subgrid scale information.

In [16,17] Hughes introduces a general procedure for finite element methods applied to multiscale problems. Extra degrees of freedom, representing the smaller scales, are eliminated in order to get a coarse grid method. This elimination in the earlier work of Hughes is analogous to the Schur complement (15) that we also use.

We shall consider a similar multiscale finite element method developed by Hou and collaborators, [14,15]. In the one dimensional case this method can be seen as a projection method of the type discussed in this paper.

Consider the two-point boundary value problem (113, 114). Approximate the equations by a standard Galerkin method with piecewise linear basis functions on the grid

$$x_j = jh, \qquad j = 0, 1, \ldots, J, \qquad Jh = 1. \tag{115}$$

This results in the linear system

$$LU = F, \tag{116}$$

where the stiffness matrix $L = \{\ell_{ij}\}$, with $1 \le i, j \le J-1$, is symmetric tridiagonal, $\ell_{ij} = 0$ for $|i-j| > 1$ and

$$\ell_{j,j+1} = \ell_{j+1,j} = -\int_{x_j}^{x_{j+1}} g(x/\varepsilon)dx, \qquad \ell_{jj} = -\int_{x_{j-1}}^{x_{j+1}} g(x/\varepsilon)dx. \tag{117}$$

Moreover, $U = (u_1, \dots, u_{J-1})^T$ and $F = (f_1, \dots, f_{J-1})^T$ where

$$u_j \approx u_\varepsilon(x_j), \qquad f_j = \int_{x_{j-1}}^{x_{j+1}} f(x)\varphi_j(x)dx. \tag{118}$$

Here, $\varphi_j(x)$ denotes the standard hat function centered at x_j. The step size h is chosen such that the ε-scale is resolved on the grid, $h < \varepsilon$.

There is also a coarser grid associated with the method,

$$X_k = kH, \qquad k = 1, 2, \dots, K, \qquad KH = 1, \qquad H = mh. \tag{119}$$

In the multiscale finite element mehtod the fine grid discretization is used to derive the coarse grid basis functions ψ_k. These functions are defined as Galerkin apprixmations of

$$L_\varepsilon \psi_k = 0, \qquad X_{k-1} < x < X_{k+1}, \quad x \neq X_k, \tag{120}$$

with boundary conditions, $\psi_k(X_j) = \delta_{kj}$. From [15] we get that the solution on the coarse and fine grids are the same in this special one dimensional case.

By appropriately defining the projection operator P we can identify this multiscale method with the different steps in our general framework. Let P be given by

$$PU = (u_m, u_{2m}, \dots, u_{(K-1)m})^T \tag{121}$$

and then Q is defined by

$$QU = (u_1, \dots, u_{m-1}, u_{m+1}, \dots, u_{2m-1}, u_{2m+1}, \dots, u_{J-1})^T. \tag{122}$$

Following (13) the matrix L will be transformed into

$$\mathcal{W}L\mathcal{W}^* = \begin{pmatrix} A & B \\ C & D \end{pmatrix} \tag{123}$$

with A block diagonal: $A = \text{diag}(A_k)$ for $k = 1, \dots, K$. Each A_k is a tridiagonal submatrix of L,

$$A_k = \{\ell_{ij}\}, \qquad km \leq i, j, < (k+1)m. \tag{124}$$

The inverse of A in the definition of the projection generated homogenized operator $\bar{L} = D - CA^{-1}B$, consist of the inverses of the separate A_k submatrices. This corresponds directly to solving for the different coarse grid basis functions ψ_k as in (120).

The final coarse space problem, $\bar{L}PU = \bar{F}$, is identical to the discrete linear system for the multiscale method with the basis functions ψ_k.

In the two dimensional case there is not as direct a correspondence. The matrix A in (123) is not block diagonal. The discrete linear system of the multiscale finite element method will only be an approximation of $\bar{L}$. The approximation is, however, very good due to a well chosen way of defining the basis functions, that is based on oversampling, see [14,15].

References

1. Andersson, U., Engquist, B., Ledfelt, G., Runborg, O. (1999) A contribution to wavelet-based subgrid modeling. Appl. Comput. Harmon. Anal. **7**, 151–164
2. Axelsson, O. (1994) Iterative Solution Methods. Cambridge University Press
3. Bensoussan, A., Lions, J.-L., Papanicolaou, G. (1978) Asymptotic Analysis for Periodic Structures. North-Holland
4. Beylkin, G., Brewster, M. (1995) A multiresolution strategy for numerical homogenization. Appl. Comput. Harmon. Anal. **2**, 327–349
5. Beylkin, G., Coifman, R., Rokhlin, V. (1991) Fast wavelet transforms and numerical algorithms I. Comm. Pure Appl. Math. **44**, 141–183
6. Beylkin, G., Coult, N. (1998) A multiresolution strategy for reduction of elliptic PDEs and eigenvalue problems. Appl. Comput. Harmon. Anal. **5**, 129–155
7. Beylkin, G., Dunn, J., Gines, D. L. (1998) LU factorization of non-standard forms and direct multiresolution solvers. Appl. Comput. Harmon. Anal. **5**, 156–201
8. Chan, T., Mathew, T. (1992) The interface probing technique in domain decomposition. SIAM J. Matrix Anal. Appl. **13**(1), 212–238
9. Concus, C., Golub, G. H., Meurant, G. (1985) Block preconditioning for the conjugate gradient method. SIAM J. Sci. Stat. Comp. **6**, 220–252
10. Daubechies, I. (1991) Ten Lectures on Wavelets. SIAM
11. Dorobantu, M., Engquist, B. (1998) Wavelet-based numerical homogenization. SIAM J. Numer. Anal. **35**(2), 540–559
12. Engquist, B., Runborg, O. (2001) Wavelet-Based Numerical Homogenization with Applications. To appear.
13. Gilbert, A. C. (1998) A comparison of multiresolution and classical one-dimensional homogenization schemes. Appl. Comput. Harmon. Anal. **5**(1), 1–35
14. Hou, T. Y., Wu, X.-H. (1997) A multiscale finite element method for elliptic problems in composite materials and porous media. J. Comput. Phys. **134**(1), 169–189
15. Hou, T. Y., Wu, X.-H., Cai, Z. (1999) Convergence of a multiscale finite element method for elliptic problems with rapidly oscillating coefficients. Math. Comp. **68**(227), 913–943
16. Hughes, T. J. R. (1995) Multiscale phenomena: Green's functions, the dirichlet-to neumann formulation, subgrid, scale models, bubbles and the origins of stabilized methods. Comput. Methods Appl. Mech. Engrg. **127**, 387–401
17. Hughes, T. J. R., Feijóo, G. R., Mazzei, L., Quicy, J.-B. (1998) The variational multiscale method – a paradigm for computational mechanics. Comput. Methods Appl. Mech. Engrg. **166**, 3–24
18. Knapek, S. (1999) Matrix-dependent multigrid-homogenization for diffusion problems. SIAM J. Sci. Stat. Comp. **20**(2), 515–533
19. Leon, D. D. (2000) Wavelet Operators Applied to Multigrid Methods. PhD thesis, Department of Mathematics, UCLA
20. Neuss, N. (1995) Homogenisierung und Mehrgitter. PhD thesis, Fakultät Mathematik Universität Heidelberg
21. Neuss, N., Jäger, W., Wittum, G. (2001) Homogenization and multigrid. Computing **66**(1), 1–26

22. Steinberg, B. Z., McCoy, J. J., Mirotznik, M. (2000) A multiresolution approach to homogenization and effective modal analysis of complex boundary value problems. SIAM J. Appl. Math. **60**(3), 939–966

Derivation of the Diphasic Biot's Law for an Elastic Solid Matrix Containing Isolated Fluid Drops

Thierry Clopeau[1], Jose Luis Ferrín[2], and Andro Mikelić[1]

[1] UFR Mathématiques, CNRS UMR 5585, Bât. 101, Université Claude Bernard Lyon 1, 43, Bd. du onze novembre, 69622 Villeurbanne Cedex, France.
[2] Departamento de Matemática Aplicada, Facultad de Matemáticas, Universidad de Santiago de Compostela, 15782 Santiago de Compostela, Spain.

Abstract. We consider a porous medium consisting of a deformable pore structure of the characteristic size ε. The solid skeleton is supposed to be elastic and the pores contain a viscous and incompressible fluid. Moreover, we consider that the solid part is connected but the fluid part is not connected. We analyse the case when the contrast of property number, and then the adimensional viscosity coefficients, are of order ε^2. By homogenization we undertake a rigorous derivation of the diphasic effective behavior already observed in papers by M. Biot.

Contents

1 Introduction

In this paper we study the case which corresponds to the effective diphasic macroscopic behavior of the fluid part and elastic matrix. We undertake a rigorous derivation of the diphasic Biot's law, describing small deformations of a deformable porous matrix containing an incompressible and viscous fluid. Original equations of Biot (see Biot [5–7]) concentrate on this particular situation and his heuristic modeling assures a kind of Darcy law for the difference between effective velocities of the solid and fluid part. Asymptotic modeling of this case was undertaken by Auriault , see chapter 8 in [4], Burridge and Keller [8], Levy [12], Nguetseng [17], Sanchez-Palencia [18], pages 158-190, and it is customary to set "the dimensionless viscosity" to be

$\mu\varepsilon^2$ and then to study the 2-scale asymptotic expansion. The meaning of $\mu\varepsilon^2$ is that the normal stress of the elastic matrix is of the same order as the fluid pressure.

Rigorous derivation of the Biot's equations from the first principles (linear conservation laws) by homogenization was undertaken in the paper by Clopeau, Ferrín, Gilbert and Mikelić [9]. They considered the case where both phases were connected. Here we consider the case of an elastic solid skeleton containing isolated viscous fluid drops. It leads to much simpler effective model.

This paper is organized as follows: We start from the first principles, i.e. from the coupled system containing the linear elasticity system for the solid part and the incompressible Stokes system for the viscous fluid. Then we derive a priori estimates for the displacements and pressure and, by using appropriate results from the theory of the two-scale convergence we pass to the limit obtaining a two-scale homogenized problem. Then by separation of the scales we get an effective system of equations, depending only on the slow scale. Finally we compare the homogenized equations with the original Biot's system. The porous medium we propose to study is obtained by a periodic arrangement of the pores. The formal description goes along the following lines:

Firstly we define the geometrical structure inside the unit cell $\mathcal{Y} =]0,1[^n$, $n = 2,3$. Let $\mathcal{Y}_f$ (the fluid part) be a closed subset of $\mathcal{Y}$, $\mathcal{Y}_f \subset\subset \mathcal{Y}$, and $\mathcal{Y}_s = \mathcal{Y} \backslash \mathcal{Y}_f$ (the solid part). Now we make the periodic repetition of $\mathcal{Y}_f$ all over $\mathbb{R}^n$ and set $\mathcal{Y}_f^k = \mathcal{Y}_f + k$, $k \in \mathbb{Z}^n$. Obviously the set $E_f = \bigcup_{k\in\mathbb{Z}^n} \mathcal{Y}_f^k$ is a closed subset of $\mathbb{R}^n$ and $E_s = \mathbb{R}^n \backslash E_f$ in an open set in $\mathbb{R}^n$. Following Allaire [2] we make the following assumptions on $\mathcal{Y}_s$ and E_s:

- $\mathcal{Y}_s$ is an open connected set of strictly positive measure, with a Lipschitz boundary and $\mathcal{Y}_f$ has strictly positive measure in $\bar{\mathcal{Y}}$ as well.
- E_s and the interior of E_f are open sets with the boundary of class $C^{0,1}$, which are locally located on one side of their boundary. We note that E_s is connected, but E_f is not connected.

Now we see that $\Omega =]0, L[^n$ is covered with a regular mesh of size ε, each cell being a cube $\mathcal{Y}_i^\varepsilon$, with $1 \le i \le N(\varepsilon) = |\Omega|\varepsilon^{-n}[1 + o(1)]$. Each cube $\mathcal{Y}_i^\varepsilon$ is homeomorphic to $\mathcal{Y}$, by linear homeomorphism Π_i^ε, being composed of translation and an homothety of ratio $1/\varepsilon$. We define

$$\mathcal{Y}_{s_i}^\varepsilon = (\Pi_i^\varepsilon)^{-1}(\mathcal{Y}_s) \qquad \text{and} \quad \mathcal{Y}_{f_i}^\varepsilon = (\Pi_i^\varepsilon)^{-1}(\mathcal{Y}_f).$$

For sufficiently small $\varepsilon > 0$ we consider the sets

$$T_\varepsilon = \{k \in \mathbb{Z}^n / \mathcal{Y}_{f_k}^\varepsilon \subset \Omega\}$$
$$K_\varepsilon = \{k \in \mathbb{Z}^n / \mathcal{Y}_{f_k}^\varepsilon \bigcap \partial\Omega \neq \emptyset\}$$

and define

$$\Omega_f^\varepsilon = \bigcup_{k\in T_\varepsilon} \mathcal{Y}_{f_k}^\varepsilon, \quad S^\varepsilon = \partial\Omega_f^\varepsilon, \quad \Omega_s^\varepsilon = \Omega \setminus \Omega_f^\varepsilon.$$

Obviously, $\partial\Omega_s^\varepsilon = \partial\Omega \cup S^\varepsilon$. The domains Ω_s^ε and Ω_f^ε represent, respectively, the solid and fluid parts of a porous medium Ω. For simplicity we suppose $L/\varepsilon \in \mathbb{N}$.

2 Microscopic model

We consider in the solid structure the linear elasticity system

$$\rho_s \frac{\partial^2 u^\varepsilon}{\partial t^2} - \operatorname{div}(\sigma^{s,\varepsilon}) = \rho_s F \text{ in } \Omega_s^\varepsilon \times]0,T[, \tag{1}$$

$$\sigma^{s,\varepsilon} = AD(u^\varepsilon) \tag{2}$$

whereas, in the fluid part we consider the Stokes system

$$\rho_f \frac{\partial^2 u^\varepsilon}{\partial t^2} - \operatorname{div}(\sigma^{f,\varepsilon}) = \rho_f F \text{ in } \Omega_f^\varepsilon \times]0,T[, \tag{3}$$

$$\sigma^{f,\varepsilon} = -p^\varepsilon I + 2\mu\varepsilon^2 D\left(\frac{\partial u^\varepsilon}{\partial t}\right), \tag{4}$$

$$\operatorname{div} \frac{\partial u^\varepsilon}{\partial t} = 0, \tag{5}$$

where $\frac{\partial u^\varepsilon}{\partial t}$ is the fluid velocity, p^ε is the fluid pressure and D is the symmetrized gradient. We note that the Lagrangean coordinates are used for the structure and Eulerian for the fluid. Since we suppose small Reynolds numbers, the inertial term in the Navier-Stokes system reduces to the non-stationary Stokes system. Next, Ω_s^ε is the reference configuration for the solid structure and the interface between two media changes with the evolution of the structure. This change of the position of the interface is quadratic in displacements and their derivatives and, in accordance with the assumptions of the linear elasticity and of a small flow, we neglect it. At the interface between fluid and solid parts we have continuity of displacements and continuity of forces, namely

$$[u^\varepsilon] = 0 \quad \text{and} \quad \sigma^{s,\varepsilon} \cdot \nu = \sigma^{f,\varepsilon} \cdot \nu \text{ on } \Gamma_\varepsilon \times]0,T[. \tag{6}$$

For large deformation or/and large Reynolds numbers the linearization made to obtain these boundary conditions is not valable any more and one should keep an evolutive interface depending on the solution. For more details we refer to the articles by Grandmont and Maday [13] and [14] and to the references therein. At the outer boundary we suppose periodicity, i.e.

$$\{u^\varepsilon, p^\varepsilon\} \text{ are } L-\text{periodic} \tag{7}$$

and for simplicity, that there was no flow and no deformations for $t = 0$, i.e.

$$u^\varepsilon(x,0) = \frac{\partial u^\varepsilon}{\partial t}(x,0) = 0 \quad \text{in } \Omega. \tag{8}$$

The variational formulation which corresponds to (1)-(8) is given by: Find $u^\varepsilon \in H^1(0,T;H^1_{\text{per}}(\Omega)^n)$ with $\dfrac{d^2u^\varepsilon}{dt^2} \in L^2(0,T;L^2(\Omega)^n)$ and $p^\varepsilon \in L^2(0,T;L^2(\Omega_f^\varepsilon))$ such that

$$\frac{d^2}{dt^2}\int_\Omega \rho^\varepsilon u^\varepsilon(t)\varphi\,dx + \frac{d}{dt}\int_{\Omega_f^\varepsilon} 2\mu\varepsilon^2 D(u^\varepsilon(t)) : D(\varphi)\,dx$$

$$+\int_{\Omega_s^\varepsilon} AD(u^\varepsilon(t)) : D(\varphi)\,dx - \int_{\Omega_f^\varepsilon} p^\varepsilon\,\mathrm{div}\varphi\,dx = \int_\Omega \rho^\varepsilon F\varphi\,dx,$$

$$\forall\varphi\in H^1_{\text{per}}(\Omega)^n, \qquad (a.e.)\text{ in }]0,T[, \tag{9}$$

where

$$\rho^\varepsilon = \rho_f\chi_{\Omega_f^\varepsilon} + \rho_s\chi_{\Omega_s^\varepsilon} \tag{10}$$

and

$$u^\varepsilon(0) = \frac{\partial u^\varepsilon}{\partial t}(0) = 0, \tag{11}$$

$$\mathrm{div}\frac{\partial u^\varepsilon}{\partial t} = 0 \text{ in } \Omega_f^\varepsilon\times]0,T[. \tag{12}$$

3 A priori estimates for (9)–(12)

After taking $\varphi = \frac{\partial u^\varepsilon}{\partial t}$ as the test function in (9) we obtain

$$\|\frac{\partial u^\varepsilon}{\partial t}\|_{L^\infty(0,T;L^2(\Omega)^n)} + \|D(u^\varepsilon)\|_{L^\infty(0,T;L^2(\Omega_s^\varepsilon)^{n^2})} \le C, \tag{13}$$

$$\|D\left(\frac{\partial u^\varepsilon}{\partial t}\right)\|_{L^2(0,T;L^2(\Omega_f^\varepsilon)^{n^2})} + \|D(u^\varepsilon)\|_{L^\infty(0,T;L^2(\Omega_f^\varepsilon)^{n^2})} \le \frac{C}{\varepsilon}. \tag{14}$$

If $F\in H^k(0,T;L^2(\Omega)^n)$ then

$$\|\frac{\partial^{k+1}u^\varepsilon}{\partial t^{k+1}}\|_{L^\infty(0,T;L^2(\Omega)^n)} + \|D\left(\frac{\partial^k u^\varepsilon}{\partial t^k}\right)\|_{L^\infty(0,T;L^2(\Omega_s^\varepsilon)^{n^2})} \le C\,, \tag{15}$$

$$\|D\left(\frac{\partial^k u^\varepsilon}{\partial t^k}\right)\|_{L^\infty(0,T;L^2(\Omega_f^\varepsilon)^{n^2})} \le \frac{C}{\varepsilon}. \tag{16}$$

For simplicity we suppose $F\in H^2(0,T;L^2(\Omega)^n)$. We extend the pressure field p^ε to $\Omega^\varepsilon\times]0,T[$ by

$$\tilde p^\varepsilon(x,t) = \begin{cases} p^\varepsilon(x,t) - \frac{1}{|\Omega|}\int_{\Omega_f^\varepsilon} p^\varepsilon(x,t)\,dx, & x\in\Omega_f^\varepsilon,\\ -\frac{1}{|\Omega|}\int_{\Omega_f^\varepsilon} p^\varepsilon(x,t)dx, & x\in\Omega_s^\varepsilon. \end{cases} \tag{17}$$

Then $\int_{\Omega} \tilde{p}^{\varepsilon} \operatorname{div}\varphi = \int_{\Omega_f^{\varepsilon}} p^{\varepsilon} \operatorname{div}\varphi$, $\forall \varphi \in H^1_{\text{per}}(\Omega)^n$, and

$$\|\tilde{p}^{\varepsilon}\|_{H^1(0,T;L^2_0(\Omega))} + \|\nabla \tilde{p}^{\varepsilon}\|_{H^1(0,T;H^{-1}_{\text{per}}(\Omega)^n)} \le C. \tag{18}$$

4 Passing to the limit $\varepsilon \to 0$

In order to prove the main convergence result of this paper we use the notion of *two-scale convergence* which was introduced in [16] and developed further in [3]. The main interest of this convergence are the compactness properties, saying that from each bounded sequence one can extract a subsequence two-scale convergent.

Definition 1. The sequence $\{w^{\varepsilon}\} \subset L^2(\Omega)$ is said to two-scale converge to a limit $w \in L^2(\Omega \times \mathcal{Y})$ iff for any $\sigma \in C^{\infty}(\Omega; C^{\infty}_{\text{per}}(\mathcal{Y}))$ ("per" denotes 1-periodicity) one has

$$\lim_{\varepsilon \to 0} \int_{\Omega} w^{\varepsilon}(x)\sigma(x, \frac{x}{\varepsilon})\, dx = \int_{\Omega} \int_{\mathcal{Y}} w(x,y)\sigma(x,y)\, dy\, dx.$$

As in our case we have two different estimates for gradients in the solid and in the fluid part, the classical way to proceed is by extending the deformation from Ω_f^{ε} to Ω and then passing to the limit $\varepsilon \to 0$. The recent results on homogenization of Neumann problems in perforated domains of general type (see [1] and also book Jikov, Kozlov, Oleinik [11]) suggest that it is enough to suppose the Lipschitz regularity for the pore boundaries. The a priori estimates (13)-(18) allow us to pass to the limit as $\varepsilon \to 0$ in the variational formulation (9). This is possible because of the following result:

Theorem 1. *(see Clopeau et al. [9]) There are subsequences such that, for all $t \in [0,T]$,*

$$u^{\varepsilon} \to u^0(x,t) + \chi_{\mathcal{Y}_f}(y)v(x,y,t) \text{ in the 2-scale sense,} \tag{19}$$

$$\chi_{\Omega_s^{\varepsilon}} \nabla u^{\varepsilon} \to \chi_{\mathcal{Y}_s}(y)[\nabla_x u^0(x,t) + \nabla_y u^1(x,y,t)] \text{ in the 2-scale sense,} \tag{20}$$

$$\varepsilon \chi_{\Omega_f^{\varepsilon}} \nabla u^{\varepsilon} \to \chi_{\mathcal{Y}_f}(y)\nabla_y v(x,y,t) \text{ in the 2-scale sense,} \tag{21}$$

$$\tilde{p}^{\varepsilon} \to \tilde{p}^0(x,y,t) \text{ in the 2-scale sense,} \tag{22}$$

with

$u^0 \in H^3(0,T; L^2(\Omega)^n) \cap H^2(0,T; H^1_{\text{per}}(\Omega)^n)$,

$u^1 \in H^3(0,T; L^2(\Omega; H^1(\mathcal{Y}_s)/\mathbb{R})^n)$,

$$\begin{cases} v \in H^3(0,T; L^2(\Omega \times \mathcal{Y})^n) \cap H^2(0,T; L^2(\Omega; H^1_{\text{per}}(\mathcal{Y}))^n), \quad v = 0 \text{ in } \mathcal{Y}_s, \\ \operatorname{div}_y \frac{\partial v}{\partial t}(x,y,t) = 0 \text{ in } \Omega \times \mathcal{Y}_f \times]0,T[\text{ and} \\ \operatorname{div}_x \int_{\mathcal{Y}_f} \frac{\partial v}{\partial t}(x,y,t)dy \in L^2(\Omega \times]0,T[) \end{cases}$$

$$\begin{cases} \tilde{p}^0(x,y,t) = \chi_{\mathcal{Y}_f}(y)p(x,t) + B(t), \quad p \in H^1(]0,T[; L^2(\Omega)) \\ B(t) = -\frac{|\mathcal{Y}_f|}{|\mathcal{Y}_s||\Omega|} \int_{\Omega} p(x,t)\, dx \end{cases}$$

and

$$u^0(x,0) = \frac{\partial u^0}{\partial t}(x,0) = 0 \quad in\ \Omega,$$
$$u^1(x,y,0) = 0, \quad in\ \Omega \times \mathcal{Y}_s,$$
$$v(x,y,0) = \frac{\partial v}{\partial t}(x,y,0) = 0 \quad in\ \Omega \times \mathcal{Y}_f.$$

By using standard procedures from the theory of 2-scale convergence it is straight-forward to pass to the limit in the system (1)-(8). For details we refer to Clopeau *et al.* [9]. We obtain the following two-scale system given for the effective displacement u^0, the effective relative pore displacement between phases v, the correction $D_y(u^1)$ and the effective and second pressures p and π, respectively

$$\rho_f \frac{\partial^2 u^0}{\partial t^2} + \rho_f \frac{\partial^2 v}{\partial t^2} - \mu \Delta_y \frac{\partial v}{\partial t} + \nabla_x p^0 + \nabla_y \pi = \rho_f F \text{ in } \Omega \times \mathcal{Y}_f \times]0,T[, \quad (23)$$

$$- \operatorname{div}_y \{A(D_x(u^0) + D_y(u^1))\} = 0 \text{ in } \Omega \times \mathcal{Y}_s \times]0,T[, \quad (24)$$

$$\rho_f \frac{\partial^2}{\partial t^2} \left\{ |\mathcal{Y}_f| u^0 + \int_{\mathcal{Y}_f} v\, dy \right\} + \rho_s |\mathcal{Y}_s| \frac{\partial^2 u^0}{\partial t^2} + |\mathcal{Y}_f| \nabla_x p$$
$$- \operatorname{div}_x \left\{ \int_{\mathcal{Y}_s} A(D_x(u^0) + D_y(u^1))\, dy \right\} = \rho F, \quad (25)$$

$$\operatorname{div}_x \left\{ |\mathcal{Y}_f| \frac{\partial u^0}{\partial t} + \int_{\mathcal{Y}_f} \frac{\partial v}{\partial t} dy \right\} = \int_{\mathcal{Y}_s} \operatorname{div}_y \frac{\partial u^1}{\partial t}\, dy \ \text{ in } \Omega \times]0,T[\quad (26)$$

$$A(D_x(u^0) + D_y(u^1))\nu = -p\,\nu \text{ on } \Omega \times (\partial\mathcal{Y}_s \backslash \partial\mathcal{Y}) \times]0,T[, \quad (27)$$
$$v = 0 \text{ on } \Omega \times (\partial\mathcal{Y}_f \backslash \partial\mathcal{Y}) \times]0,T[, \quad (28)$$

$$\{u^0, u^1, v, p\} \text{ are } L\text{-periodic in } x, \quad (29)$$
$$\{u^1, v\} \text{ are 1-periodic in } y, \quad (30)$$
$$u^0(x,0) = \frac{\partial u^0}{\partial t}(x,0) = 0 \quad \text{in } \Omega, \quad (31)$$
$$v(x,y,0) = \frac{\partial v}{\partial t}(x,y,0) = 0 \quad \text{in } \Omega \times \mathcal{Y}_f, \quad (32)$$

where $\rho = |\mathcal{Y}_f|\rho_f + |\mathcal{Y}_s|\rho_s$.

Theorem 2. *(see Clopeau et al. [9]) Let $\{u^0, u^1, v, p\}$ be given by (19)-(22). Then they define the unique variational solution for (23)-(32).*

5 Derivation of the effective equations

The system (23)-(32) is the complete upscaled problem but it is too complicated to be used directly for numerical calculations. Hence, we seek $\{u^1, v\}$ in a particular form, that we make precise below, with the aim of eliminating the fast scale y. It is eliminated by using the ansatz

$$u^1(x,y,t) = p(x,t)w^0(y) + \sum_{i,j} \left(D_x(u^0(x,t))\right)_{ij} w^{ij}(y), \tag{33}$$

where functions $\{w^{ij}\}$ and w^0 are given, respectively, by

$$\begin{cases} \operatorname{div}_y \left\{A\left(\frac{e_i\otimes e_j+e_j\otimes e_i}{2} + D_y(w^{ij})\right)\right\} = 0 & \text{in } \mathcal{Y}_s, \\ A(\frac{e_i\otimes e_j+e_j\otimes e_i}{2} + D_y(w^{ij}))\nu = 0 & \text{on } \partial\mathcal{Y}_s\backslash\partial\mathcal{Y}, \\ \int_{\mathcal{Y}_s} w^{ij}(y)\,dy = 0, \quad w^{ij} \text{ is 1-periodic} \end{cases} \tag{34}$$

and

$$\begin{cases} -\operatorname{div}_y \left\{AD_y(w^0)\right\} = 0 & \text{in } \mathcal{Y}_s, \\ AD_y(w^0)\nu = -\nu & \text{on } \partial\mathcal{Y}_s\backslash\partial\mathcal{Y}, \\ \int_{\mathcal{Y}_s} w^0(y)\,dy = 0, \quad w^0 \text{ is 1-periodic.} \end{cases} \tag{35}$$

These auxiliary functions define the following effective coefficients:

$$A^H_{klij} := \left(\int_{\mathcal{Y}_s} A\left(\frac{e_i\otimes e_j + e_j\otimes e_i}{2} + D_y(w^{ij})\right)\right)_{kl}, \tag{36}$$

$$\mathcal{B}^H := \int_{\mathcal{Y}_s} AD_y(w^0), \tag{37}$$

verifying

$$\int_{\mathcal{Y}_s} \operatorname{div}_y w^{ij}(y)\,dy = -\int_{\mathcal{Y}_s} AD_y(w^0) : D_y(w^{ij})\,dy = \mathcal{B}^H_{ij} \tag{38}$$

and

$$\int_{\mathcal{Y}_s} \operatorname{div}_y w^0\,dy = -\int_{\mathcal{Y}_s} AD_y(w^0) : D_y(w^0) < 0. \tag{39}$$

Equations (36) and (37) define tensors A^H and $\mathcal{B}^H$ which are positive definite and symmetric. On the other side, for v and π we get

$$v = 0, \tag{40}$$

$$\pi = \sum_j y_j \left\{\rho_f F_j(x,t) - \frac{\partial p^0}{\partial x_j}(x,t) - \rho_f \frac{\partial^2 u_j^0}{\partial t^2}(x,t)\right\}. \tag{41}$$

Hence, we are left with the following system for the unknowns u^0 and p

$$\rho I \frac{\partial^2 u^0}{\partial t^2} - \operatorname{div}_x \{A^H D_x(u^0)\} + (|\mathcal{Y}_f| I - \mathcal{B}^H)\nabla_x p = \rho F \quad \text{in } \Omega \times]0,T[, \tag{42}$$

$$-\frac{\partial^2 p}{\partial t^2} \int_{\mathcal{Y}_s} \operatorname{div}_y w^0 \, dy + \operatorname{div}_x \left\{ (|\mathcal{Y}_f| I - \mathcal{B}^H) \frac{\partial^2 u^0}{\partial t^2} \right\} = 0 \quad \text{in } \Omega \times]0,T[, \tag{43}$$

$$\{u^0, p\} \text{ are } L\text{-periodic in } x, \tag{44}$$

$$u^0(x,0) = \frac{\partial u^0}{\partial t}(x,0) = 0 \quad \text{in } \Omega. \tag{45}$$

We refer to Clopeau *et al.* [9] for the proof that $\{u^0, p\}$ is the unique solution for the system (42)-(45). This new system does not contain the fast scale and it is much easier to study. As we shall see it corresponds to the diphasic Biot's law, with the same displacement in the fluid and solid part. In fact we get an overall linear elastic, but anisotropic behavior.

6 Convergence and corrector results

Theorem 3. *Let $\{u^\varepsilon, p^\varepsilon\}$ be the solution for (1)-(8) and $\{u^0, p\}$ the solution for (42)-(45). The tensors A^H and $\mathcal{B}^H$ are defined by (36) and (37), respectively. They are symmetric and positive definite. Then u^ε converges towards u^0 and p^ε converges towards $\chi_{\mathcal{Y}_f}(y)p$ in 2-scale sense. Furthermore, $u^\varepsilon \rightharpoonup u^0$ weakly in $H^3(0,T;H^1(\Omega)^n)$.*

Moreover, we can obtain corrector results leading to strong convergences of the microscopic variables.

Theorem 4. *(see Clopeau et al. [9]) Let u^0, u^1 and v be defined as above. Then we have*

$$\sqrt{\rho^\varepsilon} u^\varepsilon(x,t) - \sqrt{\rho^0} u^0(x,t) \to 0 \quad \text{in } C([0,T]; L^2(\Omega)^n), \tag{46}$$

$$\begin{aligned} &\chi_{\Omega_s^\varepsilon} A^{1/2} D_x(u^\varepsilon(x,t)) - \chi_{\mathcal{Y}_s} A^{1/2} D_x(u^0(x,t)) \\ &\quad - \chi_{\mathcal{Y}_s} A^{1/2} \varepsilon D_x(u^1(x, \frac{x}{\varepsilon}, t)) \to 0 \quad \text{in } C([0,T]; L^2(\Omega)^{n^2}). \end{aligned} \tag{47}$$

where $\rho^0 = \rho_f \chi_{\mathcal{Y}_f} + \rho_s \chi_{\mathcal{Y}_s}$.

7 Concluding remarks

It remains to compare the system (42)-(45) with the Biot's system obtained with the assumption of moderate Reynolds numbers and considering the dissipative effects. Our results are to be compared with the classical references, such as the original work by Biot [5–7] and the formal asymptotic expansions used in Burridge and Keller [8]. By introducing the effective displacement of the porous medium as $\boldsymbol{u} = u^0$, we get

$$p(x,t) = \frac{1}{\int\limits_{\mathcal{Y}_s} \mathrm{div}_y w^0} \mathrm{div}_x \left\{(|\mathcal{Y}_f| I - \mathcal{B}^H)\boldsymbol{u}\right\}, \tag{48}$$

which implies that the effective medium is compressible, even when we had started from an incompressible fluid, as noticed by M. A. Biot. Formula (48) coincides with the formula proposed by Biot for a general anisotropic material. For the effective stress tensor we have obtained

$$\tau(x,t) = -\frac{1}{\int\limits_{\mathcal{Y}_s} \mathrm{div}_y w^0} (|\mathcal{Y}_f| I - \mathcal{B}^H)\, \mathrm{div}_x \left\{(|\mathcal{Y}_f| I - \mathcal{B}^H)\boldsymbol{u}\right\} + A^H D_x(\boldsymbol{u}). \tag{49}$$

We also note that the equation (42) express the total momentum of the mixed fluid-solid system and agrees with the corresponding Biot's equation (see Biot [6]).

References

1. Acerbi E., Chiado Piat V., Dal Maso G., Percivale D. (1992) An extension theorem from connected sets and homogenization in general periodic domains. Nonlinear Anal., Theory Methods Appl **18**, 481–496.
2. Allaire G. (1989) Homogenization of the Stokes flow in a connected porous medium. Asymptotic Anal. **2**, 203–222.
3. Allaire G. (1992) Homogenization and two-scale convergence. SIAM J. Math. Anal. **23.6**, 1482–1518
4. Hornung U. (1997) Homogenization and Porous Media. Interdisciplinary Applied Mathematics, Springer
5. Biot M.A. (1956) Theory of propagation of elastic waves in a fluid-saturated porous solid. I. Lower frequency range and II. Higher frequency range. J. Acoust. Soc. Am. **28**(2), 168–178 and 179–191
6. Biot M.A. (1962) Generalized theory of acoustic propagation in porous dissipative media. J. Acoust. Soc. Am. **34**, 1254–1264
7. Biot M.A. (1962) Mechanics of deformation and acoustic propagation in porous media. J. Appl. Phys. **33**, 1482–1498
8. Burridge R., Keller J.B. (1981) Poroelasticity equations derived from microstructure. J. Acoust. Soc. Am. **70**, 1140–1146
9. Clopeau Th., Ferrín J.L., Gilbert R.P., Mikelić A. (2001) Homogenizing the acoustic properties of the seabed: Part II. Math. Comput. Modelling **33** (8–9), 821–841

10. Fasano A., Mikelić A., Primicerio M. (1998) Homogenization of flows through porous media with grains. Advances in Mathematical Sciences and Applications **8**, 1–31
11. Jikov V.V., Kozlov S.M., Oleinik O.A. (1994) Homogenization of Differential Operators and Integral Functionals. Springer
12. Levy T. (1979) Propagation waves in a fluid-saturated porous elastic solid. Internat. J. Engrg. Sci. **17**, 1005–1014
13. Grandmont C., Maday Y. (1998) Analyse et méthodes numériques pour la simulation de phénomènes d'interaction fluide-structure. ESAIM: Proc. **3**, 101–117
14. Grandmont C., Maday Y. (2000), Existence for an unsteady fluid-structure interaction problem. M2AN Math. Model. Numer. Anal. **34**, 609–636
15. Mikelić A. (1994) Mathematical derivation of the Darcy-type law with memory effects, governing transient flow through porous medium. Glas. Mat. Ser. III **29** (49), 57–77
16. Nguetseng G. (1989) A general convergence result for a functional related to the theory of homogenization. SIAM J. Math. Anal. **20**, 608–623
17. Nguetseng G. (1990) Asymptotic analysis for a stiff variational problem arising in mechanics. SIAM J. Math. Anal. **20.3**, 608–623
18. Sanchez-Palencia E. (1980) Non-Homogeneous Media and Vibration Theory. Lecture Notes in Physics **129**, Springer

Mesoscopic Models of Reaction-Diffusion Processes with Exclusion Mechanism

Flavius Guiaş

Institut für Angewandte Mathematik, Im Neuenheimer Feld 294,
D–69120 Heidelberg, Germany

Abstract. We study a class of Markov jump processes describing reaction-diffusion phenomena in a dynamic medium. The model consists basically of a bounded domain divided into cells, each of them being the collection of a given number of sites. These sites are in one of two possible phases of the medium, being either empty, or occupied by particles. The cells are considered as homogenous, the spatial structure of the sites being ignored. The model describes chemical reactions and interactions with the medium, while the individual particles can also perform jumps between neighbouring cells. We consider an exclusion mechanism which prohibits adsorption or jump of particles in a saturated cell. We analyze the macroscopic behaviour of this class of processes in the limit when the cell size tends to 0 and the number of sites per cell tends to infinity, giving conditions for convergence towards the solution of a deterministic reaction-diffusion system.

Contents

1 Introduction

The model considered in this paper is closely related to a class of stochastic models for reaction-diffusion-type phenomena, known in the literature as 'density dependent models' or 'high density particle models'. Briefly, these models consist in the division of a bounded domain into cells by a regular mesh, while a large number of particles is considered inside each cell. Particles can perform random walks among neighbouring cells, or can react with a partner inside the same cell. In the works [1,7,2,3] the asymptotic properties of this class of Markov processes are analyzed, and convergence results towards a deterministic law (a reaction-diffusion system) are proved if the cell size tends to 0 and the number of particles tends to infinity. Of course, the results hold only under a suitable adjustment of the speed of convergence to 0 of the cell size and of the order of convergence to infinity of the number of particles. In [5] an infinite-dimensional system of reaction-diffusion type is considered, namely the *coagulation-fragmentation equations*, and a convergence result of the same type as mentioned above is proved. In studying the macroscopic limits of this type of models, one can take advantage of probabilistic techniques, as well as of methods from the analysis. For this reason, this approach is often more convenient than an approach based on a lattice gas model, in which the particles perform random walks and where a reaction mechanism is present in the case when two particles meet on the same site. In the present work we consider a density-dependent particle model, with the modification that the number of particles in every cell is limited to a maximal number. However, in the limit it will tend to infinity. We consider particles which can perform random walks between the cells, while they also obey to a birth-death and a binary reaction mechanism, the only restriction being the limited maximal number of particles in each cell. The products of the binary reactions are not among the considered type of particles, and are eliminated instantaneously from the cell.

A further aspect is the modelling of a surface or a volume with crystalline structure as a *dynamic medium* which varies in time, exhibiting local phase changes caused by the mutual interaction between the medium and the particle system. The resolution of this model does not go down to the level of the atomic sites of the crystal surface, but considers the cells as being homogenous patches on a *mesoscopic scale*, in which the spatial distribution of the reactants and of the crystal phases is ignored. The motivation for this approach comes from modelling oxidation of **CO** on the crystalline surface of a catalyst, as treated e.g. in [8]. The simplified model presented in our approach is intended to be a first step in the rigorous mathematical modelling by stochastic processes of this type of phenomena.

2 Construction of the stochastic model

Let $\Omega \subset \mathbb{R}^d$, $d \leq 3$, be a bounded domain with smooth boundary, which we divide into cubic cells of size ε. Let $\mathbb{G}_\varepsilon$ denote the collection of all cells which are *completely included* in Ω, and let $N_C = |\mathbb{G}_\varepsilon|$ denote their total number. Define

$$\Omega_\varepsilon := \bigcup_{G \in \mathbb{G}_\varepsilon} G \subset \Omega. \tag{1}$$

Consider particles which correspond to one of the reactants involved in the process. We associate to each cell a number of sites; each site can be either empty, or occupied by one particle. Moreover, each site is in one of the two possible phases of the medium. The sites are not a further subdivision of a single cell and their spatial structure is of no importance in this case. Due to this simplification, we call this type of model a *mesoscopic model.* Denote the set of all neighbours of a cell G which are contained in Ω_ε by $\mathbf{N}(G)$, and let $n(G) = |\mathbf{N}(G)|$. Let $n_G(t), m_G(t)$, denote respectively the number of particles and the number of sites in a reference phase of the medium (denoted from now on as "phase 1") considered in the cell G at time t. By multiplication with a scaling factor h, i.e. by defining $x_G^{h,\varepsilon}(t) = h \cdot n_G(t)$, we obtain the corresponding concentration. The factor h should be seen as the inverse of the 'total number of sites' in a cell. This implies that the concentration of the reactant cannot take values above 1. In a similar way we define $y_G^{h,\varepsilon}(t) = h \cdot m_G(t)$, which is the concentration of the sites in phase 1 in the cell G at time t. The pair $z_G^{h,\varepsilon}(t) := (x_G^{h,\varepsilon}(t), y_G^{h,\varepsilon}(t))$ is thus an element of the state space $\mathbb{E}$, which can be defined either as the finite-dimensional space of pairs of functions defined on Ω, piecewise constant on the cells in $\mathbb{G}_\varepsilon$, and zero in $\Omega \setminus \Omega_\varepsilon$, or as the space of $N_C \times 2$ matrices with real entries. In the first case, the values of the functions on the cell boundaries are of no importance, since we will work in an L^2-setting. We will make selective use of these two viewpoints, depending on the context. For simplicity, where is no danger of confusion, the dependence of the state space on ε and h will be suppressed, since we will never consider two processes on two different state spaces at the same time. Denote further by E_G^x the $N_C \times 2$ matrix which has all elements equal to 0, except the element corresponding to the pair (G, x) which equals 1. Define further E_G^y in a similar way. The notation $\chi_A(\cdot)$ will stand as usually for the characteristic function of the set A. Consider some polynomial functions with positive coefficients $d(\cdot), a(\cdot)$ defined on [0,1] and $q^1(\cdot,\cdot), q^2(\cdot,\cdot)$ defined on $[0,1]^2$. Let $z = (x, y) \in \mathbb{E}$ be the current state of the Markov process. The possible transitions are the following:

1. adsorption of new particles in an arbitrary cell G, at a rate proportional to the number of free sites, i.e.:

$$z \longrightarrow z + h \cdot E_G^x \quad \text{at rate } h^{-1} \cdot a(y_G) \cdot (1 - x_G) \tag{2}$$

2. disappearance of particles from an arbitrary cell G, i.e.:

$$z \longrightarrow z - h \cdot E_G^x \quad \text{at rate } h^{-1} \cdot d(y_G) \cdot x_G \tag{3}$$

   This transition can occur as a consequence of desorption or reaction phenomena. Since we consider only one reactant and since the reaction products are leaving instantaneously the catalysator surface, we can describe the mentioned two processes with the same type of transition. Note that, as in the case of adsorption, the rates depend also on the medium, i.e. on the proportion of the two crystal phases inside the cell.
3. diffusion of particles inside the domain, i.e.:

$$\begin{aligned} &z \longrightarrow z - h \cdot E_G^x + h \cdot E_{G'}^x \text{ at rate} \\ &\varepsilon^{-2} h^{-1} \cdot \chi_{[0,1)}(x_{G'}) \cdot x_G, \text{ where } G' \in \mathbf{N}(G). \end{aligned} \tag{4}$$

   The interpretation of this transition is the following: a particle jumps from the cell G to the neighbour cell G' with the given rate, under the condition that there still are free sites available in the target cell. This is possible only if $x_{G'} < 1$. We have introduced in this way an *exclusion mechanism* into the model.

4. change of the crystal phase

$$\text{a)}\quad z \longrightarrow z - h \cdot E_G^y \text{ at rate } h^{-1} \cdot q^1(x_G, y_G) \cdot y_G \tag{5}$$

$$\text{b)}\quad z \longrightarrow z + h \cdot E_G^y \text{ at rate } h^{-1} \cdot q^2(x_G, y_G) \cdot (1 - y_G) \tag{6}$$

This transition models the change of phase of a site in the cell G (from the total number of h^{-1} sites in the cell), which can flip between phase 1 and phase 2. The terms $q^i(\cdot,\cdot)$, $i = 1, 2$, denote the flipping rates of one single site in phase i. The rates depend on both the concentration of the reactant and the proportion of the two phases. This model contains the possibility that only one site can flip at a time, but the results will hold also if we allow the flipping of a random number of sites, under the condition that the mean (expectation) of this random number is finite. However, for simplicity we will restrict ourselves only at the situation described above in 4.

The infinitesimal generator of the process $z(t)$ is given by:

$$(\Lambda_{h,\varepsilon} g)(z) = \sum_{\tilde{z}} (g(\tilde{z}) - g(z))\, r_{z\to\tilde{z}} \tag{7}$$

for any bounded, measurable fuction g defined on $\mathbb{E}$, where the value of $r_{z\to\tilde{z}}$ runs through all previously described transition rates from the state z to the new state $\tilde{z}$. The parameter of the exponentially distributed waiting time before jumping from the state z to the state $\tilde{z}$ is given by:

$$\lambda_{h,\varepsilon}(z) = \sum_{\tilde{z}} r_{z\to\tilde{z}} \tag{8}$$

The observation that the number of possible states of the process is finite implies that the waiting time parameter function $\lambda_{h,\varepsilon}$ is bounded. Then, according to [4], pp.162-164 the jump process $z(t)$ is defined a.s. for all $t \geq 0$, i.e. the jumps do not accumulate. Define for all $r > 0$ and $z = (x, y) \in \mathbb{E}$:

$$p_{G,r}^x(z) = \beta_r(x_G) \tag{9}$$

$$p_{G,r}^y(z) = \beta_r(y_G) \tag{10}$$

where the functions β_r are defined on $\mathbb{R}$ for all $r > 0$ by:

$$\beta_r(w) = \begin{cases} w & \text{for } |w| \leq r \\ 2r \cdot sgn(w) - r^2 \cdot w^{-1} & \text{for } |w| > r. \end{cases} \tag{11}$$

$p_{G,r}^x, p_{G,r}^y$ are thus bounded C^1-functions. From our construction of the process we have ensured that $x_G, y_G \leq 1$, so the functions $p_{G,1}^x, p_{G,1}^y$ are nothing else but the projection of the component x_G, respectively y_G. The application of Dynkin's formula ([4], p.162) for these functions implies:

$$x(t) = x(0) + \int_0^t \{\Delta_\varepsilon^* x(s) - d(y(s)) \cdot x(s) + a(y(s)) \cdot (1 - x(s))\} \, ds + M_x^{h,\varepsilon}(t) \tag{12}$$

$$y(t) = y(0) + \int_0^t \{q^2(x(s), y(s)) \cdot (1 - y(s)) - q^1(x(s), y(s)) \cdot y(s)\} \, ds + M_y^{h,\varepsilon}(t) \tag{13}$$

where the terms $M_x^{h,\varepsilon}(t), M_y^{h,\varepsilon}(t)$ are martingales with the respect to the filtration generated by the process. The operator Δ_ε^* corresponds to the diffusion-type transitions of the reactant, under the exclusion mechanism taken into consideration. We have:

$$\begin{aligned} \Delta_\varepsilon^* x_G &= \varepsilon^{-2} \sum_{G' \in \mathbf{N}(G)} \{\chi_{[0,1)}(x_G) \cdot x_{G'} - \chi_{[0,1)}(x_{G'}) \cdot x_G\} \\ &= \chi_G \Delta_\varepsilon x_G - \Delta_\epsilon \chi_G x_G \end{aligned} \tag{14}$$

The operator denoted by Δ_ε, is defined by:

$$\Delta_\varepsilon x_G = \sum_{J \in \mathbb{G}_\varepsilon, J' \in \mathbf{N}(J)} (-h \cdot E_J^* + h \cdot E_{J'}^*)_G \cdot \varepsilon^{-2} h^{-1} \cdot x_J \tag{15}$$

where E_J^* denotes the characteristic function of the cell $J \in \mathbb{G}_\varepsilon$, and will turn out to be the finite-difference approximation of the Laplace operator on an ε-mesh corresponding to Neumann boundary conditions. Let us present some further notations and auxiliary results concerning this operator. Let H^ε denote the Hilbert subspace of $L^2(\Omega)$ consisting of step functions which are constant on the cells $G \in \mathbb{G}_\varepsilon$ and vanish on $\Omega \setminus \Omega_\varepsilon$. For any element $f \in H^\varepsilon$, its norm is given by the L^2-norm:

$$\|f\|_{H^\varepsilon}^2 = \sum_{G \in \mathbb{G}_\varepsilon} |G| \cdot f_G^2 = \varepsilon^d \sum_{G \in \mathbb{G}_\varepsilon} f_G^2 = \|f\|_{L^2(\Omega)}^2 \tag{16}$$

Where it is no confusion, by $\| \cdot \|_{L^2}$ we will mean always the norm in $L^2(\Omega)$. Define also the projections $\pi_\varepsilon : L^2(\Omega) \to H^\varepsilon$ by:

$$(\pi_\varepsilon f)_G = \varepsilon^{-d} \int_G f(x) dx \tag{17}$$

for any $G \in \mathbb{G}_\varepsilon$. It is easy to see that π_ε are continuous linear operators with norm 1, which implies that the internal approximation of $L^2(\Omega)$ by the spaces H^ε is stable. It can be easily seen that the approximation is also convergent, i.e.

$$\|\pi_\varepsilon f - f\|_{L^2} \to 0 \tag{18}$$

as $\varepsilon \to 0$, for any $f \in L^2(\Omega)$. Let Δ denote the L^2-realization of the Laplace operator under Neumann boundary conditions, i.e. with the domain

$$H = \left\{ u \in H^2(\Omega) : \frac{\partial u}{\partial \nu} = 0 \text{ on } \partial\Omega \right\} \tag{19}$$

On the space H we have the norm:

$$\|u\|_H = \|u\|_{L^2(\Omega)} + \|\Delta u\|_{L^2(\Omega)}, \tag{20}$$

which is equivalent with the H^2-norm (see [5]).

3 Statement of the result

The main result of this paper is the convergence of the familiy of stochastic processes to the solution of a deterministic reaction-diffusion system. Let Ω be a bounded domain in $\mathbb{R}^d, d \leq 3$ with smooth boundary and consider the polynomial functions with positive coefficients $d(\cdot), a(\cdot)$ defined on [0,1] and $q^1(\cdot,\cdot), q^2(\cdot,\cdot)$ defined on $[0,1]^2$. Denote by $x(t,\cdot), y(t,\cdot)$ the solution of the following deterministic system:

$$\begin{aligned} \frac{dx}{dt} &= \Delta x(t) - d(y(t)) \cdot x(t) + a(y(t)) \cdot (1 - x(t)) \\ \frac{dy}{dt} &= -q^1(x(t), y(t)) \cdot y(t) + q^2(x(t), y(t)) \cdot (1 - y(t)) \quad \text{in } \Omega \\ \frac{\partial x}{\partial \nu} &= 0 \quad \text{on } \partial\Omega \end{aligned} \tag{21}$$

The intial conditions $x(0)$ and $y(0)$ are assumed to be sufficiently regular functions on Ω, positive and bounded strictly from above by 1. Global existence and uniqueness results of the above system are standard (see e.g. [9]). However, for our purpose it suffices to show the existence and uniqueness of a *mild solution* of the system. This will be pointed out subsequently. The regularity is only needed in order to apply a comparison principle for showing that the components of the solution stay in the range $[0, 1]$, if the initial conditions are in this range (in the case of our stochastic process, this is already ensured by construction). Moreover, due to the dissipativity of the diffusion operator, the component $x(t)$ will stay uniformly bounded away from 1.

Theorem 1. *If $h\varepsilon^{-2-d} \to 0$ as $h, \varepsilon \to 0$ and if*

$$\|x^{h,\varepsilon}(0) - x(0)\|_{L^\infty} + \|y^{h,\varepsilon}(0) - y(0)\|_{L^\infty} \longrightarrow 0$$

in probability for $h, \varepsilon \to 0$, then the family of stochastic processes $z^{h,\varepsilon}(t) = (x^{h,\varepsilon}(t), y^{h,\varepsilon}(t))$ satisfies the convergence property:

$$\sup_{t \leq T} \left[\|x^{h,\varepsilon}(t) - x(t)\|_{L^2} + \|y^{h,\varepsilon}(t) - y(t)\|_{L^2} \right] \longrightarrow 0$$

in probability as $h, \varepsilon \to 0$, for any $T > 0$ fixed.

4 The proof

For a better overview, we will divide the proof into several lemmata.

Lemma 1. *Let $S(t)$ be the contraction semigroup generated by the operator Δ with domain H on $L^2(\Omega)$. We have then: (i) If $u \in L^\infty(\Omega)$, then $S(t)u \in L^\infty(\Omega)$ and $||S(t)u||_{L^\infty} \leq ||u||_{L^\infty}$. (ii) The operator Δ_ε generates a contraction semigroup S_ε on H^ε in the L^p-norm, for all $p \in [2, \infty]$. Moreover, if $p < \infty$, the semigroup is analytic. (iii) The semigroup $S_\varepsilon(t)$ on the space H^ε (considered as a subspace of $L^2(\Omega)$) satisfies the approximation property:*

$$||S_\varepsilon(t)\pi_\varepsilon f - S(t)f||_{L^2(\Omega)} \to 0 \quad as \quad \varepsilon \to 0$$

uniformly in t on bounded intervals and for any $f \in L^2(\Omega)$. (iv) If $q(\cdot,\cdot)$ is a polynomial and u^1, u^2, v^1, v^2 are L^∞-functions defined on Ω with norms ≤ 1, then

$$||q(u_1,u_2) - q(v_1,v_2)||_{L^2} \leq M\left(||u_1 - v_1||_{L^2} + ||u_2 - v_2||_{L^2}\right)$$

where the constant M depends only on the coefficients and of the degree of the polynomial q.

Proof. For (i)-(iii) see [5], Lemma 2.2 and Lemma 3.1. (iv) The proof is a simple computation, which we will perform only in the case of monomials. The extension to general polynomials is straightforward. We have:

$$\begin{aligned} &||u_1^i u_2^j - v_1^i v_2^j||_{L^2} \leq ||u_1^i u_2^j - v_1^i u_2^j||_{L^2} + ||v_1^i u_2^j - v_1^i v_2^j||_{L^2} \\ &\leq ||u_1^i - v_1^i||_{L^2}||u_2^j||_{L^\infty} + ||u_2^j - v_2^j||_{L^2}||v_1^i||_{L^\infty} \\ &\leq M_1\left(||u_1 - v_1||_{L^2} + ||u_2 - v_2||_{L^2}\right) \end{aligned} \tag{22}$$

Since all L^∞-norms of the considered functions are bounded from above by 1, the constant M_1 depends only on i and j. In the case of general polynomials, we simply sum up the results of the above computations and get the statement of the lemma. □

Fix some time moment $T > 0$. Consider the Banach space $Z_T := \{z = (x(t), y(t)) \in L^\infty(\Omega) \times L^\infty(\Omega)\}$ with the norm

$$||z||_{Z_T} = \Big\| ||x(t)||_{L^\infty(\Omega)} + ||y(t)||_{L^\infty(\Omega)} \Big\|_{L^\infty[0,T]}$$

Define on Z_T the operator A by $(Az)(t) := ((Az)^x(t), (Az)^y(t))$, where

$$(Az)^x(t) = S(t)x(0) + \int_0^t S(t-s)\left[-d(y(s))x(s) + a(y(s))(1 - x(s))\right] ds$$

$$(Az)^y(t) = y(0) + \int_0^t \left[-q^1(x(s),y(s))y(s) + q^2(x(s),y(s))(1 - y(s))\right] ds \tag{23}$$

Lemma 2. *For $r \leq 1$, the operator A defined above has a unique fixed point in the open ball $B_r(z(0))$ of the space Z_{T_1}, provided $T_1 = K^{-1}(1 + \|z(0)\|_{Z_{T_1}})^{-1}r$. K is a sufficiently large constant, depending only on the reaction functions and on the domain Ω. This is the mild (local) solution on the time interval $[0, T_1]$ of the system (21).*

Proof. The result follows from a straightforward application of Banach's fixed point theorem, making use also of Lemma 1 (i). It is similar to [5] Theorem 1.1. The existence result is only local, since we did not impose positivity of the initial conditions. □

Lemma 3. *(L^2-estimate of the martingales) For a given $T > 0$ and for small ε, the martingales introduced in (12) and (13) satisfy the estimates:*

$$E_{z(0)}\left[\|M_x^{h,\varepsilon}(t)\|_{H^\varepsilon}^2\right] \leq CT \cdot h \cdot \varepsilon^{-2} \tag{24}$$

$$E_{z(0)}\left[\|M_y^{h,\varepsilon}(t)\|_{H^\varepsilon}^2\right] \leq CT \cdot h \cdot \varepsilon^{-2} \tag{25}$$

where the conditional expectation is taken with respect to the initial condition $z(0)$ of the process $z(\cdot)$ and where C is a constant depending only on Ω and of the coefficients of the transition rates.

Proof. We will prove only the first estimate, the computations for the second one being completely similarly (even simpler, since for this component the diffusion is absent). For any bounded C^1-function g and for $z(t) = (x(t), y(t)) \in \mathbb{E}$ we have the standard identity (see e.g. [6]):

$$E_{z(0)}\left[g(M_x^{h,\varepsilon}(t)\right] = \int_0^t E_{z(0)}\Big[\sum_{z \to \tilde{z}} \Big(g(\tilde{x} - x + M_x(s)) - g(M_x(s)) \\ -(\tilde{x} - x)g'(M_x^{h,\varepsilon}(s))\Big) r_{z \to \tilde{z}}\Big] ds$$

where the notation $z \to \tilde{z} = (\tilde{x}, \tilde{y})$ has the same meaning as in (7). By taking $g = (p_{G,1}^x)^2$, we have then:

$$E_{z(0)}\left[\|M_x^{h,\varepsilon}(t)\|_{H^\varepsilon}^2\right] = \varepsilon^d \sum_{G \in \mathbb{G}_\varepsilon} E_{z(0)}\left[M_{x,G}^{h,\varepsilon}(t)^2\right] =$$

$$= \varepsilon^d \sum_{G \in \mathbb{G}_\varepsilon} \int_0^t \Big\{h^2 h^{-1} d(y_G) x_G + h^2 h^{-1} a(y_G)(1 - x_G) \\ + \sum_{J' \in \mathbf{N}(J)} [(-hE_J^* + hE_{J'}^*)_G]^2 \varepsilon^{-2} h^{-1} \chi_{[0,1)}(x_{J'}) x_J(s)\Big\} ds$$

$$= \varepsilon^d h \sum_{G \in \mathbb{G}_\varepsilon} \int_0^t \Big\{d(y_G) x_G + a(y_G)(1 - x_G) \\ + \varepsilon^{-2} x_G(s) \sum_{G' \in \mathbf{N}(G)} \chi_{[0,1)}(x_{G'}) + \varepsilon^{-2} \chi_{[0,1)}(x_G) \sum_{G' \in \mathbf{N}(G)} x_{G'}(s)\Big\} ds \tag{26}$$

Since $0 \leq x_G \leq 1$ for all G (ensured by definition), relation (26) implies immediately that the statement of the lemma is valid if we choose a sufficiently small ε. □

Proof of Theorem 1 Throughout this proof, by *const* we will denote a generic constant, which is independent on h, ε, t and on the considered stochastic or deterministic processes. It may depend on the domain Ω and on the polynomial reaction functions. We will show first that the statement holds locally in time, more precisely on the time interval $[0, T_1]$ given by Lemma 2 for

$$r = 1 - ||x(0)||_{L^\infty} \tag{27}$$

Define the stopping time

$$\tau^{h,\varepsilon} = \inf\{t \geq 0 : ||x^{h,\epsilon}(t)||_{L^\infty} = 1\} \wedge T_1 \tag{28}$$

That is, we stop the process as soon as we reach saturation in one of the cells or at the end of the considered time interval.. Before this time moment, the exclusion property of the process does not show up, and we have just simple, discrete diffusion of the particles. For simplicity, we will denote from now on this stoping time as τ. From identity (12) we have thus:

$$\begin{aligned} x^{h,\varepsilon}(t\wedge\tau) - M_x^{h,\varepsilon}(t\wedge\tau) &= x^{h,\varepsilon}(0) + \int_0^{t\wedge\tau} \Big\{\Delta_\varepsilon(x^{h,\varepsilon}(s) - M_x^{h,\varepsilon}(s)) \\ &+\Delta_\varepsilon M_x^{h,\varepsilon}(s) + a(y^{h,\varepsilon}(s)(1 - x^{h,\varepsilon}(s)) + d(y^{h,\varepsilon}(s))x^{h,\varepsilon}(s)\Big\}ds \end{aligned} \tag{29}$$

By variation of constants we have:

$$\begin{aligned} x^{h,\varepsilon}(t\wedge\tau) &= S_\varepsilon(t\wedge\tau)x^{h,\varepsilon}(0) + \int_0^{t\wedge\tau} S_\varepsilon(t\wedge\tau - s)\Big\{a(y^{h,\varepsilon}(s))(1 - x^{h,\varepsilon}(s)) \\ &+d(y^{h,\varepsilon}(s))x^{h,\varepsilon}(s) + \Delta_\varepsilon M_x^{h,\varepsilon}(s)\Big\}ds + M_x^{h,\varepsilon}(t\wedge\tau) \end{aligned} \tag{30}$$

Let $z = (x, y)$ be the unique fixed point of the operator A defined in (23), that is:

$$\begin{aligned} x(t\wedge\tau) = S(t\wedge\tau)x(0) &+ \int_0^{t\wedge\tau} S(t\wedge\tau - s)\big[-d(y(s))x(s) \\ &+a(y(s))(1 - x(s))\big]ds \end{aligned} \tag{31}$$

$$y(t\wedge\tau) = y(0) + \int_0^{t\wedge\tau} \left[-q^1(x(s), y(s))y(s) + q^2(x(s), y(s))(1 - y(s))\right] ds \tag{32}$$

Subtracting the equations (30) and (31), respectively (13) and (32) and taking L^2-norms,we obtain:

$$||x^{h,\varepsilon}(t\wedge\tau) - x(t\wedge\tau)||_{L^2} \leq A_1^x + A_2^x + A_3^x \tag{33}$$

$$||y^{h,\varepsilon}(t\wedge\tau) - y(t\wedge\tau)||_{L^2} \leq A_0^y + A_1^y + A_2^y + A_3^y \tag{34}$$

where

$$A_1^x = \|S_\varepsilon(t\wedge\tau)x^{h,\varepsilon}(0) - S(t\wedge\tau)x(0)\|_{L^2} \tag{35}$$

$$A_2^x = \int_0^{t\wedge\tau} \Big\| S_\varepsilon(t\wedge\tau - s)\left[a(y^{h,\varepsilon}(s))(1-x^{h,\varepsilon}(s)) + d(y^{h,\varepsilon}(s))x^{h,\varepsilon}(s)\right]$$

$$-S(t\wedge\tau - s)\left[-d(y(s))x(s) + a(y(s))(1-x(s))\right]\Big\|_{L^2} ds \tag{36}$$

$$A_3^x = \sup_{t\le T_1}\left\| \int_0^{t\wedge\tau} S_\varepsilon(t\wedge\tau - s)\Delta_\varepsilon M_x^{h,\varepsilon}(s)ds + M_x^{h,\varepsilon}(t\wedge\tau)\right\|_{L^2} \tag{37}$$

and

$$A_y^0 = \|y^{h,\varepsilon}(0) - y(0)\|_{L^2} \tag{38}$$

$$A_y^1 = \int_0^{t\wedge\tau} \|-q^1(x^{h,\varepsilon}(s), y^{h,\varepsilon}(s))y^{h,\varepsilon}(s) + q^1(x(s),y(s))y(s)\|_{L^2} ds \tag{39}$$

$$A_y^2 = \int_0^{t\wedge\tau} \|q^2(x^{h,\varepsilon}(s), y^{h,\varepsilon}(s))(1-y^{h,\varepsilon}(s)) - q^2(x(s),y(s))(1-y(s))\|_{L^2} ds \tag{40}$$

$$A_y^3 = \sup_{t\le T_1} \|M_y^{h,\varepsilon}(t\wedge\tau)\|_{L^2} \tag{41}$$

Let us estimate these terms, by using the properties of the operator semigroups stated in Lemma 1.

$$\begin{aligned}
A_x^1 &\le \|S_\varepsilon(t\wedge\tau)x^{h,\varepsilon}(0) - S_\varepsilon(t\wedge\tau)\pi_\varepsilon x(0)\|_{L^2} \\
&\quad + \|S_\varepsilon(t\wedge\tau)\pi_\varepsilon x(0) - S(t\wedge\tau)x(0)\|_{L^2} \\
&\le \|x^{h,\varepsilon}(0) - \pi_\varepsilon x(0)\|_{L^2} + \|S_\varepsilon(t\wedge\tau)\pi_\varepsilon x(0) - S(t\wedge\tau)x(0)\|_{L^2} \\
&\le \|x^{h,\varepsilon}(0) - x(0)\|_{L^2} + \|x(0) - \pi_\varepsilon x(0)\|_{L^2} \\
&\quad + \|S_\varepsilon(t\wedge\tau)\pi_\varepsilon x(0) - S(t\wedge\tau)x(0)\|_{L^2} \\
&\le \|x^{h,\varepsilon}(0) - x(0)\|_{L^2} + c_1(\varepsilon)
\end{aligned} \tag{42}$$

where, according to (18) and Lemma 1 (iii), $c_1(\varepsilon) \to 0$, as $\varepsilon \to 0$. Let us make the following notations:

$$D_{h,\varepsilon}(s) = a(y^{h,\varepsilon}(s))(1-x^{h,\varepsilon}(s)) + d(y^{h,\varepsilon}(s))x^{h,\varepsilon}(s) \tag{43}$$

$$D(s) = a(y(s))(1-x(s)) - d(y(s))x(s) \tag{44}$$

We have then:

$$\begin{aligned}
A_x^2 &\le \int_0^{t\wedge\tau} \|S_\varepsilon(t\wedge\tau - s)\left[D_{h,\varepsilon}(s) - \pi_\varepsilon D(s)\right]\|_{L^2} ds \\
&\quad + \int_0^{t\wedge\tau} \|(S_\varepsilon(t\wedge\tau - s)\pi_\varepsilon - S(t\wedge\tau - s))D(s)\|_{L^2} ds
\end{aligned}$$

$$\leq \int_0^{t\wedge\tau} \|D_{h,\varepsilon}(s) - \pi_\varepsilon D(s)\|_{L^2} ds$$
$$+ \int_0^{t\wedge\tau} \|(S_\varepsilon(t\wedge\tau - s)\pi_\varepsilon - S(t\wedge\tau - s))D(s)\|_{L^2} ds$$
$$\leq \int_0^{t\wedge\tau} \|D_{h,\varepsilon}(s) - D(s)\|_{L^2} ds + \int_0^{T_1} \|D(s) - \pi_\varepsilon D(s)\|_{L^2} ds$$
$$+ \int_0^{T_1} \|(S_\varepsilon(t\wedge\tau - s)\pi_\varepsilon - S(t\wedge\tau - s))D(s)\|_{L^2} ds$$
$$\leq \int_0^{t\wedge\tau} \|D_{h,\varepsilon}(s) - D(s)\|_{L^2} ds + c_2(\varepsilon) \tag{45}$$

where, according to (18), Lemma 1 (iii), and the dominated convergence theorem, $c_2(\varepsilon) \to 0$, as $\varepsilon \to 0$. By Lemma 1 (iv), we have:

$$\|D_{h,\varepsilon}(s) - D(s)\|_{L^2} \leq const \left(\|x^{h,\varepsilon}(s) - x(s)\|_{L^2} + \|y^{h,\varepsilon}(s) - y(s)\|_{L^2}\right) \tag{46}$$

which yields:

$$A_x^2 \leq const \int_0^{t\wedge\tau} \left(\|x^{h,\varepsilon}(s) - x(s)\|_{L^2} + \|y^{h,\varepsilon}(s) - y(s)\|_{L^2}\right) ds + c_2(\varepsilon) \tag{47}$$

We will discuss now the term A_3^x using the framework considered e.g. in [7]. Let $\mathcal{H}$ be a separable Hilbert space and M an $\mathcal{H}$-valued martingale which is square integrable, right continuous and has left limits (cadlag). Consider on $\mathcal{H}$ a C_0-semigroup of operators $U(t)$ which satisfies $\|U(t)\|_{L(\mathcal{H})} \leq \exp(\mu t)$, and let A be its infinitesimal generator. By partial integration we have:

$$\int_0^t U(t-s)AM(s)ds = -\int_0^t \frac{d}{ds}\left(U(t-s)\right)M(s)ds = U(t)M(0) - M(t)$$
$$+ \int_0^t U(t-s)dM(s) \tag{48}$$

The term $\int_0^t U(t-s)dM(s)$ is called *stochastic convolution-type integral* and can be regarded as a Stieltjes integral of the evolution operators $U(t-s)$ with respect to $M(\cdot)$. Moreover, cf. [7] p.179 it satisfies the following Doob-like inequality:

$$P\left(\sup_{t\leq T}\left\|\int_0^t U(t-s)dM(s)\right\|_{\mathcal{H}} \geq \delta\right) \leq \frac{4\exp(4\mu T)}{\delta^2} E\left[\|M(T)\|_{\mathcal{H}}^2\right] \tag{49}$$

for all $0 \leq t \leq T$ and $\delta > 0$. Returning to our problem where $\mu = 0$, we define

$$Y_{h,\varepsilon}(t) := \int_0^{t\wedge\tau} S_\varepsilon(t\wedge\tau - s)\Delta_\varepsilon M_x^{h,\varepsilon}(s)ds + M_x^{h,\varepsilon}(t\wedge\tau)$$
$$= \int_0^{t\wedge\tau} S_\varepsilon(t\wedge\tau - s)dM_x^{h,\varepsilon}(s) \tag{50}$$

By (49) and Lemma 3 we have thus for any $\delta > 0$:

$$P(A_3^x \geq \delta) = P(\sup_{t\leq T_1} \|Y_{h,\varepsilon}(t)\|_{H^\epsilon} \geq \delta) \leq 4\delta^{-2} const T_1 h\varepsilon^{-2} \tag{51}$$

By Lemma 1 (iv), we have:

$$A_y^1 + A_y^2 \leq const \int_0^{t\wedge\tau} \left(\|x^{h,\varepsilon}(s) - x(s)\|_{L^2} + \|y^{h,\varepsilon}(s) - y(s)\|_{L^2}\right) ds \tag{52}$$

For $\delta > 0$, we have by Doob's inequality for the submartingale $\|M_y^{h,\epsilon}(t\wedge\tau)\|_{H^\epsilon}^2$ and by Lemma 3:

$$P(A_y^3 \geq \delta) \leq \frac{1}{\delta^2} E[\|M_y^{h,\epsilon}(T_1)\|_{H^\epsilon}^2] \leq 4\delta^{-2} const T_1 h\varepsilon^{-2} \tag{53}$$

Summing up the relations (33) and (34) and using the estimates (42), (47), (52) we obtain:

$$\begin{aligned} \|x^{h,\varepsilon}(t\wedge\tau) - x(t\wedge\tau)\|_{L^2} + \|y^{h,\varepsilon}(t\wedge\tau) - y(t\wedge\tau)\|_{L^2} &\leq c_1(\varepsilon) + c_2(\varepsilon) \\ +A_x^3 + A_y^3 + \|x^{h,\varepsilon}(0) - x(0)\|_{L^2} + \|y^{h,\varepsilon}(0) - y(0)\|_{L^2} & \\ +const \int_0^{t\wedge\tau} \left(\|x^{h,\varepsilon}(s) - x(s)\|_{L^2} + \|y^{h,\varepsilon}(s) - y(s)\|_{L^2}\right) ds & \end{aligned} \tag{54}$$

Gronwall's inequality implies:

$$\begin{aligned} \|x^{h,\varepsilon}(t\wedge\tau) - x(t\wedge\tau)\|_{L^2} + \|y^{h,\varepsilon}(t\wedge\tau) - y(t\wedge\tau)\|_{L^2} \leq \Big(c_1(\varepsilon) + c_2(\varepsilon) \\ +A_x^3 + A_y^3 + \|x^{h,\varepsilon}(0) - x(0)\|_{L^2} + \|y^{h,\varepsilon}(0) - y(0)\|_{L^2}\Big) e^{const T_1} \end{aligned} \tag{55}$$

The convergence to 0 of $c_1(\varepsilon), c_2(\varepsilon)$, relations (51), (53) and the hypotheses of the theorem imply that

$$\sup_{t\leq T_1} \left(\|x^{h,\varepsilon}(t\wedge\tau) - x(t\wedge\tau)\|_{L^2} + \|y^{h,\varepsilon}(t\wedge\tau) - y(t\wedge\tau)\|_{L^2}\right) \longrightarrow 0$$

in probability as $h, \varepsilon \to 0$. We will show now that the stopping time τ will be with high probability equal to T_1, meaning that in the limit the saturation is not reached. Let us analyze the event $\{\tau < T_1\}$, that is, the situation when the saturation is reached before the end of the time interval. Considering now in relation (12) $t = \tau$, taking L^∞-norms and using also Lemma 1 (ii), we have:

$$\begin{aligned} 1 = \|x^{h,\varepsilon}(\tau)\|_{L^\infty} &\leq \|x^{h,\varepsilon}(0)\|_{L^\infty} + \|Y_{h,\epsilon}(\tau)\|_{L^\infty} + \int_0^\tau \|D_{h,\epsilon}(s)\|_{L^\infty} ds \\ &\leq \|x(0)\|_{L^\infty} + \|x^{h,\varepsilon}(0) - x(0)\|_{L^\infty} + \varepsilon^{-d/2}\|Y_{h,\epsilon}(\tau)\|_{L^2} + K\tau \end{aligned} \tag{56}$$

Where the constant K is defined in Lemma 2. Together with (27), this implies

$$KT_1 > K\tau \geq r - \|x^{h,\varepsilon}(0) - x(0)\|_{L^\infty} - \varepsilon^{-d/2}\|Y_{h,\epsilon}(\tau)\|_{L^2} \tag{57}$$

Using the relation between T_1 and r from Lemma 2 we get:

$$K\|z(0)\|_{Z_{T_1}} T_1 \leq \|x^{h,\varepsilon}(0) - x(0)\|_{L^\infty} + \varepsilon^{-d/2}\|Y_{h,\epsilon}(\tau)\|_{L^2} \tag{58}$$

Denote the event of this inequality by $\mathcal{E}$. We have shown that

$$\{\tau < T_1\} \Longrightarrow \mathcal{E}$$

and thus

$$P\{\tau < T_1\} \leq P\{\mathcal{E}\}$$

But, by the hypothesis of the theorem together with (51),

$$P\{\mathcal{E}\} \to 0 \text{ as } h, \varepsilon \to 0$$

This shows that the convergence property holds on the entire time interval $[0, T_1]$. Since we considered a solution of (21) which, by the assumption on the reaction functions, stays uniformly bounded away from 1, we have $\|x(t)\|_{L^\infty} \leq 1 - \kappa$ for all $t \in [0, T_1]$, for some positive constant κ. This implies that $r \geq \kappa$ and that $1 - \|x(T_1)\|_{L^\infty} \geq \kappa > 0$. We can now repeat the construction from Lemma 2 starting form the initial value $(x(T_1), y(T_1))$ and obtain a prolongation of the solution on the interval $[T_1, T_2]$ with $T_2 = T_1 + K^{-1}(1 + \|z(T_1)\|_{Z_{T_2}})^{-1}(1 - \|x(T_1)\|_{L^\infty})$. We have then $T_2 - T_1 \geq (K/3)^{-1}\kappa$. Applying this procedure repeatedly, we obtain a solution on any arbitrary finite interval $[0, T]$, since the local existence intervals do not shrink. In every step we repeat the argumentation for the stochastic convergence which we made for the interval $[0, T_1]$ and cover thus the whole interval $[0, T]$, finishing the proof of the theorem. □

References

[1] Arnold, L., Theodosopulu, M. (1980) Deterministic limit of the stochastic model of chemical reactions with diffusion. Adv.Appl.Prob. **12**, 367-379

[2] Blount, D. (1994) Density dependent limits for a nonlinear reaction-diffusion model. Ann.Probab. **22** No.4, 2040-2070

[3] Blount, D. (1996) Diffusion limits for a nonlinear density-dependent space-time population model. Ann.Probab. **24** No.2, 639-659

[4] Ethier, S., Kurtz, T.G. (1986) Markov Processes: Characterization and Convergence. J.Wiley & Sons, New York

[5] Guiaş, F. (2001) Convergence properties of a stochastic model for coagulation–fragmentation processes with diffusion. Stochastic.Anal.Appl. **19** No.2, 245-278

[6] Guiaş, F. (1997) A Monte Carlo approach to the Smoluchowski equations. Monte Carlo Meth.Appl. **3** No.4, 313-326

[7] Kotelenez, P. (1986) Law of large numbers and central limit theorem for linear chemical reactions with diffusion. Ann.Probab. **14** No.1, 173-193

[8] Reichert, C., Starke, J., Eiswirth, M. (2001) Stochastic Modelling of CO-Oxidation on Platinum Surfaces and Deterministic Limit. (Preprint)

[9] Smoller, J. (1983) Shock Waves and Reaction-Diffusion Equations. Springer

Two–Scale Models for Liquid–Solid Phase Transitions in Binary Material with Equiaxed Microstructure

Christof Eck and Peter Knabner

Institute for Applied Mathematics, University Erlangen–Nürnberg, Martensstr. 3, 91058 Erlangen, Germany

Abstract. In many liquid–solid phase transitions, a specific dendritic microstructure of the phase interface is observed. In this contribution we present two–scale models capable to describe the evolution of equiaxed microstructure. The models are based either on a sharp interface model or on a phase field model for phase transitions in binary alloys. In both cases, a formal asymptotic expansion in terms of a scale parameter ε for the microstructure is carried out, with solute diffusivity scaling proportional to ε^2. In the limit $\varepsilon \to 0$ we obtain a two scale–model consisting of a macroscopic heat transport equation and, at each point of the macroscopic domain, of a local cell problem modeling the microsegregation and the evolution of the microstructure.

Contents

1 Introduction

In many liquid–solid phase transition processes as e.g. the casting of metals we observe a specific dendritic (tree–like) microstructure of the solid–liquid interface. This microstructure is a consequence of an instability of a regular phase interface with respect to small perturbations. There are mainly two morphologies of this microstructure, the equiaxed one, where small solid kernels nucleate in the melt and grow in all directions; and the columnar one, where columnar crystals grow in the opposite direction of the heat flow; in both cases we observe side–branching in dendrites of second or third order (see Fig. 1). Models capable to describe the evolution of dendritic crystals are either sharp interface models with curvature undercooling and possibly kinetic undercooling, [8] and [12], or phase field models [2]. Numerical methods for the solution of these problems are based on interface tracking [10], variational algorithms [1], level set methods [11] and discretizations of the phase field model [6]. However, all these methods are computationally challenging even for one or few crystals and two dimensions; and it is not possible to compute the microstructure evolution in a whole macroscopic domain.

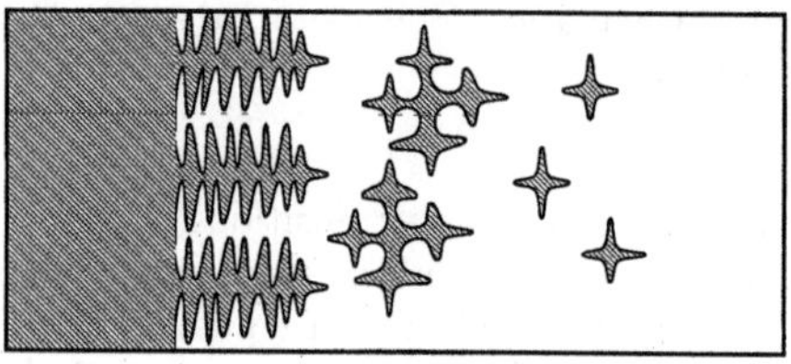

Fig. 1. Microstructure in liquid–solid phase transitions

In the present paper we develop two–scale models for phase transitions with equiaxed microstructure in binary alloys with small solute diffusion. The derivation of the models starts from a sharp interface model analyzed in [8] or a phase field model introduced in [4]. In both cases, a formal asymptotic expansion in terms of the microstructure scale ε is carried out, where the solute diffusivity is scaled proportional to ε^2. In the limit $\varepsilon \to 0$; a two–scale model is obtained, consisting of a macroscopic heat equation and, at each point of the macroscopic domain, of a local cell problem modeling the microsegregation and the evolution of the microstructure.

2 The sharp interface model

In the sharp interface model the domain under consideration $\Omega \subset \mathbb{R}^N$ with space dimension $N \in \mathbb{N}$ is divided for every time t into the domains $\Omega_\ell(t)$ of liquid material and $\Omega_s(t)$ of solid material. Both domains are separated by the phase interface $\Gamma_I(t)$. We denote the variables by temperature T, concentration c_k of one of the components of the material in liquid ($k = \ell$) and solid ($k = s$) material and velocity v_s of the phase interface, measured in direction of the outer normal n_s of the solid domain. The material properties are described by density of internal energy $u_k = u_k(T, c_k)$, chemical potential $\mu_k = \mu_k(T, c_k)$, heat conductivity $K^{(k)}$, solute diffusivity $D^{(k)}$, curvature multiplier σ, kinetic multiplier β, and a function $b(T, \mu)$ describing the deviation from equilibrium melting point. Then the model is given by the equations

$$\partial_t u_k - \nabla \cdot (K^{(k)} \nabla T) - \nabla \cdot \left(D^{(k)} \frac{\partial u_k}{\partial c_k} \nabla c_k \right) = f, \tag{1}$$

$$\partial_t c_k - \nabla \cdot \left(D^{(k)} \nabla c_k\right) = 0 \tag{2}$$

defined for $t \in I_{\mathcal{T}} = [0, \mathcal{T}]$ and $x \in \Omega_k(t)$, $k = \ell, s$; supplemented by boundary conditions on the phase interface $\Gamma_I(t)$,

$$\mu_\ell(T, c_\ell) = \mu_s(T, c_s), \tag{3}$$

$$\left[K\nabla T \cdot n + D\frac{\partial u}{\partial c}\nabla c \cdot n\right] = -[u]v_s, \tag{4}$$

$$[D\nabla c \cdot n] = -[c]v_s, \tag{5}$$

$$b(T, \mu) = -\sigma\kappa - \beta v_s, \tag{6}$$

by boundary conditions on the boundary of the whole domain $\Gamma = \partial\Omega$

$$K\nabla T \cdot n + \alpha T = f_\Gamma, \tag{7}$$

$$D\nabla c \cdot n = 0, \tag{8}$$

and by initial conditions

$$T(t = 0, x) = T^{(0)}(x) \text{ and } c(t = 0, x) = c^{(0)}(x), \quad x \in \Omega$$

for temperature and concentration and

$$\Omega_\ell(t = 0) = \Omega_\ell^{(0)}$$

for the initial partition into solid and liquid domain. Here, κ denotes the mean curvature, taken positive for Ω_s being convex, and the bracket $[\cdot]$ denotes the solid–liquid difference of the corresponding quantity at the phase interface, $[K\nabla T \cdot n] := K^{(\ell)}\nabla T_\ell \cdot n_\ell + K^{(s)}\nabla T_s \cdot n_s$ and $[u] := u_s - u_\ell$. The functions u_k and μ_k are assumed to be sufficiently smooth, but they may be different in solid and liquid. Equation (6) is a generalized Stefan condition with Gibbs–Thomson undercooling and kinetic undercooling adapted to the case of a binary mixture. The function b is specified from the phase diagram; in the simplest case of a linear model it is given by $b(T, \mu) = C_1 T + C_2 \mu + C_3$ with coefficients C_1, C_2 and C_3; in a thermodynamically consistent model b is defined as solid–liquid difference of the Legendre transform of the entropy. An existence result for this model is given in [8] for the case of vanishing kinetic coefficient $\beta = 0$.

3 Homogenization of the sharp–interface model

In order to perform the homogenization, we assume an idealized microstructure consisting of equiaxed crystals whose midpoints are located at the sites of a uniform grid. The scale of the microstructure is given by a small parameter $\varepsilon > 0$ signifying the spacing of this grid. This situation arises in the case of *instantaneous* nucleation, where small solid kernels of a given density and a given size nucleate at a given undercooling. The assumption of periodic distribution of the initial kernels is, of course, not very realistic, but it is frequently made in homogenization theory in order to derive macroscopic models which are then also used for more general situations. The initial conditions of the problem have to reflect this instantaneous periodic nucleation, they are given by

$$T_\varepsilon(t=0,x) = T^{(0)}(x,\tfrac{x}{\varepsilon}), \tag{9}$$

$$c_\varepsilon(t=0,x) = c^{(0)}(x,\tfrac{x}{\varepsilon}) \tag{10}$$

with functions $T^{(0)}$, $c^{(0)} : \Omega \times Y \to \mathbb{R}$ being Y–periodic with respect to the second variable, and an initial solid domain

$$\Omega_s^{(0)} := \Omega \cap \bigcup_{k\in\mathbb{Z}^N} \varepsilon\left(\{k\}\oplus Y_s^{(0)}\right)$$

with unit solid domain $Y_s^{(0)} \subset Y$. Here, T_ε and c_ε denote solutions of the problem with scale parameter ε.

In the limit procedure $\varepsilon \to 0$ several physical parameters must be scaled in dependence of ε. In order to maintain an equilibrium between surface- and volume energy, the density of surface energy σ is scaled proportional to ε,

$$\sigma = \varepsilon\sigma_0.$$

The solute diffusivity is scaled proportional to ε^2,

$$D^{(k)} = \varepsilon^2 D_0^{(k)},$$

this is justified by the fact that solute diffusivity is usually much smaller than heat conductivity. Moreover, in many situations the microstructure is generated by a competition between solute diffusion and surface energy; hence, in a two scale model the solute diffusivity has to be scaled with some power of ε. The power ε^2 chosen here leads to a decoupling of the roles of T and c, it turns out that T will be an essentially macroscopic quantity, while c will be a microscopic variable. The kinetic coefficient is scaled proportional to ε^{-1},

$$\beta = \varepsilon^{-1}\beta_0.$$

The homogenization is carried out by the method of asymptotic expansions. We assume that the functions T_ε, c_ε and the velocity v_ε of the phase interface admit the asymptotic representations

$$\begin{aligned} T_\varepsilon(t,x) &= T_0\big(t,x,\tfrac{x}{\varepsilon}\big) + \varepsilon T_1\big(t,x,\tfrac{x}{\varepsilon}\big) + \varepsilon^2 T_2\big(t,x,\tfrac{x}{\varepsilon}\big) + \cdots, \\ c_\varepsilon(t,x) &= c_0\big(t,x,\tfrac{x}{\varepsilon}\big) + \varepsilon c_1\big(t,x,\tfrac{x}{\varepsilon}\big) + \varepsilon^2 c_2\big(t,x,\tfrac{x}{\varepsilon}\big) + \cdots, \\ v_\varepsilon(t,x) &= \varepsilon v_0\big(t,x,\tfrac{x}{\varepsilon}\big) + \varepsilon^2 v_1\big(t,x,\tfrac{x}{\varepsilon}\big) + \cdots \end{aligned} \tag{11}$$

valid for $\varepsilon \to 0$. These expansions are plugged into problem (1)–(6). The gradient of a function $f(x, \frac{x}{\varepsilon})$ depending on a "slow" variable x and a "fast" variable $y = \frac{x}{\varepsilon}$ is given by $\nabla f = \nabla_x f + \frac{1}{\varepsilon}\nabla_y f_{|y=\frac{x}{\varepsilon}}$ or, shortly, $\nabla = \nabla_x + \frac{1}{\varepsilon}\nabla_y$. This leads to a sequence of problems of different orders.

The **problem of 0th order** consists of the terms in equation (1) of order ε^{-2}, of the continuity of the temperature of order ε^0 and of equation (4) of order ε^{-1}. Since these terms contain only derivatives with respect to the microscopic space variable y (representing $\frac{x}{\varepsilon}$), the problem of 0th order can be written as a family of problems defined on the (up–scaled) unit cell Y depending on the macroscopic space variable x as a parameter:

$$\begin{aligned} -\nabla_y \cdot \left(K^{(k)}\nabla_y T_0\right) &= 0 \quad \text{in } Y_k(t,x),\ k = \ell, s, \\ [K\nabla_y T_0 \cdot n_y] &= 0 \quad \text{on } \Gamma_I(t,x), \\ [T_0] &= 0 \quad \text{on } \Gamma_I(t,x), \end{aligned}$$

where T_0 satisfies periodic boundary conditions on ∂Y. Here, Y_k for $k = s$ and $k = \ell$ denote the solid and liquid part of the unit cell Y, and n_y is the unit normal vector on their common boundary Γ_I, pointing outward of Y_s. Let us remark that Y_k depends on time (due to the evolution of the phase transition) and on the macroscopic space variable (since the solid crystals may be different at different macroscopic positions $x \in \Omega$). The solutions of the problem of 0th order are constant functions. Hence, T_0 is independent of y, $T_0 = T_0(t,x)$.

The **problem of 1st order** consists of the terms in equation (1) of order ε^{-1}, of the continuity of the temperature of order ε^1 and of equation (4) of order ε^0,

$$\begin{aligned} -\nabla_y \cdot \left(K^{(k)}(\nabla_y T_1 + \nabla_x T_0)\right) &= 0 \quad \text{in } Y_k(t,x),\ k = \ell, s, \\ [K(\nabla_y T_1 + \nabla_x T_0)\cdot n_y] &= 0 \quad \text{on } \Gamma_I(t,x), \\ [T_1] &= 0 \quad \text{on } \Gamma_I(t,x) \end{aligned}$$

and T_1 satisfies periodic boundary conditions on ∂Y. The solution of this problem can be represented by

$$T_1(t,x,y) = \sum_{j=1}^{N} H_j(t,x,y)\partial_{x_j} T_0(t,x), \tag{12}$$

where H_j, $j = 1, \ldots, N$, are solutions of the *local cell problem*

$$\begin{aligned} -\nabla_y \cdot \left(K^{(k)}\nabla_y H_j\right) &= \nabla_y \cdot \left(K^{(k)} e_j\right) \quad \text{in } Y_k(t,x),\ k = \ell, s, \\ [K\nabla_y H_j \cdot n_y] &= -[K e_j \cdot n_y] \quad \text{on } \Gamma_I(t,x) \end{aligned} \tag{13}$$

satisfying periodic boundary conditions on ∂Y. Here, e_j denotes the j–th unit vector in $\mathbb{R}^N$. If the heat conductivity is constant and its value in liquid and solid is equal, $K^{(\ell)} = K^{(s)}$, then $T_1(t,x,y)$ is also independent of y.

The **problem of 2nd order** consists of the terms of order ε^0 in (1), (2), (3), (6), of order ε^1 in (4), (5) and of order ε^2 in the continuity of the temperature,

$$\begin{aligned}
\partial_t u_k(T_0, c_0^{(k)}) - \nabla_y \cdot \left(K^{(k)} \nabla_y T_2\right) - \nabla_x \cdot \left(K^{(k)} (\nabla_x T_0 + \nabla_y T_1)\right) & \\
- \nabla_y \cdot \left(K^{(k)} \nabla_x T_1 + D_0^{(k)} \frac{\partial u_k}{\partial c_k} \nabla_y c_0^{(k)}\right) &= f \text{ in } \Omega \times Y_k(t), \quad (14)\\
\left[\left(K(\nabla_y T_2 + \nabla_x T_1) + D_0 \frac{\partial u}{\partial c} \nabla_y c_0\right) \cdot n_y\right] & \\
+ [u(T_0, c_0)] v_0 &= 0 \text{ on } \Omega \times \Gamma_I(t), \quad (15)\\
\partial_t c_0^{(k)} - \nabla_y \cdot \left(D_0^{(k)} \nabla_y c_0^{(k)}\right) &= 0 \text{ in } \Omega \times Y_k(t), \quad (16)\\
\mu_\ell(T_0, c_0^{(\ell)}) &= \mu_s(T_0, c_0^{(s)}) \text{ on } \Omega \times \Gamma_I(t), \quad (17)\\
[D_0 \nabla_y c_0 \cdot n_y] + [c_0] v_0 &= 0 \text{ on } \Omega \times \Gamma_I(t), \quad (18)\\
b(T_0, \mu_0) &= -\sigma_0 \kappa_0 - \beta_0 v_0 \text{ on } \Omega \times \Gamma_I(t). \quad (19)
\end{aligned}$$

Integrating equation (14) with respect to $y \in Y$ and using the interfacial boundary condition (15) yields the *macroscopic heat equation*

$$\partial_t u^*_{c_0, Y_s}(T_0) - \nabla \cdot \left(K^* \nabla T_0\right) = f \quad \text{in } \Omega, \tag{20}$$

with the averaged density of internal energy

$$u^*_{c_0, Y_s}(T_0) = \sum_{k=\ell,s} \int_{Y_k} u_k(T_0, c_0^{(k)})\, dy,$$

the effective heat conductivity

$$K^*_{ij} = \sum_{k=\ell,s} \int_{Y_k} \left(K_{ij}^{(k)} + \sum_{r=1}^{N} K_{ir}^{(k)} \partial_{y_r} H_j\right) dy$$

and a possibly averaged heat source f. Equations (16) – (19) represent a family of generalized Stefan problems for concentration c_0 and the partition into solid domain Y_s and liquid domain Y_ℓ, depending on the parameter $x \in \Omega$ and on the local macroscopic temperature evolution $T(\cdot, x)$ as given data. They can be interpreted as *local cell problems* for the microsegregation c_0 and the evolution of the dendritic microstructure,

$$\partial_t c_k(T_0; \mu_0) - \nabla_y \cdot \left(\widetilde{D}_0^{(k)} \nabla_y \mu_0\right) = 0 \text{ in } Y_k(t),\ k = \ell, s, \tag{21}$$

$$\left[\widetilde{D}_0 \nabla_y \mu_0 \cdot n_y\right] + \left[c(T_0; \mu_0)\right] v_0 = 0 \text{ on } \Gamma_I(t), \tag{22}$$

$$b(T_0, \mu_0) = -\sigma_0 \kappa_0 - \beta_0 v_0 \text{ on } \Gamma_I(t). \tag{23}$$

In order to illustrate the structure as a generalized Stefan problem, the chemical potential μ is used as variable. This requires a modification of the diffusivity to $\widetilde{D}^{(k)} = D^{(k)} \frac{\partial c_k}{\partial \mu}$.

4 The phase field model

The homogenization approach will be applied to a phase field model for binary mixtures introduced in [4]. It is not the most general model available, in fact it can be characterized as the most simple model capable to describe the release of latent heat and the miscibility gap (the difference of concentration in solid and liquid phase at the phase interface). The model is given by the differential equations

$$\partial_t T + L\partial_t \Phi - \nabla \cdot (K\nabla T) = f, \tag{24}$$

$$\partial_t c - \nabla \cdot (D\nabla c) - \nabla \cdot (DM(c)\nabla \Phi) = 0, \tag{25}$$

$$\alpha\xi^2 \partial_t \Phi - \xi^2 \Delta\Phi + p'(\Phi) + q(T, c) = 0. \tag{26}$$

The variables are temperature T, concentration c and phase field Φ, where $\Phi = -1$ stands for solid and $\Phi = +1$ for liquid material. Moreover, L is half of the latent heat, $K = \{K_{ij}\}_{i,j=1}^N$ the heat conductivity and $D = \{D_{ij}\}_{i,j=1}^N$ the solute diffusivity. The term $\nabla \cdot (DM(c)\nabla\Phi)$ with $M : [0, 1] \to \mathbb{R}$ generates a transition layer in the concentration from a transition layer in the phase field and thereby realizes the miscibility gap mentioned above. The parameter ξ describes the thickness of the phase transition layer between solid and liquid material, and α is a relaxation parameter. The function p represents a double–well potential, e.g. $p(\Phi) = \frac{1}{8}\left(\Phi^2 - 1\right)^2$, and $q : \mathbb{R}^2 \to \mathbb{R}$ is defined by $q(T, c) = \frac{2\xi}{3\sigma}[s](T - T_1 c - T_2(1 - c))$ with surface tension σ, solid–liquid difference of entropy $[s]$ and equilibrium melting points T_1 and T_2 of the two pure substances.

5 Homogenization of the phase field model

As in the case of the sharp–interface model it is assumed that the initial data exhibit a small scale microstructure having a size proportional to the scale parameter $\varepsilon > 0$. For the unit cell $Y = [0, 1]^N$ the initial data are given by

$$T_0^{(\varepsilon)}(x) = T^{(0)}(x, \tfrac{x}{\varepsilon}),$$

$$c_0^{(\varepsilon)}(x) = c^{(0)}(x, \tfrac{x}{\varepsilon}) \text{ and}$$

$$\Phi_0^{(\varepsilon)}(x) = \Phi^{(0)}(x, \tfrac{x}{\varepsilon})$$

with functions $T^{(0)}, c^{(0)}, \Phi^{(0)} : \Omega \times Y \to \mathbb{R}$ being Y–periodic with respect to the second variable $y \in Y$. In order to obtain a well–defined homogenized problem it is necessary to scale some given data in dependence of ε. Here it is assumed

$$\xi = \varepsilon\xi_0,$$

$$\alpha = \varepsilon^{-2}\alpha_0,$$

$$D = \varepsilon^2 D^{(0)}.$$

The required scaling of ξ is obvious; if the size of a solid crystal varies proportional to ε, then the width of the solid–liquid transition layer must be bounded by const $\cdot\,\varepsilon$, too. The scaling of the solute diffusivity is the same as in the sharp interface model above. The relaxation parameter α is scaled such that the total factor $\xi^2\alpha$ of the time derivative in the phase field equation remains constant; in this case the time scale of the problem is independent of ε.

In order to study the limit $\varepsilon \to 0$ the existence of an asymptotic expansion

$$\begin{aligned} T_\varepsilon(t,x) &= T_0\big(t,x,\tfrac{x}{\varepsilon}\big) + \varepsilon T_1\big(t,x,\tfrac{x}{\varepsilon}\big) + \varepsilon^2 T_2\big(t,x,\tfrac{x}{\varepsilon}\big) + \cdots, \\ c_\varepsilon(t,x) &= c_0\big(t,x,\tfrac{x}{\varepsilon}\big) + \varepsilon c_1\big(t,x,\tfrac{x}{\varepsilon}\big) + \varepsilon^2 c_2\big(t,x,\tfrac{x}{\varepsilon}\big) + \cdots, \\ \Phi_\varepsilon(t,x) &= \Phi_0\big(t,x,\tfrac{x}{\varepsilon}\big) + \varepsilon \Phi_1\big(t,x,\tfrac{x}{\varepsilon}\big) + \varepsilon^2 \Phi_2\big(t,x,\tfrac{x}{\varepsilon}\big) + \cdots \end{aligned} \tag{27}$$

is supposed. The asymptotic expansions and the formal relation $\nabla = \nabla_x + \frac{1}{\varepsilon}\nabla_y$ are put into the differential equations (24)–(26). Then the coefficients of different powers of ε are collected, starting from the lowest order.

The **problem of 0th order** consists of the terms of order ε^{-2} in the heat equation (24); these are

$$\begin{aligned} -\nabla_y \cdot \big(K\nabla_y T_0\big) \;&=\; 0 \text{ on } Y, \\ T_0 \text{ is } Y\text{–periodic with respect to } y. \end{aligned}$$

The solutions of this problem are constant, hence $T_0(t,x,y) = T_0(t,x)$ is independent of y.

The **problem of 1st order** is given by the terms of order ε^{-1} of the heat equation,

$$\begin{aligned} -\nabla_y \cdot \big(K(\nabla_y T_1 + \nabla_x T_0)\big) \;&=\; 0 \text{ on } Y, \\ T_1 \text{ is } Y\text{–periodic with respect to } y. \end{aligned}$$

In our simple case with K independent of y (and in particular of the phase field), the solution T_1 is also a constant function, hence $T_1 = T_1(t,x)$ is independent of y. For K depending on the fast variable y one would get the usual local cell problem leading to a homogenized effective heat conductivity.

The **problem of 2nd order** consists of the term of order ε^0 in the heat equation, the diffusion equation and the phase field equation. This gives

$$\partial_t T_0 + L\partial_t \Phi_0 - \nabla_y \cdot \big(K\nabla_y T_2\big) - \nabla_x \cdot \big(K\nabla_x T_0\big) = f, \tag{28}$$

$$\partial_t c_0 - \nabla_y \cdot \big(D^{(0)}\nabla_y c_0\big) - \nabla_y \cdot \big(D^{(0)} M(c_0)\nabla_y \Phi_0\big) = 0, \tag{29}$$

$$\alpha_0\xi_0^2\partial_t\Phi_0 - \xi_0^2\Delta_y\Phi_0 + p'(\Phi_0) + q(T_0,c_0) = 0 \tag{30}$$

on $I_T \times \Omega \times Y$, supplemented by periodic boundary conditions

$$T_2, c_0 \text{ and } \Phi_0 \text{ are } Y\text{–periodic with respect to } y. \tag{31}$$

Equations (29) and (30) do not contain any derivatives with respect to x. Hence they can be interpreted as a set of differential equations defined on $I_T \times Y$ for every parameter $x \in \Omega$. Equation (28) is transformed into a macroscopic equation for $T_0 = T_0(t,x)$ by integration with respect to $y \in Y$. Due to the periodic boundary conditions of T_2 the $\nabla_y\cdot$–term disappears and the following problem is obtained:

$$\partial_t T_0 + L\partial_t \overline{\Phi}_0 - \nabla_x \cdot \big(K\nabla_x T_0\big) = f \tag{32}$$

with solid volume fraction $\overline{\Phi}_0(t,x) := \int_Y \Phi_0(t,x,y)\,dy$.

The result of the formal asymptotic expansion is a two–scale model consisting of the macroscopic heat transport equation (32) and, for each point $x \in \Omega$ of the macroscopic domain, of a local cell problem (29)–(30), supplemented by the appropriate periodic boundary conditions. The local cell problem describes the evolution of the phases and the microscopic solute concentration. Both problems are coupled; the release of latent heat in the heat equation is given by the macroscopic average of the local microscopic phase field Φ; and the temperature $T_0(t,x)$ appears in the function $q(T_0, c_0)$ of the phase field equation (30) and describes in particular the melting point.

6 A numerical example

In order to illustrate the numerical application of the two–scale model presented here, we consider a simple sharp interface model defined by internal energy $u_k(T,c) = T + L\delta_{k\ell}$ with latent heat L specified below and Kronecker symbol $\delta_{k\ell} = \begin{cases} 1, & k=\ell, \\ 0, & k \neq \ell, \end{cases}$ heat conductivity $K = 1$, chemical potential $\mu = \mu_k(c_k) = c_k + \delta_{k\ell}\lambda$, with constant miscibility gap $\lambda = 0.5$, solute diffusivity $D = \varepsilon^2$, function $b(T,\mu) = T - \mu$, anisotropic curvature multiplier $\sigma(\varphi_n) = \overline{\sigma}\big(1 - \frac{4}{45}\sin(4\varphi_n) - \frac{1}{189}\sin(8\varphi_n)\big)$ with φ_n denoting the angle between the normal vector of the interface and the x–axis and $\overline{\sigma} = 0.002\varepsilon$, no kinetic undercooling ($\beta = 0$) and initial data $T^{(0)} = 0$, $\mu^{(0)} = 0.5$. The problem is solved on a rectangular macroscopic domain $\Omega = [0;2] \times [0;0.5]$, c.f. Fig. 2. At initial time $t_0 = 0$, small solid kernels of radius $r_0 = 0.05\varepsilon$ are assumed to have just formed at the grid points of a uniform square grid with a density of $4/\varepsilon^2$; this gives a local cell problem defined on a square with edge length 2. Remember that the two–scale model is an approximation for a whole family of phase transition problems depending on scale parameter ε with increasing accuracy for decreasing ε. The domain is cooled by the Neumann boundary condition $K\frac{\partial T}{\partial n} = -5$ on part of boundary Γ_4, while the remaining boundary is isolated.

The space discretizations for both the macroscopic and the microscopic problems is done by a finite element method on a uniform rectangular grid with bilinear shape functions; we use 8×3 elements for the macroscopic

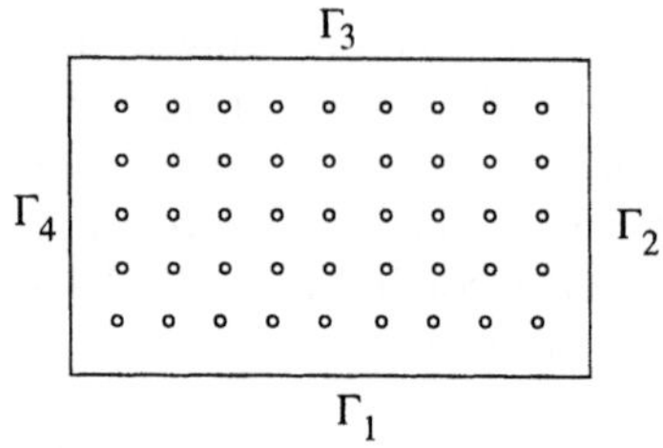

Fig. 2. Model problem for two–scale method

problem and 400 × 400 nodes for the microscopic problem. The time discretization is done by an implicit Euler scheme with time step $\Delta t = 0.0001$. The problems are coupled in an explicit way; first, all the local cell problems are solved for the temperature of the previous time step, then the heat equation is solved with the new release of latent heat computed from the cell problems. For the solution of the cell problems we use an algorithm developed by R. Almgren, [1]. The problem is essentially one–dimensional; in the graphs below we illustrate the properties of an array of 8 crystals in x–direction.

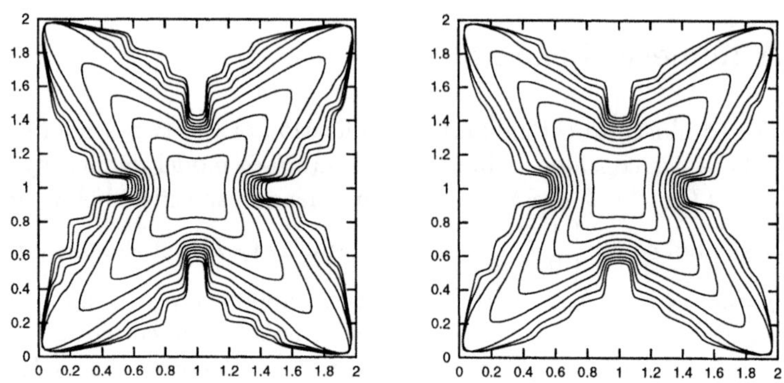

Fig. 3. Evolution of left and right crystal for latent heat $L = 10$

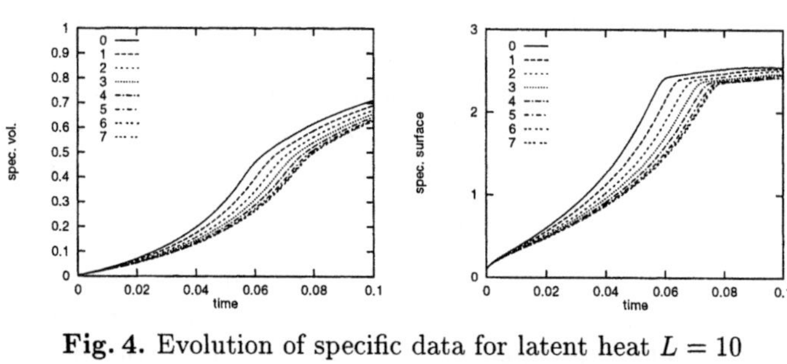

Fig. 4. Evolution of specific data for latent heat $L = 10$

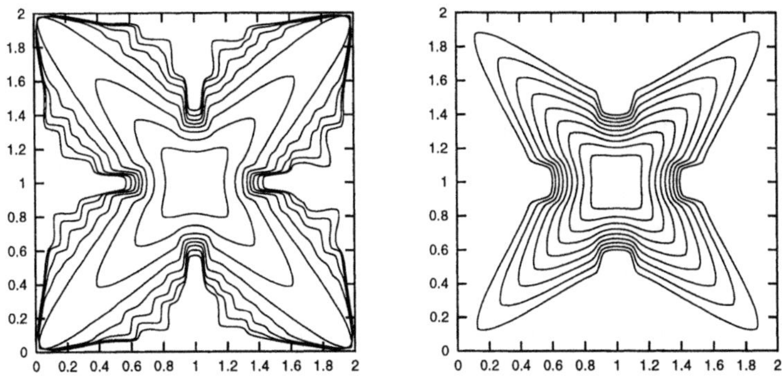

Fig. 5. Evolution of left and right crystal for latent heat $L = 5$

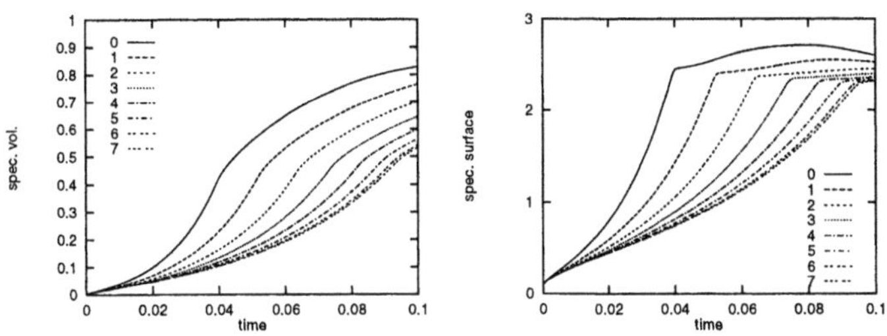

Fig. 6. Evolution of specific data for latent heat $L = 5$

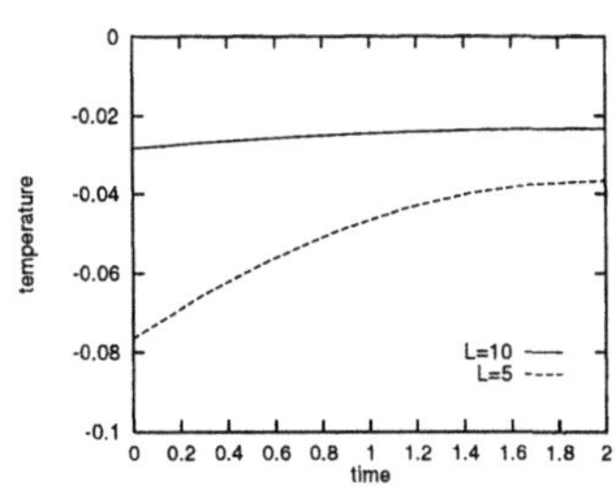

Fig. 7. Macroscopic temperature for latent heat $L = 10$ and $L = 5$ at time $t = 0.1$

In order to show the influence of macroscopic heat transport on the evolution of the solid crystals, we use two different values $L = 10$ and $L = 5$ for the latent heat. Fig. 3 shows the evolution of the first crystal (that one at the cooled boundary Γ_4) and the last crystal (that one at boundary part Γ_2) for latent heat $L = 10$. In Fig. 4 the evolution of the macroscopic specific volume and specific surface is depicted. Fig. 5 and 6 show the corresponding graphs for latent heat $L = 5$. In this case, we observe a much bigger dependence of the microscopic data on the location of the crystal. The reason can be seen in the graph of the temperature distribution at a fixed time $t = 0.1$ in Fig. 7. For large value of the latent heat, the growth of the first crystal releases enough latent heat to almost compensate the cooling condition on the boundary, this leads to an almost uniform temperature field, and the main driving force of the crystal growth is the initial solutal undercooling, which is the same for all crystals. For small latent heat, the growth of the first crystals is limited by solute diffusion, hence they cannot release enough latent heat to consume up all the boundary cooling and the variation of the temperature field as well as that of the crystals is much bigger.

7 Conclusion

Based on two different models for the microscopic scale, the sharp interface model and the phase field model, we propose two two–scale models for liquid–solid phase transitions with equiaxed microstructure. In both cases, the model consists of a macroscopic energy transport equation to be solved on the given domain, and of local cell problems to be solved on the local periodicity cell for every point of the macroscopic domain. The cell problems have the structure of generalized isothermal phase transition problems for binary material. The derivation presented here has only a formal character, because it is a prior not guaranteed that asymptotic expansions of the type (11) or (27) exist. In order to be mathematically rigorous, it is necessary to justify the homogenization result. This can be done e.g. by comparing the differences $u_\varepsilon - u_0^\varepsilon$, where u_ε for $u = T, c, \Phi$ signifies the solution of the original model with scale parameter ε and $u_0^\varepsilon = u_0(t, x, \frac{x}{\varepsilon})$ is a reconstruction of the solution to the two–scale model. The proof of an estimate for $\sum_{u=T,c} \|u_\varepsilon - u_0^\varepsilon\|$ with a suitable norm $\|\cdot\|$ is probably a very hard task in the case of the sharp interface model, because these models are difficult to analyze; and uniqueness results usually are available only for *small* initial data and/or a *small* time interval. In the case of the phase field model a rigorous justification seems to be possible. Since phase field models converge to sharp interface models for vanishing parameter ξ (at least for suitable scalings of other parameters), it could be possible to get a justification for the homogenization of the sharp interface model via the corresponding phase field model and its limit.

References

1. Almgren, R. (1993) Variational algorithms and pattern formation in dendritic solidification. J. Comput. Physics **106**, 337–354
2. Caginalp, G. (1986) An analysis of a phase field model of a free boundary. Arch. Ration. Mech. Anal. **92**, 205–245
3. Caginalp, G., Xie, W. (1993) Phase–field and sharp interface alloy models. Physical Rev. E **48**, 1897–1909

4. Caginalp, G., Xie, W. (1998) An analysis of phase field alloys and transition layers. Arch. Rat. Mech. Anal. **142**, 293–329
5. Eck, Ch., Knabner, P., Korotov, S. (2001) A two–scale method for the computation of solid–liquid phase transitions with dendritic microstructure. Preprint No. 278, Institute for Applied Mathematics, University Erlangen–Nürnberg
6. Kobayashi, R. (1993) Modeling and numerical simulation of dendritic crystal growth. Physica D **63**, 410–423
7. Luckhaus, S. (1990) Solutions for the two–phase Stefan problem with the Gibbs–Thomson law for the melting temperature. Eur. J. Appl. Math. **1**, 101–111
8. Luckhaus, S. (1993) Solidification of alloys and the Gibbs–Thomson law. Preprint, University of Bonn
9. Roosen, A.R., Taylor, J.E. (1994) Modeling crystal growth in a diffusion field using fully faceted interfaces. J. Comput. Physics **114**, 113–128
10. Schmidt, A. (1996) Computation of three–dimensional dendrites with finite elements. J. Comput. Physics **125**, 293–312
11. Sethian, J.A., Strain, J. (1992) Crystal growth and dendritic solidification. J. Comput. Physics **98**, 231–253
12. Visintin, A. (1996) Models of Phase Transitions. Birkhäuser, Boston

Asymptotic Behaviour of the Incompressible Newtonian Flow Through Thin Constricted Fracture

Olivier Gipouloux[1,2] and Eduard Marušić–Paloka[3]

[1] Calcul Scientifique et Modélisation Mathématique, Université de Saint Etienne, 23 Rue du Docteur P. Michelon 42023 Saint Etienne, France
[2] Laboratoire de Mécanique et d'Acoustique, 31 Chemin Joseph Aiguier 13402 Marseille, France
[3] Department of Mathematics, University of Zagreb, Bijenička 30, 10000 Zagreb, Croatia

Abstract. We study the fluid flow through a thin fracture with prescribed pressure drop. We suppose that the fracture has a constriction such that the flow is not purely Poiseuille. We find the corrector for the Poiseuille flow due to the constriction, we prove the corresponding error estimate and we illustrate the theoretical results by some numerical experiments.

Contents

1 Introduction

The goal of this paper is to study the asymptotic behaviour of the flow in a thin fracture Ω_ε, with respect to the thickness of the fracture ε. For the sake of simplicity we assume that the flow is two dimensional. We are interested in the situation when the flow is governed by the prescribed pressure drop between the edges of the fracture. If the fracture has plain-parallel walls, the situation is simple and the answer is well known. Indeed, the 2D flow through a simple fracture $\Omega_\varepsilon =]0,\ell[\ ;\times\ ]0,\varepsilon[$, governed by the pressure drop, can be described by the Poiseuille formula:

$$\begin{cases} \boldsymbol{u}^\varepsilon = -\frac{1}{2\mu}\, y(\varepsilon - y)\ (p_\ell - p_0)\ \boldsymbol{e}_1 \\ p^\varepsilon = \frac{x}{\ell}\ (p_\ell - p_0)\ + p_\ell\ . \end{cases} \tag{1}$$

where p_0 and p_ℓ are the prescribed pressures at $x = 0$ and $x = \ell$ respectively. Such $(\boldsymbol{u}^\varepsilon, p^\varepsilon)$ is an exact solution of the Navier-Stokes system:

$$\begin{cases} -\mu\Delta\boldsymbol{u}^\varepsilon + (\boldsymbol{u}^\varepsilon\cdot\nabla)\boldsymbol{u}^\varepsilon + \nabla p^\varepsilon = 0\ ,\ \ \operatorname{div}\boldsymbol{u}^\varepsilon = 0\ \text{ in }\ \Omega_\varepsilon \\ \boldsymbol{u}^\varepsilon \times \boldsymbol{e}_1 = 0\ \text{ for }\ x = 0,\ell\ ,\ \boldsymbol{u}^\varepsilon = 0\ \text{ for }\ y = 0,\varepsilon \\ p^\varepsilon = p_i\ \text{ for }\ x = i \in \{0,\ell\} \end{cases} \tag{2}$$

If the fracture is constricted:

$$\Omega_\varepsilon = \{(x,y) \in \mathbf{R}^2\ ;\ 0 < x < \ell\ ,\ \varepsilon h^-(x) < y < \varepsilon h^+(x)\ \}\ ,$$

situation si much more complicated and the Poiseuille flow takes the form:

$$\begin{cases} \boldsymbol{u}^0_\varepsilon = -\frac{1}{2\mu}\,(y - \varepsilon h^-)(\varepsilon h^+ - y)\ \frac{(p_\ell - p_0)}{(h^+ - h^-)^3\int_0^\ell \frac{ds}{[h^+(s)-h^-(s)]^3}}\ \boldsymbol{e}_1 \\ p^0_\varepsilon = \frac{\int_0^x \frac{ds}{[h^+(s)-h^-(s)]^3}}{\int_0^\ell \frac{ds}{[h^+(s)-h^-(s)]^3}}\ (p_\ell - p_0)\ + p_\ell\ . \end{cases} \tag{3}$$

However, this is not an exact solution of the Navier-Stokes system

$$\begin{cases} -\mu\Delta\boldsymbol{u}^\varepsilon + (\boldsymbol{u}^\varepsilon\cdot\nabla)\boldsymbol{u}^\varepsilon + \nabla p^\varepsilon = 0\ ,\ \ \operatorname{div}\boldsymbol{u}^\varepsilon = 0\ \text{ in }\ \Omega_\varepsilon \\ \boldsymbol{u}^\varepsilon \times \boldsymbol{e}_1 = 0\ \text{ for }\ x = 0,\ell\ ,\ \boldsymbol{u}^\varepsilon = 0\ \text{ for }\ y = \varepsilon h^-, \varepsilon h^+ \\ p^\varepsilon = p_i\ \text{ for }\ x = i \in \{0,\ell\} \end{cases} \tag{4}$$

but only (as we shall prove in the last section) its approximation satisfying[1]

$$|\Omega_\varepsilon|^{-1/2}\ |\boldsymbol{u}^\varepsilon - \boldsymbol{u}^0_\varepsilon|_{L^2(\Omega_\varepsilon)} \le C\varepsilon^3 \tag{5}$$

$$|\Omega_\varepsilon|^{-1/2}\ |p^\varepsilon - p^0_\varepsilon|_{L^2(\Omega_\varepsilon)} \le C\varepsilon\ . \tag{6}$$

Indeed

$$\operatorname{div}\boldsymbol{u}^0_\varepsilon = -\frac{(p_\ell - p_0)}{2\mu\int_0^\ell \frac{ds}{[h^+(s)-h^-(s)]^3}}\ \frac{\partial}{\partial x}\left(\frac{(y-\varepsilon h^-)(\varepsilon h^+ - y)}{(h^+ - h^-)^3}\right) \neq 0,$$

so that the incompressibility condition is satisfied only in mean, i.e.

$$\frac{\partial}{\partial x}\left(\int_{\varepsilon h^-}^{\varepsilon h^+} \boldsymbol{u}^0_\varepsilon\ dy\right) = 0\ \text{ in }\ ]0,1[\ .$$

That means that the flow profile does not have a perfectly parabolic shape. In this paper we compute correctors to the Poiseuille formula. We find a recurrent algorithm giving explicit formulas (26)-(28) for computing the whole asymptotic expansion of the solution, reducing the problem to the 1-dimensional Reynolds equation (29). We prove the error estimate for such expansion.

[1] In fact the approximation for the first component of $\boldsymbol{u}^\varepsilon$ is better and we have $|\Omega_\varepsilon|^{-1/2}\ |u^\varepsilon_1 - \boldsymbol{u}^0_\varepsilon\cdot\boldsymbol{e}_1|_{L^2(\Omega_\varepsilon)} \le C\varepsilon^4$.

1.1 Some bibliographic remarks

The asymptotic behaviour of the fluid flow in thin domains has been extensively studied in the literature. Let us mention some references related to our problem. The flow through networks of thin pipes and channels having constant cross-section with different boundary conditions was considered in [1], [8] (Dirichlet's boundary condition) and [6] (pressure boundary condition). In [7] complete asymptotic expansion for the flow in an infinite pipe governed by an exterior force was found. Behaviour of the flow through periodically (rapidly) constricted fracture, was studied in [2] via homogenization.

2 Solvability of the ε- problem

Our first goal is to prove the existence of the solution for such problem since, in general, the Navier-Stokes system has the solution with pressure boundary condition only for small data. It is well known that the Stokes system with pressure boundary condition has a unique solution, while for the Navier-Stokes system one needs to prescribe either the, so called, dynamic pressure $p + \frac{1}{2}|\boldsymbol{u}|^2$ or to restrict to the case of small data p_i (see e.g. [3], [5]). In the first case one can use the compactness method, since prescribing the dynamic pressure assures the energy equality. In the second case, due to the lack of compactness, the only tool that remains is the contraction method. We can write our problem (2) in the variational form and eliminate the pressure:
Find $\boldsymbol{u}^\varepsilon \in V_\varepsilon = \{\boldsymbol{\phi} \in H^1(\Omega_\varepsilon)^3 \; ; \; \boldsymbol{\phi} = 0 \quad \text{for} \quad y = h^+, h^- \; , \; \operatorname{div} \boldsymbol{\phi} = 0 \text{ in } \Omega_\varepsilon \; , \; \boldsymbol{\phi} \times \boldsymbol{e}_1 = 0 \; \text{ for } \; x = 0, \ell \,\}$ such that

$$\mu \int_{\Omega_\varepsilon} \nabla \boldsymbol{u}^\varepsilon \nabla \boldsymbol{\phi} + \int_{\Omega_\varepsilon} (\boldsymbol{u}^\varepsilon \nabla) \boldsymbol{u}^\varepsilon \, \boldsymbol{\phi} + (p_\ell - p_0) \int_{\varepsilon h^-(0)}^{\varepsilon h^+(0)} \boldsymbol{\phi}(0, y) \cdot \boldsymbol{e}_1 \, dy = 0 \quad (7)$$

for any $\boldsymbol{\phi} \in V_\varepsilon$.

We notice that the relevant quantity is the pressure drop $d = p_\ell - p_0$. We shall suppose that the pressure drop is reasonably small (see (8)). Since the pipes are thin the no-slip boundary condition then implies that the characteristic velocity is not too large and that the inertial term is small compared to the viscous term (i.e. that the Reynolds number is small). Using the contraction method we prove:

Theorem 1. *If $|d|$ and the fracture diameter $\rho_\varepsilon = \varepsilon|h^+ - h^-|_{L^\infty(0,\ell)}$ are such that*

$$\rho_\varepsilon^{3/2} \, |d| < \frac{\mu^2}{17 \, (1 + 4\rho_\varepsilon^2 |1 - 1/\ell|)^{1/2}} \; , \quad (8)$$

then the problem (7) has a solution. Furthermore, that solution is the only solution laying in the ball $B_\varepsilon = \{\boldsymbol{\phi} \in V_\varepsilon \; ; \; \|\nabla \boldsymbol{\phi}\|_{L^2(\Omega_\varepsilon)} \leq \frac{\mu}{8\rho_\varepsilon} \; \}$.

For the proof, we need the sharp estimates for the Sobolev and trace constants:

Lemma 1. *For any* $\phi \in H^1(\Omega_\varepsilon)$ *such that* $\phi = 0$ *for* $y = \varepsilon h^+, \varepsilon h^-$ *the following inequalities hold*

$$|\phi|_{L^2(\Omega_\varepsilon)} \leq 2\rho_\varepsilon |\frac{\partial \phi}{\partial y}|_{L^2(\Omega_\varepsilon)} \tag{9}$$

$$|\phi|_{L^4(\Omega_\varepsilon)} \leq 2\sqrt{\rho_\varepsilon}|\nabla\phi|_{L^2(\Omega_\varepsilon)} \tag{10}$$

$$|\phi(k,\,\cdot\,)|_{L^2(\varepsilon h^-(k),\varepsilon h^+(k))} \leq (1+4\rho_\varepsilon^2|1-1/\ell|)^{1/2}|\nabla\phi|_{L^2(\Omega_\varepsilon)} \quad , \; k = 0, \ell. \tag{11}$$

Proof of lemma 1 follows by direct integration (see e.g. [2]).
Proof of theorem 1. We use the Banach fixed point theorem. Let $T : B_\varepsilon \to B_\varepsilon$ be the mapping defined by $T(\boldsymbol{v}) = \boldsymbol{u}$, where $\boldsymbol{u}$ is the solution of the variational problem:
Find $\boldsymbol{u} \in V_\varepsilon$ such that

$$\mu \int_{\Omega_\varepsilon} \nabla\boldsymbol{u}\nabla\boldsymbol{\phi} + \int_{\Omega_\varepsilon} (\boldsymbol{v}\nabla)\boldsymbol{u}\,\boldsymbol{\phi} + (p_\ell - p_0)\int_{\varepsilon h^-(0)}^{\varepsilon h^+(0)} \boldsymbol{\phi}(0,y)\cdot \boldsymbol{e}_1\, dy = 0 \tag{12}$$

for any $\boldsymbol{\phi} \in V_\varepsilon$.

Such T is well defined on B_ε since the quadratic form

$$a(\boldsymbol{u},\boldsymbol{\phi}) = \mu \int_{\Omega_\varepsilon} \nabla\boldsymbol{u}\nabla\boldsymbol{\phi} + \int_{\Omega_\varepsilon} (\boldsymbol{v}\nabla)\boldsymbol{u}\,\boldsymbol{\phi}$$

is elliptic:

$$a(\boldsymbol{u},\boldsymbol{u}) \geq (\mu - 4\rho_\varepsilon\,||\nabla\boldsymbol{v}||_{L^2(\Omega_\varepsilon)})\,||\nabla\boldsymbol{u}||^2_{L^2(\Omega_\varepsilon)} \geq \frac{\mu}{2}\,||\nabla\boldsymbol{u}||^2_{L^2(\Omega_\varepsilon)}\,.$$

To prove that $T(B_\varepsilon) \subset B_\varepsilon$ it is sufficient to notice that, due to (11)

$$\left|(p_\ell - p_0)\int_{\varepsilon h^-(0)}^{\varepsilon h^+(0)} \boldsymbol{\phi}(0,y)\cdot\boldsymbol{e}_1\,dy\right| \leq\, |d|\,\sqrt{\rho_\varepsilon}(1+4\rho_\varepsilon^2|1-1/\ell|)^{1/2}\,||\nabla\boldsymbol{\phi}||_{L^2(\Omega_\varepsilon)}$$

leading to $T(\boldsymbol{v}) \in B_\varepsilon$. In fact $|\nabla T(v)|_{L^2(\Omega_\varepsilon)} \leq \frac{2\mu}{17}$. Finally, we prove that T is contraction. Indeed, subtracting the equation (12) for $\boldsymbol{v}$ and for $\boldsymbol{w}$ and taking $\boldsymbol{\phi} = \boldsymbol{v} - \boldsymbol{w}$ we obtain

$$\begin{aligned}\frac{\mu}{2}||\nabla(T(\boldsymbol{v}) - T(\boldsymbol{w}))||^2_{L^2(\Omega_\varepsilon)} &= \left|\int_{\Omega_\varepsilon} [(\boldsymbol{v}-\boldsymbol{w})\nabla]T(\boldsymbol{v})\,(T(\boldsymbol{v}) - T(\boldsymbol{w}))\right| \leq \\ &\leq 4\rho_\varepsilon\,||\nabla(\boldsymbol{v}-\boldsymbol{w})||_{L^2(\Omega_\varepsilon)}\,||\nabla T(\boldsymbol{v})||_{L^2(\Omega_\varepsilon)}||\nabla(T(\boldsymbol{v}) - T(\boldsymbol{w}))||_{L^2(\Omega_\varepsilon)} \\ &\leq \frac{8\mu}{17}\,||\nabla(\boldsymbol{v}-\boldsymbol{w})||_{L^2(\Omega_\varepsilon)}\,||\nabla(T(\boldsymbol{v}) - T(\boldsymbol{w}))||_{L^2(\Omega_\varepsilon)}\;\;.\end{aligned}$$

Since $\boldsymbol{v} \in B_\varepsilon$ we get

$$||\nabla(T(\boldsymbol{v}) - T(\boldsymbol{w}))||_{L^2(\Omega_\varepsilon)} \leq \frac{16}{17}\,||\nabla(\boldsymbol{v}-\boldsymbol{u})||_{L^2(\Omega_\varepsilon)}\;,$$

proving that T is contractive. The Banach contraction theorem implies the existence and uniqueness of $\boldsymbol{u}^\varepsilon$. □

Remark 1. We notice that the solution $\boldsymbol{u}^\varepsilon \in B_\varepsilon$, i.e. that

$$|\nabla \boldsymbol{u}^\varepsilon|_{L^2(\Omega_\varepsilon)} \leq \frac{\mu}{8\rho_\varepsilon} \ . \tag{13}$$

In fact, as we shall see in the last section, the optimal estimate is

$$|\nabla \boldsymbol{u}^\varepsilon|_{L^2(\Omega_\varepsilon)} \leq C\varepsilon \ . \tag{14}$$

3 Formal computation of the asymptotic expansion

To avoid the boundary layers at $x = 0, \ell$ we suppose that the fracture is not constricted at its extremities, i.e. that there exists some $\delta > 0$ such that $h^+ = 1$ and $h^- = 0$ for $0 \leq x < \delta$ and $1 - \delta < x \leq 1$.

We seek an asymptotic expansion of the solution in the form

$$\boldsymbol{u}^\varepsilon \approx \varepsilon^2 \boldsymbol{u}^0(x, \frac{y}{\varepsilon}) + \varepsilon^3 \boldsymbol{u}^1(x, \frac{y}{\varepsilon}) + \cdots \tag{15}$$

$$p^\varepsilon \approx p^0(x) + \varepsilon p^1(x, \frac{y}{\varepsilon}) + \varepsilon^2 p^2(x, \frac{y}{\varepsilon}) + \varepsilon^3 p^3(x, \frac{y}{\varepsilon}) + \cdots \ , \tag{16}$$

where

$$\boldsymbol{u}^0(x, z) = -\frac{(p_\ell - p_0)\,(z - h^-(x))(h^+(x) - z)}{2\mu(h^+(x) - h^-(x))^3 \int_0^\ell \frac{ds}{[h^+(s) - h^-(s)]^3}} \boldsymbol{e}_1 \tag{17}$$

$$p^0(x) = \frac{\int_0^x \frac{ds}{[h^+(s) - h^-(s)]^3}}{\int_0^\ell \frac{ds}{[h^+(s) - h^-(s)]^3}} (p_\ell - p_0) \ + p_\ell \ , \tag{18}$$

is the Poiseuille solution.

For $(\boldsymbol{u}^1, p^2)$ we get the following problem

$$-\mu \frac{\partial^2 u_1^1}{\partial z^2} + \frac{\partial p^1}{\partial x} = 0 \ , \quad -\mu \frac{\partial^2 u_2^1}{\partial z^2} + \frac{\partial p^2}{\partial z} = 0 \tag{19}$$

$$\frac{\partial u_1^0}{\partial x} + \frac{\partial u_2^1}{\partial z} = 0 \ , \quad \frac{\partial}{\partial x}\left(\int_{h^-}^{h^+} u_1^1 \, dz\right) = 0 \ . \tag{20}$$

The above system can be solved by taking $u_1^1 = p^1 = 0$ and

$$u_2^1 = \frac{p_\ell - p_0}{2\mu \int_0^\ell \frac{ds}{(h^+ - h^-)^3}} \frac{\partial}{\partial x} (h^+ - h^-)^{-3} \int_{h^-}^{z} (\tau - h^-)\,(h^+ - \tau)\, d\tau = \tag{21}$$

$$= \frac{p_\ell - p_0}{2\mu \int_0^\ell \frac{ds}{(h^+ - h^-)^3}} \cdot$$

$$\cdot \frac{\partial}{\partial x} \frac{\frac{1}{2}z^2(h^+ + h^-) - z h^+ h^- - \frac{1}{3}z^3 + \frac{1}{2}h^+(h^-)^2 - \frac{1}{6}(h^-)^3}{(h^+ - h^-)^3}$$

$$p^2 = \frac{p_\ell - p_0}{2 \int_0^\ell \frac{ds}{(h^+ - h^-)^3}} \frac{\partial}{\partial x} \frac{(z - h^-)\,(h^+ - z)}{(h^+ - h^-)^3} + q^2(x) \ . \tag{22}$$

We can proceed the above computation up to any order of accuracy (provided that h^+ and h^- are smooth enough). In general the system satisfied by $(\boldsymbol{u}^m, p^m)$ reads

$$
\begin{aligned}
&-\mu\frac{\partial^2 u_1^{m-2}}{\partial x^2} - \mu\frac{\partial^2 u_1^m}{\partial z^2} + \frac{\partial p^m}{\partial x} + \\
&+\sum_{j=0}^{m-4}\left(u_1^j\frac{\partial u_1^{m-j-4}}{\partial x} + u_2^j\frac{\partial u_1^{m-j-3}}{\partial z}\right) + u_2^{m-3}\frac{\partial u_1^0}{\partial z} = 0 \qquad (23)\\
&-\mu\frac{\partial^2 u_2^{m-2}}{\partial x^2} - \mu\frac{\partial^2 u_2^m}{\partial z^2} + \frac{\partial p^{m+1}}{\partial z} + \sum_{j=0}^{m-4}\left(u_1^j\frac{\partial u_2^{m-j-4}}{\partial x} + u_2^j\frac{\partial u_2^{m-j-3}}{\partial z}\right) = 0\\
&\frac{\partial u_1^{m-1}}{\partial x} + \frac{\partial u_2^m}{\partial z} = 0 \qquad (24)\\
&\frac{\partial}{\partial x}\int_{h^-}^{h^+} u_1^m \, dz = 0 \ , \qquad (25)
\end{aligned}
$$

where we accept the notation $\boldsymbol{u}^j = 0$ and $p^j = 0$ if $j < 0$. It can be solved by taking

$$
\begin{aligned}
u_1^m &= \int_{h^-}^{z}(z-\tau)\left[\frac{1}{\mu}\frac{\partial p^m}{\partial x} - \frac{\partial^2 u_1^{m-2}}{\partial x^2} + \right.\\
&\left. + \sum_{j=0}^{m-4}\left(u_1^j\frac{\partial u_1^{m-j-4}}{\partial x} + u_2^j\frac{\partial u_1^{m-j-3}}{\partial z}\right) + u_2^{m-3}\frac{\partial u_1^0}{\partial z}\right] d\tau \qquad (26)\\
u_2^m &= -\frac{\partial}{\partial x}\int_{h^-}^{z} u_1^{m-1}\, d\tau \qquad (27)\\
p^m &= \pi^m(x,z) + q^m(x) \qquad (28)\\
\pi^m &= \mu\frac{\partial}{\partial x}\left(\frac{\partial}{\partial x}\int_{h^-}^{z} u_2^{m-3} - u_1^{m-2}\right) + \sum_{j=0}^{m-5}\left(u_1^j\frac{\partial u_2^{m-j-5}}{\partial x} + u_2^j\frac{\partial u_2^{m-j-4}}{\partial z}\right)
\end{aligned}
$$

and q^m is to be determined from the equation (25), i.e.

$$
\begin{aligned}
&\frac{1}{3}\frac{\partial}{\partial x}\left[(h^+ - h^-)^3\frac{\partial q^m}{\partial x}\right] = \frac{\partial}{\partial x}\int_{h^-}^{h^+}(h^+ - z)^2\left\{\frac{\partial^2 u_1^{m-2}}{\partial x^2} - \right.\\
&\left. -\frac{\partial \pi^m}{\partial x} - \frac{1}{\mu}\left[\sum_{j=0}^{m-4}\left(u_1^j\frac{\partial u_1^{m-j-4}}{\partial x} + u_2^j\frac{\partial u_1^{m-j-3}}{\partial z}\right) + u_2^{m-3}\frac{\partial u_1^0}{\partial z}\right]\right\}\\
&q^m(0) = q^m(\ell) = 0 \ , \qquad (29)
\end{aligned}
$$

which can be solved by direct integration.

4 Convergence of the asymptotic expansion

We denote by

$$\boldsymbol{u}_\varepsilon^m = \sum_{j=0}^{m} \varepsilon^{j+2} \boldsymbol{u}^j(x, \frac{y}{\varepsilon}) \tag{30}$$

$$p_\varepsilon^m = p^0(x) + \sum_{j=1}^{m} \varepsilon^j p^j(x, \frac{y}{\varepsilon}) \tag{31}$$

the m-th partial sum of the asymptotic expansion (15), (16). We can now prove the following result:

Theorem 2. *Let $h^+, h^- \in C^\infty([0,\ell])$ and let $m \in \mathbf{N} \cup \{0\}$. Let $(\boldsymbol{u}^\varepsilon, p^\varepsilon)$ be the solution of the Navier-Stokes system (2) with $\boldsymbol{u}^\varepsilon$ laying in the ball B_ε and let $(\boldsymbol{u}_\varepsilon^m, p_\varepsilon^m)$ be defined by (30) and (31). Then*

$$|\Omega_\varepsilon|^{-1/2} \, |\boldsymbol{u}^\varepsilon - \boldsymbol{u}_\varepsilon^m|_{H^1(\Omega_\varepsilon)} \le C\varepsilon^{m+2} \tag{32}$$

$$|\Omega_\varepsilon|^{-1/2} \, |\boldsymbol{u}^\varepsilon - \boldsymbol{u}_\varepsilon^m|_{L^2(\Omega_\varepsilon)} \le C\varepsilon^{m+3} \tag{33}$$

$$|\Omega_\varepsilon|^{-1/2} \, |p^\varepsilon - p_\varepsilon^m|_{L^2(\Omega_\varepsilon)} \le C\varepsilon^{m+1} \tag{34}$$

For the proof we need the following technical lemma that can be proved by simple change of variables (see e.g. [4]).

Lemma 2. *Let $f \in L_0^2(\Omega_\varepsilon)$. Then the problem*

$$div\, \boldsymbol{w} = f \;\; in \;\; \Omega_\varepsilon \tag{35}$$

$$\boldsymbol{w} = 0 \;\; on \;\; \partial\Omega_\varepsilon \tag{36}$$

admits a solution that satisfies the estimate

$$|\nabla \boldsymbol{w}|_{L^2(\Omega_\varepsilon)} \le C\varepsilon^{-1} |f|_{L^2(\Omega_\varepsilon)} \;,$$

with some $C > 0$ independent from ε.

Proof of theorem 2. We begin by noticing that

$$\operatorname{div} \boldsymbol{u}_\varepsilon^m = \varepsilon^{m+2} \frac{\partial u_1^m}{\partial x} \neq 0 \;.$$

Therefore we need to define the divergence corrector. To get an optimal result we need to do it carefully. We first define

$$\psi(x,z) = \int_{h^-}^{z} \frac{\partial u_1^m}{\partial x}(x,\tau)\, d\tau \;.$$

Then we chose $\boldsymbol{w}_\varepsilon$ as the solution of the problem

$$\operatorname{div} \boldsymbol{w}_\varepsilon = -\frac{\partial \psi_\varepsilon}{\partial x} \text{ in } \Omega_\varepsilon \ , \quad \boldsymbol{w}_\varepsilon = 0 \text{ on } \partial\Omega_\varepsilon \ ,$$

where $\psi_\varepsilon(x,y) = \psi(x, y/\varepsilon)$. We notice that

$$\int_{\varepsilon h^-}^{\varepsilon h^+} \frac{\partial \psi_\varepsilon}{\partial x} = 0,$$

so that such $\boldsymbol{w}_\varepsilon$ exists. Furthermore, due to the lemma 2, $\boldsymbol{w}_\varepsilon$ can be chosen to satisfy the estimate

$$|\nabla \boldsymbol{w}_\varepsilon|_{L^2(\Omega_\varepsilon)} \le C\varepsilon^{-1} |\frac{\partial \psi_\varepsilon}{\partial x}|_{L^2(\Omega_\varepsilon)} \le C\varepsilon^{-1/2} \ .$$

It is now easy to see that the function

$$\boldsymbol{v}_\varepsilon^m = \boldsymbol{u}_\varepsilon^m - \varepsilon^{m+3}(\psi_\varepsilon \boldsymbol{e}_1 + \boldsymbol{w}_\varepsilon)$$

is divergence free and satisfies

$$-\mu \Delta \boldsymbol{v}_\varepsilon^m + \nabla p_\varepsilon^m + (\boldsymbol{v}_\varepsilon^m \cdot \nabla)\boldsymbol{v}_\varepsilon^m = R_\varepsilon^m \ . \tag{37}$$

where

$$|R_\varepsilon^m|_{L^\infty(\Omega_\varepsilon)} \le C\varepsilon^{m+1} \ . \tag{38}$$

Subtracting the above equation from (2), multiplying by $\boldsymbol{u}^\varepsilon - \boldsymbol{v}_\varepsilon^m$ and integrating over Ω_ε yields

$$\begin{aligned} &\mu |\nabla(\boldsymbol{u}^\varepsilon - \boldsymbol{v}_\varepsilon^m)|^2_{L^2(\Omega_\varepsilon)} \\ &+ \int_{\Omega_\varepsilon} \{(\boldsymbol{u}^\varepsilon \cdot \nabla)(\boldsymbol{u}^\varepsilon - \boldsymbol{v}_\varepsilon^m) + [(\boldsymbol{u}^\varepsilon - \boldsymbol{v}_\varepsilon^m) \cdot \nabla]\boldsymbol{v}_\varepsilon^m\}(\boldsymbol{u}^\varepsilon - \boldsymbol{v}_\varepsilon^m) = \\ &= \int_{\Omega_\varepsilon} R_\varepsilon^m \, (\boldsymbol{u}^\varepsilon - \boldsymbol{v}_\varepsilon^m) \le C\varepsilon^{m+2+1/2} |\nabla(\boldsymbol{u}^\varepsilon - \boldsymbol{v}_\varepsilon^m)|_{L^2(\Omega_\varepsilon)} \ . \end{aligned}$$

Two integrals on the right-hand side can be estimated as

$$\begin{aligned} \textstyle\int_{\Omega_\varepsilon} \{(\boldsymbol{u}^\varepsilon \cdot \nabla)(\boldsymbol{u}^\varepsilon - \boldsymbol{v}_\varepsilon^m) \, (\boldsymbol{u}^\varepsilon - \boldsymbol{v}_\varepsilon^m) &\le 4\rho_\varepsilon |\nabla \boldsymbol{u}^\varepsilon|_{L^2(\Omega_\varepsilon)} |\nabla(\boldsymbol{u}^\varepsilon - \boldsymbol{v}_\varepsilon^m)|^2_{L^2(\Omega_\varepsilon)} \le \\ &\le (\text{using (13)}) \le \frac{\mu}{2} |\nabla(\boldsymbol{u}^\varepsilon - \boldsymbol{v}_\varepsilon^m)|^2_{L^2(\Omega_\varepsilon)} \\ \textstyle\int_{\Omega_\varepsilon} [(\boldsymbol{u}^\varepsilon - \boldsymbol{v}_\varepsilon^m) \cdot \nabla] \, \boldsymbol{v}_\varepsilon^m \, (\boldsymbol{u}^\varepsilon - \boldsymbol{v}_\varepsilon^m) &\le C\varepsilon^2 |\nabla(\boldsymbol{u}^\varepsilon - \boldsymbol{v}_\varepsilon^m)|^2_{L^2(\Omega_\varepsilon)} \ . \end{aligned}$$

The estimates (32) and (33) are now straightforward (we notice that the norm of the divergence corrector is negligible). Consequently, we get the estimate for the velocity (14).

Let us prove (34). We normalize the pressure such that

$$\int_{\Omega_\varepsilon} (p^\varepsilon - p_\varepsilon^m) = 0 \ .$$

Due to the lemma 2, we can now define ϕ^ε such that

$$\operatorname{div} \phi^\varepsilon = p^\varepsilon - p_\varepsilon^m \text{ in } \Omega_\varepsilon \ , \quad \phi^\varepsilon = 0 \text{ on } \partial\Omega_\varepsilon$$

and

$$|\nabla \phi^\varepsilon|_{L^2(\Omega_\varepsilon)} \leq C\varepsilon^{-1} |(p^\varepsilon - p_\varepsilon^m)|_{L^2(\Omega_\varepsilon)} \ .$$

Subtracting (37) from (2), multiplying by ϕ^ε and integrating over Ω_ε leads to

$$|p^\varepsilon - p_\varepsilon^m|^2_{L^2(\Omega_\varepsilon)} \leq \left| \int_{\Omega_\varepsilon} \{(\boldsymbol{u}^\varepsilon \cdot \nabla)(\boldsymbol{u}^\varepsilon - \boldsymbol{v}_\varepsilon^m) + [(\boldsymbol{u}^\varepsilon - \boldsymbol{v}_\varepsilon^m) \cdot \nabla] \boldsymbol{v}_\varepsilon^m\} \phi^\varepsilon \right| +$$

$$+ \left| \int_{\Omega_\varepsilon} R_\varepsilon^m \ \phi^\varepsilon \right| \leq C\varepsilon^{m+2+1/2} |\nabla \phi^\varepsilon|_{L^2(\Omega_\varepsilon)} \leq C\varepsilon^{m+1+1/2} |p^\varepsilon - p_\varepsilon^m|_{L^2(\Omega_\varepsilon)}.$$

□

5 Numerical experiments

5.1 Geometry description

Let consider the computational domain $\Omega_\varepsilon = \{(x,y) \in \mathbf{R}^2;\ 0 < x < 1,\ ,\varepsilon h^-(x) < y < \varepsilon h^+(x)\}$ where (h^-, h^+) are defined by $h^+(x) = 1$ for all $0 < x < 1$ and $h^-(x) = 0$ if $0 < x \leq 0.45$ or $0.55 \leq x < 1$ and $h^-(x) = \exp(\dfrac{-1}{1-(20x-10)^2})$ for $0.45 < x < 0.55$, and we will consider values of ε between 10^{-1} and 10^{-2}. On Fig. 1 are plotted two domains for $\varepsilon = 0.1$ and $\varepsilon = 0.02$. It should be noticed that the constriction ratio in the tube is the same in the two case.

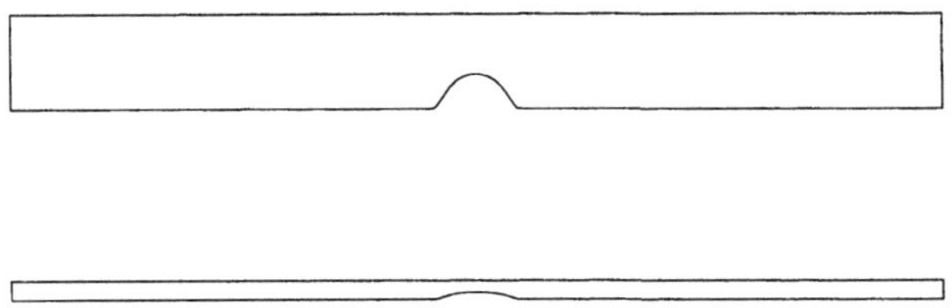

Fig. 1. Computational domains for $\varepsilon = 0.1$ and $\varepsilon = 0.02$

5.2 Numerical method

Due to the fact that it is not possible to compute effectively the solution of the problem (2) in a divergence free space of finite elements as in the weak formulation (7), we will satisfy the divergence free condition by decomposing the pressure p^ε in a macroscopic part $p(x) = (p_\ell - p_0)x + p_0$ and a local part $\tilde{p}^\varepsilon(x, y)$ which satisfy $\tilde{p}^\varepsilon(0, y) = 0$ and $\tilde{p}^\varepsilon(1, y) = 0$ and solving the problem:

$$\begin{cases} -\mu\Delta \boldsymbol{u}^\varepsilon + (\boldsymbol{u}^\varepsilon \cdot \nabla)\boldsymbol{u}^\varepsilon + \nabla\tilde{p}^\varepsilon = (p_0 - p_\ell)\boldsymbol{e}_1 \ , \ \operatorname{div} \boldsymbol{u}^\varepsilon = 0 \ \text{in} \ \Omega_\varepsilon \\ \boldsymbol{u}^\varepsilon \times \boldsymbol{e}_1 = 0 \ \text{for} \ x = 0, 1 \ , \ \boldsymbol{u}^\varepsilon = 0 \ \text{for} \ y = \varepsilon h^-, \varepsilon h^+ \\ \tilde{p}^\varepsilon = 0 \ \text{for} \ x = i \in \{0, 1\} \end{cases} \tag{39}$$

This problem is discretized by a finite element method and the divergence free condition is solved by an Augmented Lagrangian algorithm. We chose a classical finite element method such that the Brezzi–Babuska stability condition (see e.g. [9]) is satisfied: The P_2^+ element for the velocity field and the P_1 discontinuous element for the pressure are well known to give very good results for the Stokes or Navier Stokes problem (see e.g. [10]). The bubble of the velocity element is eliminated following [11]. By these choices, we are assured to satisfy the divergence free constraint, since the spatial discretization step is of order ε^2. In this algorithmic frame, an initial pressure equal to zero assures the convergence to a pressure $\tilde{p}^\varepsilon$ which satisfies the boundary conditions $\tilde{p}^\varepsilon = 0$ for $x = i \in \{0, 1\}$. This is due to the parabolic shape of the flow at the entry end the end of the tube.

5.3 Numerical results

We have considered a fluid with viscosity $\mu = 1$ and we present numerical for $p_0 - p_\ell = 1$. Computations for other values of μ and $p_0 - p_\ell$ have been performed without significant qualitative modification of the results. Then the solution of the problem (2) and the asymptotic development have been computed for four values of the thickness of the tube: $\varepsilon \in \{0.1, 0.05, 0.02, 0.01\}$. At first, on Figs. 2–4 have been plotted at left the field $\boldsymbol{u}^\varepsilon$ solution of the

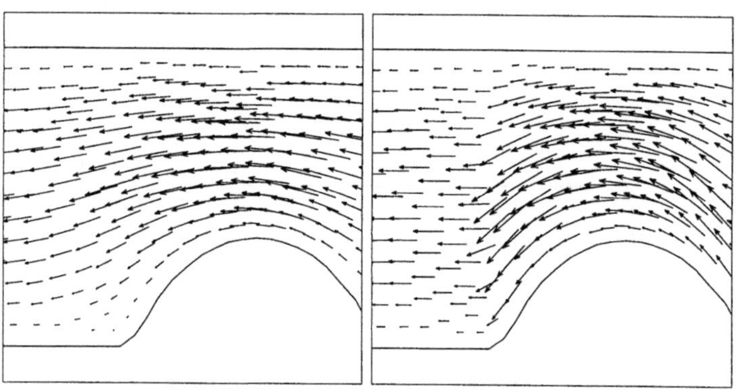

Fig. 2. Velcity field around the constriction for $\varepsilon = .1$, **(a)** $\boldsymbol{u}^\varepsilon$, **(b)** $\boldsymbol{u}_\varepsilon^1$

problem (2) and at right, the field $\boldsymbol{u}_\varepsilon^1$. One may see that the corrector on the

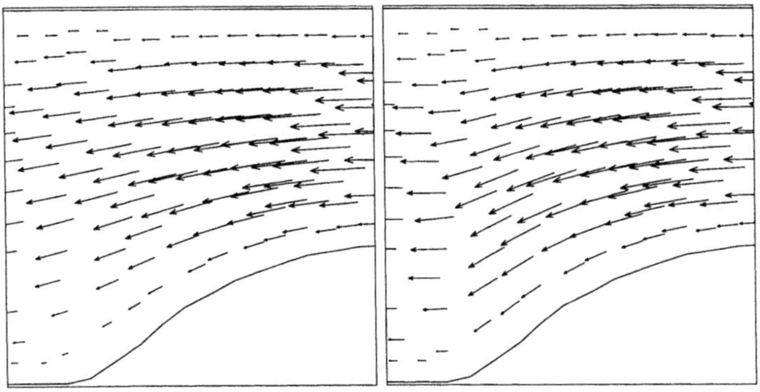

Fig. 3. Velcity field around the constriction for $\varepsilon = .05$, **(a)** $\boldsymbol{u}^{\varepsilon}$, **(b)** $\boldsymbol{u}^{1}_{\varepsilon}$

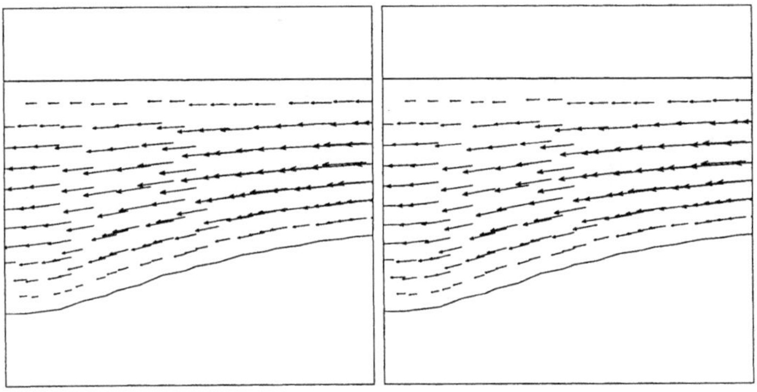

Fig. 4. Velcity field around the constriction for $\varepsilon = .02$, **(a)** $\boldsymbol{u}^{\varepsilon}$, **(b)** $\boldsymbol{u}^{1}_{\varepsilon}$

second component of the velocity is not a good corrector for $\varepsilon = 0.1$ (Fig. 2): it is active only on the constriction part of the tube. This effect disapears as ε decreases, one may not notice a perceptual difference between $\boldsymbol{u}^{\varepsilon}$ and $\boldsymbol{u}^{1}_{\varepsilon}$ for $\varepsilon \leq 0.02$ (Fig. 4)

For each value of ε , let us define the following fields:

$$\delta u_1^{\varepsilon} = \frac{|u_1^{\varepsilon} - \varepsilon^2 u_1^0|}{\|u_1^{\varepsilon}\|_{L^{\infty}(\Omega_{\varepsilon})}}, \quad \delta u_2^{\varepsilon} = \frac{|u_2^{\varepsilon} - \varepsilon^3 u_2^1|}{\|u_2^{\varepsilon}\|_{L^{\infty}(\Omega_{\varepsilon})}},$$
$$\delta \boldsymbol{u}^{\varepsilon} = \frac{\sqrt{(u_1^{\varepsilon} - \varepsilon^2 u_1^0)^2 + (u_2^{\varepsilon} - \varepsilon^3 u_2^1)^2}}{\|\sqrt{(u_1^{\varepsilon})^2 + (u_1^{\varepsilon})^2}\|_{(L^{\infty}(\Omega_{\varepsilon}))}}, \quad \delta p^{\varepsilon} = \frac{|p^{\varepsilon} - p^0|}{\|p^{\varepsilon}\|_{L^{\infty}(\Omega_{\varepsilon})}} \tag{40}$$

which describe the gap between the solution of the problem (2) and each of the components of the asymptotic development. On Figs. 5–7 have been plotted the gaps δu_1^{ε}, δu_2^{ε} and $\delta \boldsymbol{u}^{\varepsilon}$ around the constriction. Let us consider first the Fig. 5. One may see the decreasing of the size of the gap on the first component of the solution of the problem (2) and $\varepsilon^2 u^0$. We notice that the most important part of the gap δu_1^{ε} localized at the cntry (resp end) of

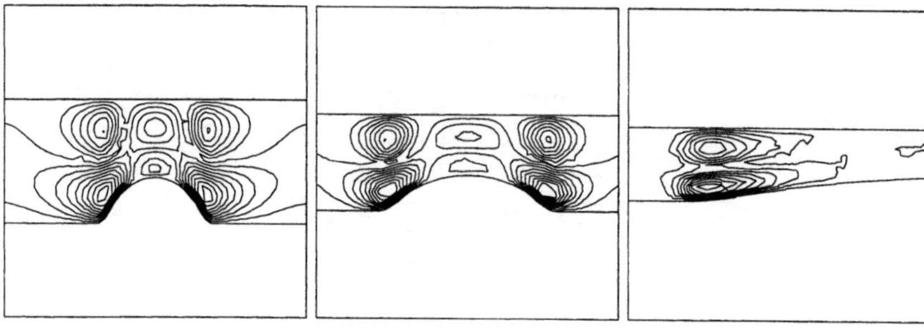

Fig. 5. Isovalues of the gap δu_1^ε between the first components of the numerical solution and the first order aymptotic expansion: (**a**) $\varepsilon = .1$, $(\delta u_1^\varepsilon)_{max} = .286$; (**b**) $\varepsilon = .05$, $(\delta u_1^\varepsilon)_{max} = .138$; (**c**) $\varepsilon = .01$, $(\delta u_1^\varepsilon)_{max} = .0153$

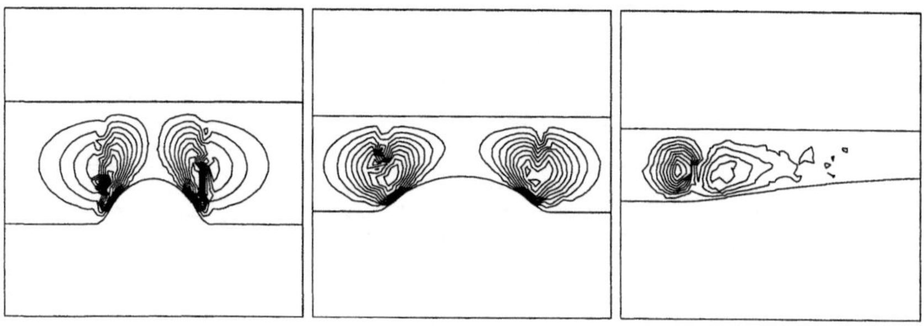

Fig. 6. Isovalues of the gap δu_2^ε between the second components of the numerical solution and the first order aymptotic expansion: (**a**) $\varepsilon = .1$, $(\delta u_2^\varepsilon)_{max} = 1.87$; (**b**) $\varepsilon = .05$, $(\delta u_2^\varepsilon)_{max} = .724$; (**c**) $\varepsilon = .01$, $(\delta u_2^\varepsilon)_{max} = .151$

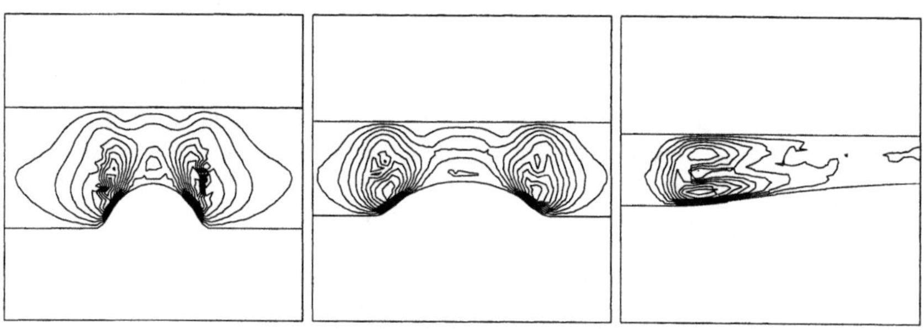

Fig. 7. Isovalues of the gap $\delta \boldsymbol{u}^\varepsilon$ between the modulus of the numerical solution and the first order aymptotic expansion: (**a**) $\varepsilon = .1$, $\delta \boldsymbol{u}^\varepsilon_{max} = .536$; (**b**) $\varepsilon = .05$, $\delta \boldsymbol{u}^\varepsilon_{max} = .189$; (**c**) $\varepsilon = .01$, $\delta \boldsymbol{u}^\varepsilon_{max} = .0151$

the constriction. Concerning the comportment of the quantity δu_2^ε, plotted on Fig. 6, one sees its decreasing in function of ε, but the value of this one is more important that for the first component one (δu_1^ε). Next, one may see that the most important part of this gap δu_2^ε is localized before the begining, or after the end of the constriction, where the correctors do not act, because in this part of the domain h^+ and h^- are constant and then $u_2^m = 0$. But in the constriction, one sees clearly the decreasing of the gap in function of ε. Finally, on Fig. 7 one may see the relative influence of each of the components of $\boldsymbol{u}_\varepsilon^1$ on $\boldsymbol{u}^\varepsilon$: Outside of the constriction, the second component $(u_\varepsilon^1)_2$ is dominant, but inside the constriction, it is the first component $(u_\varepsilon^0)_1$ which is preponderant. Concerning the pressure, one have plotted on Fig. 8 the gap δp^ε between the

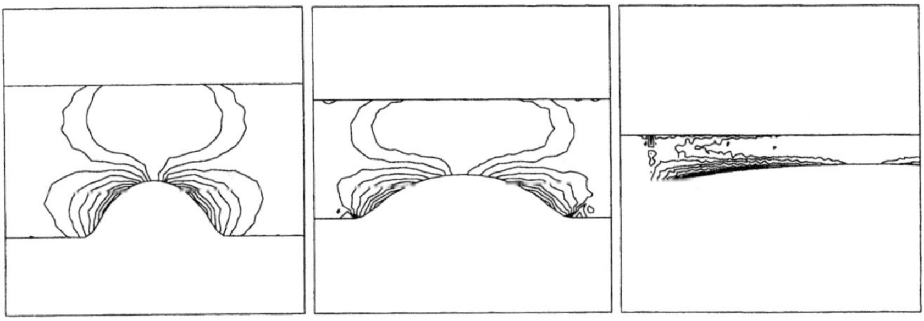

Fig. 8. Isovalues of the gap δp^ε between the pressure of the numerical solution and the first order aymptotic expansion: (**a**) $\varepsilon = .1$, $\delta p^\varepsilon_{max} = .094$; (**b**) $\varepsilon = .05$, $\delta p^\varepsilon_{max} = .033$; (**c**) $\varepsilon = .01$, $\delta p^\varepsilon_{max} = .001$

pressure p^ε and p^0. One may see that this gap is decreasing as ε decreases. Its most important part is localized on the constriction, this is due to the fact that the isolines of the pressures should be perpendicular to the non slip boundaries and that the quantity p^0 depends only on x and not on z. Then the gap is decreasing as the curvature of the constriction decreases.

Finally, we have computed the L^2 norms of the gap between the solution of the problem (2) and the first order asymptotic expansion as defined as follows:

$$\Delta \boldsymbol{u} = \|\boldsymbol{u}^\varepsilon - \boldsymbol{u}_\varepsilon^1\|_{(L^2(\Omega_\varepsilon))^2} \tag{41}$$

$$\Delta u_1 = \|u_1^\varepsilon - \varepsilon^2 u_1^0\|_{L^2(\Omega_\varepsilon)}, \quad \Delta u_2 = \|u_2^\varepsilon - \varepsilon^3 u_2^1\|_{L^2(\Omega_\varepsilon)},$$

On Fig. 9 we have plotted those quantities in function of ε. One may see the decreasing of Δu_1, Δu_2 and $\Delta \boldsymbol{u}$ in function of ε. Finally, one can deduce numerically from this graph, by computing the slope of the three straight lines, the estimate (33).

References

1. Blanc F., Gipouloux O., Panasenko G., Zine A.M. (1999) Asymptotic analysis and partial asymptotic decomposition of domain for Stokes Equation in tube structure. Math. Models Methods Appl. Sci. M^{3}AS **9**(9), 1351–1378

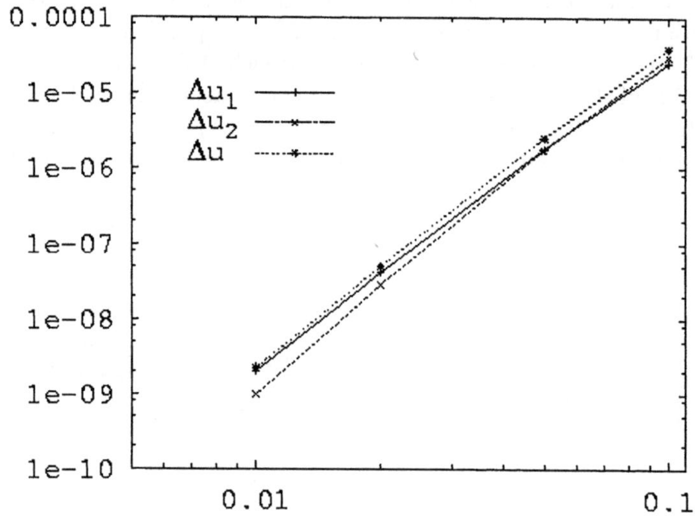

Fig. 9. Gaps Δu_1, Δu_2 and $\boldsymbol{\Delta u}$ in function of ε

2. Bourgeat A., Marušić-Paloka E. (1998) Nonlinear Effects for Flow in Periodically Constricted Channel Caused by High Injection Rate. Math. Models Methods Appl. Sci. M^3AS **9**(8), 1–25
3. Conca C., Murat F., Pironneau O. (1994) The Stokes and Navier-Stokes equations with boundary conditions involving the pressure. Japan J. Math. **20**, 263-318
4. Duvnjak A., Marušić-Paloka E. (2000) Derivation of the Reynolds equation for lubrication of a rotating shaft. Arch. Math. (Brno) **36**(4), 239-253
5. Heywood J.G., Rannacher R., Turek S. (1996) Artificial boundaries and flux and pressure conditions for the incompressible Navier-Stokes equations. Int. J. Numer. Methods Fluids **22**, 325-352
6. Marušić-Paloka E. (2001) Fluid flow through a network of thin pipes. C. R. Acad. Sci. Paris, Série II b **329**
7. Nazarov S.A., Piletskas K.I. (1991) The Reynolds Flow in a Thin Three-Dimensional Channel, Lithuanian Math. J. **30**, 366-375
8. Panasenko G. (1998) Asymptotic expansion of the solution of Navier-Stokes equation in a tube structure. C. R. Acad. Sci. Paris, Série II b, **326**, 867-872
9. Brezzi, F. (1974) On the existence, uniqueness and approximation of saddle points arising from Lagrangian multipliers. R.A.I.R.O., ser. Num. Anal., **8**
10. Brezzi, F., Fortin, M. (1991) Mixed and Hybrid Finite Element Methods. Springer
11. Fortin, M., Fortin, A. (1985) Experiments with several elements for viscous incompressible flows. Int. J. numer. methods fluids **5**, 911–928

Finite Dimensional Approximations for Elliptic Problems with Rapidly Oscillating Coefficients

Ana-Maria Matache and Christoph Schwab

Seminar for Applied Mathematics, ETH-Zentrum, CH-8092 Zürich, Switzerland

Dedicated to Professor Ivo Babuška at the 75th anniversary

Abstract. We derive and analyze finite dimensional approximations for solutions to elliptic problems with coefficients or geometries which are oscillating at a small length scale ε. We introduce problem-dependent, generalized Finite Element (FE) spaces with non-polynomial micro-shape functions that reflect the correct oscillatory behaviour of the solution. Our choice of micro-scale shape functions is motivated by representation formulas for solutions on unbounded domains, which are based on scale separation and generalized Fourier inversion formulas. Under analyticity assumptions on the input data, we prove exponential convergence of the generalized FEM based on these spaces which is robust, i.e., the rate of convergence is independent of ε.[1]

Contents

[1] Research supported by the Swiss National Science Foundation under Project Number BBW 21-58754.99 and under IHP Network 'Homogenization and Multiple Scales' "IIMS2000" of the EC.

1 Introduction

The problem that we consider is the classical elliptic problem

$$L^{\varepsilon}(x,\partial_x)u^{\varepsilon} := -\nabla\cdot(a^{\varepsilon}(x)\nabla u^{\varepsilon}) + a_0^{\varepsilon}(x)u^{\varepsilon} = f \quad \text{in } \Omega^{\varepsilon}, \tag{1}$$

$$Bu^{\varepsilon} = 0 \quad \text{on } \partial\Omega^{\varepsilon}. \tag{2}$$

Here, the set $\Omega^{\varepsilon} = \Omega^{\varepsilon}_{\infty}\cap\Omega$ is defined as the intersection of a Lipschitz domain Ω and the lattice of width ε given by

$$\Omega^{\varepsilon}_{\infty} := \varepsilon\Omega_{\infty}, \quad \Omega_{\infty} := \bigcup_{k\in\mathbb{Z}^n}(2\pi k + \widehat{Q}),$$

where $\widehat{Q}\subset[-\pi,\pi]^n$, $\partial\widehat{Q}$ Lipschitz, is the unit-cell. The case when $\widehat{Q}\neq[-\pi,\pi]^n$ and the interior boundary of $\widehat{Q}$ is merely piecewise smooth models problems where the material has holes or hard inclusions. If $\widehat{Q}=[-\pi,\pi]^n$, then $\Omega^{\varepsilon}=\Omega$. We assume throughout that the domain Ω is a bounded, connected subset of $\mathbb{R}^n$ with $\operatorname{diam}(\Omega)=1$, boundary $\partial\Omega$ and boundary operator B which may be either the trace operator or the conormal derivative operator. The problem is assumed strongly elliptic, i.e., $a^{\varepsilon}(x), a_0^{\varepsilon}(x)$ are positive.

The numerical solution of (1)–(2) by the Finite Element Method [5,6], for example, is well-known to be difficult if the mesh width H is much larger than the fine-scale ε of the solution. Therefore, one has replaced the problem (1)–(2) by so-called *effective* or averaged equations with smooth coefficients and solution. The homogenization by asymptotic analysis is by now well understood [3,4,7,17].The numerical solution of the homogenized problem with smooth coefficients allows for standard error estimates, but cannot reduce the homogenization error introduced by the coefficient averaging. To account for the fine scale in the solution, correctors must to be introduced into the homogenization process which are as complicated to compute as the original problem. In addition, two-scale asymptotic expansion of the solution u^{ε} about $\varepsilon=0$ may not be defined in polyhedral domains or for non-smooth data f in (1) [16]. Thus, attempts have been made to incorporate fine scale behavior of the solution directly into the FE space [8]. In special cases of laminated materials, analytic expressions have been derived for the shape-functions [2]. For general coefficients, shape-functions must be determined locally and numerically in parallel [9], requiring resolution of the finest scale of the coefficients throughout the domain.

In the present paper, we address the design of generalized FE spaces for periodic homogenization problems at fixed, positive scale ε. Based on a Fourier-Representation formula of the solution u^{ε} of (1) on an unbounded domain [14,15], we first prove a scale separation result for u^{ε}. The analyticity of the two-scale symbol of the differential operator is then exploited to show exponential convergence, uniform in ε, of a truncated Poisson summation formula of the Fourier integral in the representation formula. Finally, we show how our analysis leads to the design of generalized FE-spaces for homogenization problems which are used in [10–13,19].

The outline of the paper is the following: in Section 2 we consider (1)–(2) with periodic coefficients which oscillate at the length scale $\varepsilon \ll 1$ and we formulate the Finite Element discretization for such problems. Our approach is based on special, ε-dependent FE-spaces $V_N = \mathcal{V}_\varepsilon^N$ and is motivated by a refined analysis of the elliptic operator L^ε carried over in Section 3. We show that the space of admissible solutions can be enlarged and the inverse Fourier symbol of L^ε can be identified as element of suitable weighted Sobolev spaces. This allows us to give a Fourier-Bochner integral representation of the solution with integral kernel being the generalized Fourier transform of the Green function. We show that the integral kernel is completely described by a family of solutions to parameter dependent unit-cell problems and we investigate the regularity of the kernel in dependence on these parameters. Based on this refined regularity analysis we derive finite element approximation spaces $\mathcal{V}_\varepsilon^N$ in Section 4. For analytic input data the method based on the spaces $\mathcal{V}_\varepsilon^N$ will converge exponentially, independent of the length scale ε.

In Section 5 we address the construction of the generalized FE spaces $\mathcal{V}_\varepsilon^N$ on bounded domains. We discuss the approximation properties of the spaces $\mathcal{V}_\varepsilon^N$ constructed in Section 4 for solutions on bounded domains and we address the enforcement of the essential boundary conditions.

2 The homogenization problem and finite element discretization

We assume that the heterogenities are periodically distributed, i.e., $a^\varepsilon(x) = a(x/\varepsilon)$, $a_0^\varepsilon(x) = a_0(x/\varepsilon)$ and $a(\cdot) := (a_{pq}(\cdot))_{pq} \in L^\infty(\mathbb{R}^n)^{n\times n}_{\mathrm{sym}}$, $a_0(\cdot) \in L^\infty(\mathbb{R}^n)$ are 2π-periodic functions in each variable. It is also assumed that there exist positive constants $\gamma_0, \gamma_1 > 0$ such that

$$\gamma_0\, \eta^\top \eta \le \sum_{p,q=1}^{n} a_{pq}(\xi)\eta_p\eta_q \le \gamma_1\, \eta^\top \eta,\ \forall \eta \in \mathbb{R}^n \text{ a.e. } \xi \in \mathbb{R}^n \tag{3}$$

$$\gamma_0 \le a_0(\xi) \le \gamma_1 \text{ a.e. } \xi \in \mathbb{R}^n . \tag{4}$$

Exemplarily, we consider the scalar model problem (1) with homogeneous Dirichlet boundary conditions on $\Gamma_D^\varepsilon := \partial\Omega^\varepsilon \cap \partial\Omega$ and homogeneous Neumann boundary conditions on the (if non-empty) interior cavity boundaries $\Gamma_N^\varepsilon := \partial\Omega^\varepsilon \backslash \Gamma_D^\varepsilon$

$$\begin{aligned} u^\varepsilon_{\Gamma_D^\varepsilon} &= 0 \\ n \cdot \left(a\left(\frac{x}{\varepsilon}\right) \nabla u^\varepsilon \right)_{\Gamma_N^\varepsilon} &= 0. \end{aligned} \tag{5}$$

The solution u^ε of (1), (5) is an element of the space $H^1_{\Gamma^\varepsilon_D}(\Omega^\varepsilon)$ defined as

$$H^1_{\Gamma^\varepsilon_D}(\Omega^\varepsilon) := \left\{ u \in H^1(\Omega^\varepsilon) \,|\, \gamma_0 u|_{\Gamma^\varepsilon_D} = 0 \right\},$$

where γ_0 here denotes the Dirichlet trace operator. The variational form of (1), (5) reads

$$\text{Find } u^\varepsilon \in H^1_{\Gamma^\varepsilon_D}(\Omega^\varepsilon) \;:\; B^\varepsilon_{\Omega^\varepsilon}(u^\varepsilon, v) = (f, v)_{L^2(\Omega^\varepsilon)} \quad \forall v \in H^1_{\Gamma^\varepsilon_D}(\Omega^\varepsilon), \tag{6}$$

in which the bilinear form $B^\varepsilon_{\Omega^\varepsilon}(\cdot,\cdot) : H^1_{\Gamma^\varepsilon_D}(\Omega^\varepsilon) \times H^1_{\Gamma^\varepsilon_D}(\Omega^\varepsilon) \to \mathbb{R}$ is given by

$$B^\varepsilon_{\Omega^\varepsilon}(u, v) = \int\limits_{\Omega^\varepsilon} \left(a\left(\frac{x}{\varepsilon}\right) \nabla u^\varepsilon(x) \right) \cdot \nabla v(x) + a_0\left(\frac{x}{\varepsilon}\right) u^\varepsilon(x) v(x)\, dx.$$

Let $V_N \subset H^1_{\Gamma^\varepsilon_D}(\Omega^\varepsilon)$ be any subspace of dimension $N = \dim(V_N) < \infty$. The Finite Element Method (FEM) for (6)

$$\text{Find } u^\varepsilon_N \in V_N \;:\; B^\varepsilon_{\Omega^\varepsilon}(u^\varepsilon, v) = (f, v)_{L^2(\Omega^\varepsilon)} \quad \forall v \in V_N, \tag{7}$$

defines a unique FE solution u^ε_N and

$$\|u^\varepsilon - u^\varepsilon_N\|_{H^1(\Omega^\varepsilon)} \le C \min_{v \in V_N} \|u^\varepsilon - v\|_{H^1(\Omega^\varepsilon)}, \tag{8}$$

where $C > 0$ is independent of ε, i.e., the FE-error is bounded by the best approximation of u^ε from V_N. Finite Element convergence of u^ε_N is therefore related to the regularity of u^ε in dependence on the scale parameter ε.

3 The homogenization problem in infinite periodic structures

In view of (8), the key to robust approximations is the design of V_N. Rather than incorporating the asymptotics of u^ε (which is not always defined, see [4,17] and the references there) into the FE-space V_N, we design V_N based on a refined regularity theory of u^ε. To this end, we address first problem (1) on the unbounded domain $\Omega^\varepsilon_\infty$. Based on a mixed variational formulation we show existence and uniqueness of the solution u^ε in exponentially weighted Sobolev spaces $H^1_{-\nu}(\Omega^\varepsilon_\infty)$, $\nu \in (0, \nu_0)$, for general right hand sides f in the dual space of $H^1_\nu(\Omega^\varepsilon_\infty)$. Further, we prove an integral representation formula for u^ε. This is a Fourier-Bochner integral representation with respect to the Banach spaces $H^1_{-\nu}(\Omega^\varepsilon_\infty)$. The starting point is the analysis of the Fourier-Bochner integral kernel introduced as the inverse Fourier symbol of the elliptic operator (1). We show in Theorem 2 that this is precisely the product of the standard Fourier waves $e^{it\cdot x}$ and periodic, highly oscillatory functions $\phi(x/\varepsilon, \varepsilon, t)$ encoding the fast scale features of the problem. The periodic functions $\phi(y, \varepsilon, t)$ arise as solutions of parameter dependent problems on the unit-cell $\widehat{Q}$. Next, we concentrate on the regularity properties of the integral kernel $e^{it\cdot x}\phi(x/\varepsilon, \varepsilon, t)$ in dependence on its parameters $\varepsilon > 0$, $t \in \mathbb{R}^n$.

We conclude this section by extending our theory to the case of strongly elliptic systems. There, the integral kernel transforms in a tensor valued function and arise as solution of a parameter dependent system of unit-cell problems.

3.1 Variational setting

Let $\varepsilon > 0$ be a length scale and let $\widehat{Q} \subset \widehat{Y} := [-\pi, \pi]^n$ be a unit cell with Lipschitz boundary. We define the infinite lattice $\Omega^\varepsilon_\infty$ of period $\varepsilon\widehat{Y}$ as

$$\Omega^\varepsilon_\infty := \varepsilon\Omega_\infty, \quad \Omega_\infty := \bigcup_{k\in\mathbb{Z}^n} (2\pi k + \widehat{Q}) \tag{9}$$

and assume $\Omega^\varepsilon_\infty$ is connected. We denote by $\widehat{\Gamma}_N := \partial\widehat{Q}\backslash\partial\widehat{Y}\widehat{\Gamma}_{\text{per}}$ and by $\widehat{\Gamma}_{\text{per}} := \partial\widehat{Q} \cap \partial\widehat{Y} (= \partial\widehat{Q}\backslash\widehat{\Gamma}_N)$.

We consider the model problem:

$$L^\varepsilon\left(\frac{x}{\varepsilon}, \partial_x\right) u^\varepsilon(x) = f(x) \text{ in } \Omega^\varepsilon_\infty \tag{10}$$

and, if $\widehat{\Gamma}_N \neq \emptyset$,

$$\underline{n} \cdot a\left(\frac{x}{\varepsilon}\right)\nabla u^\varepsilon = 0 \text{ on } \partial\Omega^\varepsilon_\infty, \tag{11}$$

where we denoted by $\underline{n}$ the unit outward normal to $\partial\Omega^\varepsilon_\infty$. Here, $L^\varepsilon(x/\varepsilon, \partial_x)$ is the second order elliptic operator given by

$$L^\varepsilon\left(\frac{x}{\varepsilon}, \partial_x\right) u(x) := -\nabla\cdot\left(a\left(\frac{x}{\varepsilon}\right)\nabla u\right) + a_0\left(\frac{x}{\varepsilon}\right) u,$$

$\varepsilon > 0$ and $f \in L^2(\Omega^\varepsilon_\infty)$.

We considered homogenous Neumann boundary conditions on the interior cell boundaries, but emphasize that the case when $\widehat{Q} = \widehat{Y}$ or the case when homogeneous Dirichlet boundary conditions on the interior boundary $\partial\widehat{Q}\backslash\partial\widehat{Y}$ are prescribed can be treated completely analogously.

Then, by the Lax-Milgram Theorem, for every $f \in L^2(\Omega^\varepsilon_\infty)$ and for every $\varepsilon > 0$, (10)–(11) admits a unique weak solution $u^\varepsilon \in H^1(\Omega^\varepsilon_\infty)$, i.e.,

$$B^\varepsilon(u^\varepsilon, v) = (f, v)_{L^2(\Omega^\varepsilon_\infty)} \quad \forall v \in H^1(\Omega^\varepsilon_\infty), \tag{12}$$

where the bilinear form $B^\varepsilon(\cdot,\cdot)$ is given by

$$B^\varepsilon(u, v) := \left(\nabla v, a\left(\frac{x}{\varepsilon}\right)\nabla u\right)_{L^2(\Omega^\varepsilon_\infty)} + \left(v, a_0\left(\frac{x}{\varepsilon}\right)u\right)_{L^2(\Omega^\varepsilon_\infty)} \quad \forall u, v \in H^1(\Omega^\varepsilon_\infty).$$

3.2 Fourier representation of the solution

Let $f = f(x)$ be independent of ε and $f \in L^2(\mathbb{R}^n)$ (in particular, f has to be defined everywhere on $\mathbb{R}^n$, not just on $\Omega^\varepsilon_\infty$). Then, the following Fourier inversion formula in $L^2(\mathbb{R}^n)$ holds

$$f(x) = \frac{1}{(2\pi)^{n/2}} \int_{\mathbb{R}^n} \hat{f}(t) e^{it\cdot x} dt,$$

where

$$\hat{f}(t) := \frac{1}{(2\pi)^{n/2}} \int_{\mathbb{R}^n} e^{-it\cdot x} f(x) dx \in L^2(\mathbb{R}^n)$$

denotes the standard Fourier transform of f.

We might therefore think of representing $u^\varepsilon(x)$ in the form

$$u^\varepsilon(x) = \frac{1}{(2\pi)^{n/2}} \int_{\mathbb{R}^n} \hat{f}(t) \psi^\varepsilon(x,t)\, dt \tag{13}$$

as a superposition of elementary solutions $\hat{f}(t)\psi^\varepsilon(x,t)$ of (10)–(11) with right hand side $\hat{f}(t)e^{it\cdot x}$, in which the kernel $\psi^\varepsilon(x,t)$ satisfies formally

$$\begin{aligned} L^\varepsilon(x/\varepsilon, \partial_x)\psi^\varepsilon(x,t) &= e^{it\cdot x} \quad \text{in } \Omega^\varepsilon_\infty \\ \text{and, if } \widehat{\Gamma}_N \neq \emptyset,\ \underline{n} \cdot a\left(\frac{x}{\varepsilon}\right) \nabla_x \psi^\varepsilon(x,t) &= 0 \quad \text{on } \partial\Omega^\varepsilon_\infty. \end{aligned} \tag{14}$$

We have to give a rigorous mathematical meaning to the integral (13). Since $f(x) = e^{it\cdot x}$ in (14) is not in $L^2(\Omega^\varepsilon_\infty)$, (14) has no solution in $H^1(\Omega^\varepsilon_\infty)$ in the sense of (12). We therefore weaken the notion of solution of (10)–(11) and allow for more general input data such as $f(x) = e^{it\cdot x}$.

To characterize precisely this weakened notion of solution of (10)–(11) in $\Omega^\varepsilon_\infty$, we introduce weighted Sobolev spaces $H^j_\nu(\Omega^\varepsilon_\infty)$ of complex-valued functions with exponential weights depending on a real parameter ν.

Definition 1. For $j = 0, 1$ and for any $\nu \in \mathbb{R}$ the weighted Sobolev spaces $H^j_\nu(\Omega^\varepsilon_\infty)$ equipped with the norm $\|\cdot\|_{j,\nu}$ are defined to be

$$H^j_\nu(\Omega^\varepsilon_\infty) = \overline{C^\infty_0(\mathbb{R}^n;\mathbb{C})\Big|_{\Omega^\varepsilon_\infty}}^{\|\cdot\|_{j,\nu}}, \tag{15}$$

where

$$\|u\|^2_{j,\nu} = \int_{\Omega^\varepsilon_\infty} \left(\sum_{|\alpha|\leq j} |D^\alpha_x u|^2 \right) e^{2\nu\|x\|}\, dx \quad (D^\alpha_x u = \partial^{\alpha_1}_{x_1} \ldots \partial^{\alpha_n}_{x_n} u,\ \forall \alpha \in \mathbb{N}^n_0). \tag{16}$$

Note that for $\nu > 0$ holds $H^1_\nu \subset H^1_0 = H^1 \subset H^1_{-\nu}$. We will prove that the solution $u^\varepsilon \in H^1(\Omega^\varepsilon_\infty)$ of (12) can be viewed as element of exponentially weighted Sobolev spaces $H^1_{-\nu}(\Omega^\varepsilon_\infty)$ for some $\nu > 0$ and has the following representation: for $x \in \Omega^\varepsilon_\infty$, there holds

$$u^\varepsilon(x) = \frac{1}{(2\pi)^{n/2}} \int^{(B)}_{\mathbb{R}^n} \hat{f}(t)\psi^\varepsilon(x,t)dt. \tag{17}$$

Remark 1. Here and in the following we write $\int^{(B)}$ to designate a Bochner integral of $H^1_{-\nu}$-valued functions.

To specify the meaning to (14), we must generalize (12) for right hand sides $f(x)$ which are not decaying at ∞. To do so, let us introduce the following sesquilinear form $\Psi(\varepsilon)[\cdot,\cdot] : H^1_{-\nu}(\Omega^\varepsilon_\infty) \times H^1_\nu(\Omega^\varepsilon_\infty) \to \mathbb{C}$:

$$\Psi(\varepsilon)[u,v] = \int_{\Omega^\varepsilon_\infty} \left\{ \sum_{p,q=1}^{n} a_{pq}\left(\frac{x}{\varepsilon}\right) \frac{\partial u}{\partial x_p}(x) \overline{\frac{\partial v}{\partial x_q}(x)} + a_0\left(\frac{x}{\varepsilon}\right) u(x)\overline{v(x)} \right\} dx. \tag{18}$$

For all $\varepsilon > 0$ and for $\nu > 0$ sufficiently small, $\Psi(\varepsilon)$ is bounded and 'coercive' with respect to $H^1_{-\nu}(\Omega^\varepsilon_\infty) \times H^1_\nu(\Omega^\varepsilon_\infty)$, in the sense that the *inf-sup* stability condition holds (note that for $\nu = 0$, $\Psi(\varepsilon)$ coincides with B^ε in (12)). There holds:

Proposition 1. *There exist positive constants ν_0, C and γ such that for all $\nu \in (0,\nu_0)$ and all $\varepsilon > 0$*

1. $|\Psi(\varepsilon)[u,v]| \le C\|u\|_{1,-\nu}\|v\|_{1,\nu}$,
2. $\inf\limits_{\|u\|_{1,-\nu}=1} \sup\limits_{\|v\|_{1,\nu}=1} |\Psi(\varepsilon)[u,v]| \ge \gamma > 0$
3. $\sup\limits_{u \in H^1_{-\nu}(\Omega^\varepsilon_\infty)} |\Psi(\varepsilon)[u,v]| > 0$ *for all* $v \in H^1_\nu(\Omega^\varepsilon_\infty)$ *and* $v \ne 0$.

The continuity of the sesquilinear form $\Psi(\varepsilon)$ stated in *1.* is obvious. The *inf-sup* condition *2.* and the injectivity property *3.* can be verified in the following way: based on the coercivity of the bilinear form B^ε for the case $\nu = 0$ a perturbation argument can be employed to prove the existence of a positive $\nu_0 > 0$ such that *2.* and *3.* hold. We emphasize that ν_0 is independent of ε and depends only on the upper and lower bounds of the matrix a and of the zero order coefficient a_0. We prove this in Section 3.5 below in a more general setting. Proposition 1 implies now [1,18]:

Proposition 2. *The properties 1, 2 and 3 of $\Psi(\varepsilon)$ imply that the variational problem: Given $f \in \left(H^1_\nu(\Omega^\varepsilon_\infty)\right)^*$, find*

$$u \in H^1_{-\nu}(\Omega^\varepsilon_\infty) \,:\, \Psi(\varepsilon)[u,v] = \langle f, v\rangle_{(H^1_\nu(\Omega^\varepsilon_\infty))^* \times H^1_\nu(\Omega^\varepsilon_\infty)},\ \forall v \in H^1_\nu(\Omega^\varepsilon_\infty), \tag{19}$$

admits a unique weak solution $u \in H^1_{-\nu}(\Omega^\varepsilon_\infty)$ *and the a-priori estimate*

$$\|u\|_{H^1_{-\nu}(\Omega^\varepsilon_\infty)} \leq (1/\gamma)\|f\|_{(H^1_\nu(\Omega^\varepsilon_\infty))^*}$$

holds.

Corollary 1. *For each* $\varepsilon > 0$, $t \in \mathbb{R}^n$ *and* $\nu \in (0, \nu_0)$, *there exists a unique weak solution* $\psi^\varepsilon(\cdot, t) \in H^1_{-\nu}(\Omega^\varepsilon_\infty)$ *of*

$$\Psi(\varepsilon)\left[\psi^\varepsilon(\cdot, t), v\right] = \langle e^{it\cdot(\cdot)}, v\rangle_{(H^1_\nu)^* \times H^1_\nu}, \quad \forall v \in H^1_\nu(\Omega^\varepsilon_\infty). \tag{20}$$

Furthermore,

$$\left\|\psi^\varepsilon(x, t)\right\|_{1,-\nu} \leq \frac{1}{\gamma\nu^{n/2}}.$$

We prove now the Fourier-Bochner integral representation formula (17) in weighted Sobolev spaces for the solution $u^\varepsilon(x)$ of (10) on $\Omega^\varepsilon_\infty$. The proof follows the ideas in [14].

Theorem 1. *Let* $f(x) \in L^2(\mathbb{R}^n)$, $\nu \in (0, \nu_0)$ *and* $\varepsilon > 0$. *Let further* $u^\varepsilon(x) \in H^1(\Omega^\varepsilon_\infty)$ *be the solution of (10) on* $\Omega^\varepsilon_\infty$. *Then, as an element of the weighted Sobolev space* $H^1_{-\nu}(\Omega^\varepsilon_\infty)$, u^ε *admits the Bochner integral representation (17).*

Proof. The proof consists of two steps. First, we reduce the proof to the case when the Fourier transform $\hat{f}$ of f has compact support. Under the assumption that $\hat{f}$ has compact support in $\mathbb{R}^n$, we show then that the integrand in (17) is strongly measurable and that the $\|\cdot\|_{1,-\nu}$-norm of the integrand is Lebesgue integrable over *supp* f.

To begin, let us consider the truncated series $\hat{f}_M(t) := \mathbf{1}_{\Omega_M := \{\|t\| \leq M\}} \hat{f}(t)$, $M = 1, 2, \ldots$, of $\hat{f}$ and denote by $u^\varepsilon_M(x) \in H^1_{-\nu}(\Omega^\varepsilon_\infty)$ the Bochner integral

$$u^\varepsilon_M(x) = \frac{1}{(2\pi)^{n/2}} \int\limits_{\|t\| \leq M} \hat{f}(t)\psi^\varepsilon(x, t)\, dt. \tag{21}$$

We need to show now that u^ε_M defined above is the solution of (10) with right hand side

$$f_M(x) := \frac{1}{(2\pi)^{n/2}} \int\limits_{\|t\| \leq M} e^{it\cdot x} \hat{f}(t)\, dt.$$

Then, by Proposition 2 and by Parseval's equality it holds

$$\begin{aligned}
\|u^\varepsilon - u^\varepsilon_M\|_{1,-\nu} &\leq \frac{1}{\gamma}\|f - f_M\|_{L^2(\Omega^\varepsilon_\infty)} \leq \frac{1}{\gamma}\|f - f_M\|_{L^2(\mathbb{R}^n)} \\
&= \frac{1}{\gamma}\|\hat{f} - \hat{f}_M\|_{L^2(\mathbb{R}^n)} \\
&= \frac{1}{\gamma}\Bigg(\int\limits_{\|t\| \geq M} |\hat{f}(t)|^2 dt\Bigg)^{1/2} \to 0, \quad \text{as } M \to \infty.
\end{aligned}$$

We prove now that, for fixed $M \in \mathbb{N}$, u_M^ε given by the integral representation (21) with $f = f_M$ is well defined and solves (10), i.e., for an arbitrary fixed test function $v \in C^\infty(\Omega_\infty^\varepsilon)$, $supp\, v \subset\subset \mathbb{R}^n$, it holds

$$\int_{\Omega_\infty^\varepsilon} \nabla u_M^\varepsilon(x) \cdot a\left(\frac{x}{\varepsilon}\right) \nabla v(x) + a_0\left(\frac{x}{\varepsilon}\right) u_M^\varepsilon(x) v(x)\, dx = \int_{\Omega_\infty^\varepsilon} f_M(x) v(x)\, dx. \quad (22)$$

We extend v by 0 to $\mathbb{R}^n \backslash \Omega_\infty^\varepsilon$. For $k = 1, 2, \ldots$ we construct a sequence $\hat{f}_M^k(t)$ of simple functions on Ω_M and a sequence $\{\Omega_M^{k,j}\}_{j=1}^{J_k}$ of measurable partitions of Ω_M with the following properties:

i) for all $k = 1, 2 \ldots$, $j = 1, \ldots, J_k$ and for $t, \tau \in \Omega_M^{k,j}$

$$\left\| \psi^\varepsilon(\cdot, t) - \psi^\varepsilon(\cdot, \tau) \right\|_{1,-\nu} \leq 2^{-k}, \quad |\hat{v}(t) - \hat{v}(\tau)| \leq 2^{-k}.$$

Note that this is feasible due to the Lipschitz continuity of the mapping

$$\mathbb{R}^n \ni t \to \psi^\varepsilon(\cdot, t) \in H^1_{-\nu}(\Omega_\infty^\varepsilon).$$

The proof of this assertion is postponed to Section 3.4, (Theorem 3), where the regularity properties of the integral kernel $\psi^\varepsilon(\cdot, \varepsilon, t)$ are collected.

ii) $\hat{f}_M^k(t)|_{\Omega_M^{k,j}} \equiv const$ and $|\hat{f}_M^k(t)| \leq |\hat{f}_M(t)|$ a.e. in Ω_M.

iii) $\hat{f}_M^k(t) \to \hat{f}_M(t)$ as $k \to \infty$ a.e. in Ω_M.

We construct now a sequence of simple functions to approximate the integrand in (21). For arbitrary $\tau_M^{k,j} \in \Omega_M^{k,j}$ we define

$$\psi_M^{k,\varepsilon}(x,t)\Big|_{\Omega_M^{k,j}} \equiv \psi^\varepsilon(x, \tau_M^{k,j}), \quad \forall j = 1, \ldots, J_k$$

$$\hat{v}^k(t)\Big|_{\Omega_M^{k,j}} \equiv \hat{v}(\tau_M^{k,j}) \quad \forall j = 1, \ldots, J_k.$$

Then, almost everywhere in Ω_M, due to the assumptions *i)* and *ii)* above it holds

$$\left\| \hat{f}_M(t)\psi^\varepsilon(x,t) - \hat{f}_M^k(t)\psi_M^{k,\varepsilon}(x,t) \right\|_{1,-\nu} \leq C|\hat{f}_M(t) - \hat{f}_M^k(t)| + |\hat{f}_M(t)|2^{-k},$$

therefore by *iii)* we have that

$$\lim_{k\to\infty} \left\| \hat{f}_M(t)\psi^\varepsilon(x,t) - \hat{f}_M^k(t)\psi_M^{k,\varepsilon}(x,t) \right\|_{1,-\nu} = 0 \quad \text{a.e. } t \in \Omega_M.$$

Hence, by Lebesgue's Theorem on dominated convergence it follows that

$$\lim_{k\to\infty} \int_{\Omega_M} \left\| \hat{f}_M(t)\psi^\varepsilon(x,t) - \hat{f}_M^k(t)\psi_M^{k,\varepsilon}(x,t) \right\|_{1,-\nu} dt = 0,$$

i.e., in $H^1_{-\nu}(\Omega^\varepsilon_\infty)$ topology

$$u^\varepsilon_M(x) = \lim_{k\to\infty} \frac{1}{(2\pi)^{n/2}} \sum_{j=1,\dots,J_k} \hat{f}^k_M(\tau^{k,j}_M)\psi^\varepsilon(x,\tau^{k,j}_M)|\Omega^{k,j}_M|.$$

Turning back to the variational formulation (22) and recalling that by Theorem 2 for a fixed wave parameter τ, $\psi^\varepsilon(\cdot,\tau)$ solves (20) with right hand side $e^{i\tau\cdot x}$ we obtain that

$$\int_{\Omega^\varepsilon_\infty} \nabla u^\varepsilon_M(x)\cdot a\left(\frac{x}{\varepsilon}\right)\nabla v(x) + a_0\left(\frac{x}{\varepsilon}\right)u^\varepsilon_M(x)v(x)\,dx$$

$$= \lim_{k\to\infty} \sum_{j=1,\dots,J_k} \hat{f}^k_M(\tau^{k,j}_M)\overline{\hat{v}(\tau^{k,j}_M)}|\Omega^{k,j}_M| = \lim_{k\to\infty}\int_{\Omega_M} \hat{f}^k_M(t)\overline{\hat{v}^k(t)}dt$$

$$= \int_{\Omega_M} \hat{f}_M(t)\overline{\hat{v}(t)}dt = \int_{\Omega^\varepsilon_\infty} f_M(x)v(x)\,dx$$

□

3.3 Scale separation

So far, $\psi^\varepsilon(x,t)$ is solution of problem (20) on an unbounded domain and is, in a sense, the generalized fundamental solution of L^ε (resp. the two-scale Fourier symbol of $(L^\varepsilon)^{-1}$ on $\Omega^\varepsilon_\infty$). A remarkable observation at this point is that the integral representation of u^ε is true for general coefficients $a^\varepsilon(x)$, $a^\varepsilon_0(x)$ in (1). In the case of periodic coefficients $a^\varepsilon(x) = a(x/\varepsilon)$, $a^\varepsilon_0(x) = a_0(x/\varepsilon)$ with $a(\cdot)$, $a_0(\cdot)$ periodic, however, we have for $\psi^\varepsilon(x,t)$ a scale separation property.

Theorem 2. *The solution $\psi^\varepsilon(\cdot,t) \in H^1_{-\nu}(\Omega^\varepsilon_\infty)$ of (20) admits the representation*

$$\psi^\varepsilon(x,t) = \psi\left(\frac{x}{\varepsilon},\varepsilon,t\right) := e^{it\cdot x}\phi(y,\varepsilon,t)|_{y=\frac{x}{\varepsilon}} \quad y = \frac{x}{\varepsilon}\in\Omega_\infty, \tag{23}$$

where $\phi(\cdot,\varepsilon,t)$ is the 2π-periodic solution of the so-called unit-cell problem

$$\begin{aligned}\mathcal{L}(\varepsilon,t;y,\partial_y)\phi &:= e^{-i\varepsilon t\cdot y}L^\varepsilon(y,\varepsilon^{-1}\partial_y)\,e^{i\varepsilon t\cdot y}\phi \\ &= -e^{-i\varepsilon t\cdot y}\sum_{p,q=1}^{n}\varepsilon^{-2}\frac{\partial}{\partial y_p}\left(a_{pq}(y)\frac{\partial}{\partial y_q}\left(\phi(y,\varepsilon,t)e^{it\cdot\varepsilon y}\right)\right) \\ &\quad + a_0(y)\phi(y,\varepsilon,t) = 1, \quad \text{in } \widehat{Q}, \\ \underline{n}\cdot a(y)(it\varepsilon+\nabla_y)\phi &= 0 \ \text{ on } \widehat{\Gamma}_N\,. \end{aligned} \tag{24}$$

Remark 2. The unit-cell solution $\phi(y,\varepsilon,t)$ is computable by FE-discretization of the variational form

$$\text{Find } \phi(\cdot) \in H^1_{\text{per}}(\widehat{Q}) \text{ such that } \Phi(\varepsilon,t)[\phi,v] = \varepsilon^2 \int_{\widehat{Q}} \overline{v(y)}dy \ \forall v \in H^1_{\text{per}}(\widehat{Q}), \quad (25)$$

where for each $\varepsilon > 0$ and $t \in \mathbb{C}^n$ we define the sesquilinear form $\Phi(\varepsilon,t)[\cdot,\cdot]$: $H^1_{\text{per}}(\widehat{Q}) \times H^1_{\text{per}}(\widehat{Q}) \to \mathbb{C}$ by

$$\Phi(\varepsilon,t)[\phi,v] := \int_{\widehat{Q}} \Big(a(y)\nabla_y \left(\phi(y)e^{it\cdot\varepsilon y}\right) \cdot \overline{\nabla_y \left(v(y)e^{it\cdot\varepsilon y}\right)} + \varepsilon^2 a_0(y)\phi(y)\overline{v(y)}\Big)\, dy. \quad (26)$$

We will refer to $x \in \mathbb{R}^n$ as the *slow* variable and to $y = x/\varepsilon$ as the *fast* variable describing the fast scale of the problem.
Proof of Theorem 2. For each $k \in \mathbb{Z}^n$ define $\widehat{Q}(\varepsilon,k)$, $\widehat{Y}(\varepsilon,k)$ as

$$\widehat{Q}(\varepsilon,k) = \varepsilon(\widehat{Q} + k\pi), \quad \widehat{Y}(\varepsilon,k) = \varepsilon(\widehat{Y} + k\pi)$$

and consider a locally finite, C^∞-partition of unity $\{\chi_k(\cdot,\varepsilon)\}_k$ associated to $\{\widehat{Y}(\varepsilon,k)\}_k$ such that $|\chi_k(x,\varepsilon)|$ and $|\nabla\chi_k(x,\varepsilon)|$ are uniformly bounded for $x \in \mathbb{R}^n$ and $k \in \mathbb{Z}^n$. This ensures that for any $v \in H^1_\nu(\Omega^\varepsilon_\infty)$, $\sum_{k\in\mathbb{Z}^n} \chi_k(\cdot,\varepsilon)v$ converges to v with respect to the topology of $H^1_\nu(\Omega^\varepsilon_\infty)$. Since by Proposition 1, $\Psi(\varepsilon)$ is a continuous sesquilinear form on $H^1_{-\nu}(\Omega^\varepsilon_\infty) \times H^1_\nu(\Omega^\varepsilon_\infty)$, for $v \in H^1_\nu(\Omega^\varepsilon_\infty)$ it holds

$$\Psi(\varepsilon)\left[\psi\left(\frac{x}{\varepsilon},\varepsilon,t\right),v\right] = \sum_{k\in\mathbb{Z}^n} \Psi(\varepsilon)\left[\psi\left(\frac{x}{\varepsilon},\varepsilon,t\right),v_k(x,\varepsilon)\right], \quad (27)$$

where $v_k(x,\varepsilon) := \chi_k(x,\varepsilon)v(x)$ is compactly supported in $\widehat{Y}(\varepsilon,k)$. We reduce now the integrals on $\widehat{Q}(\varepsilon,k)$ to integrals on the unit-cell domain $\widehat{Q}$, by appropriate scalings and translations. Let $\tilde{k} \in (2\mathbb{Z})^n$ be given by

$$\tilde{k}_j = \begin{cases} k_j,\ k_j \text{ even}, \\ k_j - 1,\ k_j \text{ odd}. \end{cases}$$

Substituting $x/\varepsilon = y + \pi\tilde{k}$, the integrals in (27) transform in

$$\begin{aligned} &\Psi(\varepsilon)\left[\psi\left(\frac{x}{\varepsilon},\varepsilon,t\right),v_k(x,\varepsilon)\right] = \\ &= \int_{\widehat{Q}(1,k-\tilde{k})} \Big\{\varepsilon^{-2}\overline{\nabla_y\left(v_k(\varepsilon(y+\pi\tilde{k}),\varepsilon)\right)} \cdot a(y)\nabla_y\left(\phi(y,\varepsilon,t)e^{i\varepsilon t\cdot(y+\pi\tilde{k})}\right) \\ &\quad + a_0(y)\phi(y,\varepsilon,t)e^{i\varepsilon t\cdot(y+\pi\tilde{k})}\overline{v_k(\varepsilon(y+\pi\tilde{k}),\varepsilon)}\Big\}\, \varepsilon^n dy. \end{aligned} \quad (28)$$

Since the support of $y \to v_k(\varepsilon(y+\pi\tilde{k}),\varepsilon)$ is contained in $\widehat{Y}(1,k-\tilde{k})$, if we extend periodically $v_k(\varepsilon(y+\pi\tilde{k}),\varepsilon)e^{-i\varepsilon t\cdot(y+\pi\tilde{k})}$ to Ω_∞ we obtain a $H^1_{\mathrm{per}}(\widehat{Q})$ function. Then, we can replace $\widehat{Q}(1,k-\tilde{k})$ by $\widehat{Q}$ in (28) and by (25) we obtain that

$$\begin{aligned}
&\Psi(\varepsilon)\left[\psi\left(\frac{x}{\varepsilon},\varepsilon,t\right),v\right] = \\
&= \sum_{k\in\mathbb{Z}^n} \varepsilon^{n-2}\Phi(\varepsilon,t)\left[\phi(y,\varepsilon,t), v_k(\varepsilon(y+\pi\tilde{k}),\varepsilon)e^{-i\varepsilon t\cdot(y+\pi\tilde{k})}\right] \\
&= \sum_{k\in\mathbb{Z}^n} \varepsilon^{n} \int_{\widehat{Q}} \overline{v_k(\varepsilon(y+\pi\tilde{k}),\varepsilon)e^{-i\varepsilon t\cdot(y+\pi\tilde{k})}}\,dy \\
&= \int_{\Omega^\varepsilon_\infty} e^{it\cdot x}\overline{v(x)}\,dx, \quad \forall\, v\in H^1_\nu(\Omega^\varepsilon_\infty),
\end{aligned}$$

so that $\psi(x/\varepsilon,\varepsilon,t)$ defined in (23) is the weak solution of (19) with respect to the functional $f = e^{it\cdot x} \in \left(H^1_\nu(\Omega^\varepsilon_\infty)\right)^*$. □

We will show how the kernels $\phi(y,\varepsilon,t), \psi(y,\varepsilon,t)$ can be used to design FE approximations of (1), (2) which encode the microstructure and coefficient regularity. A crucial role for obtaining error estimates will be played by the kernels' analyticity with respect to t.

3.4 Analyticity of the kernels ψ, ϕ

The kernel $\phi(\cdot,\varepsilon,t)$ can be continued analytically with respect to (ε,t) in a complex neighborhood $\widehat{G}\subset\mathbb{C}^{n+1}$ of $\mathbb{R}^{n+1}$, with values in $H^1_{\mathrm{per}}(\widehat{Q})$ [15]. We now show that, for every fixed $\varepsilon>0$, the mappings

$$\mathbb{R}^n \ni t \to \phi\left(\frac{\cdot}{\varepsilon},\varepsilon,t\right),\ \psi\left(\frac{\cdot}{\varepsilon},\varepsilon,t\right) \in H^1_{-\nu}(\Omega^\varepsilon_\infty)$$

can be continued analytically with respect to t in a strip neighbourhood of $\mathbb{R}^n$, and the width of the strip is independent of ε.

To prove this, let $\nu\in(0,\nu_0)$, $\nu_n = \nu/(2\sqrt{n})$ and denote by $\mathcal{D}_{\nu_n}$ the strip region of width ν_n

$$\mathcal{D}_{\nu_n} := \{t\in\mathbb{C}^n \;:\; \max_{j=1,\dots,n} |\mathrm{Im}\, t_j| < \nu_n\}.$$

We start by proving

Lemma 1. *The mapping*

$$\mathcal{D}_{\nu_n} \ni t \to G(t) := e^{it\cdot x} \in \left(H^1_\nu(\Omega^\varepsilon_\infty)\right)^* \tag{29}$$

is holomorphic in $\mathcal{D}_{\nu_n}$ with values in the Banach space $\left(H^1_\nu(\Omega^\varepsilon_\infty)\right)^*$. Moreover, $G_k(t) := (ix)^k e^{it\cdot x} \in \left(H^1_\nu(\Omega^\varepsilon_\infty)\right)^*$ is the $k\in\mathbb{N}^n$−th partial derivative *with respect to t of the $\left(H^1_\nu(\Omega^\varepsilon_\infty)\right)^*$-valued mapping $G(t)$ and its norm has the following bound*

$$\forall t \in \mathcal{D}_{\nu_n}: \quad \|G_k(t)\|_{(H^1_\nu(\Omega^\varepsilon_\infty))^*} \le C_{\nu,n}\frac{2^{|k|}k!}{\nu^{|k|}}n^{|k|/2}, \quad k \in \mathbb{N}^n, \tag{30}$$

where $k! := k_1! \dots k_n!$.

Proof. It is sufficient to show that for every $v \in H^1_\nu(\Omega^\varepsilon_\infty)$ the application

$$\mathcal{D}_{\nu_n} \ni t \to \langle G(t), v\rangle_{(H^1_\nu(\Omega^\varepsilon_\infty))^* \times H^1_\nu(\Omega^\varepsilon_\infty)} \in \mathbb{C} \tag{31}$$

is $\mathbb{C}^n$-differentiable. Let $t_0 \in \mathcal{D}_{\nu_n}$ and $p \in 1,2,\dots,n$ arbitrary, fixed, and $t \in \mathcal{D}_{\nu_n}$ such that $|t_p - t_{0p}| \le \nu_n/2$ and $t_q = t_{0q}\,\forall q \ne p$. Then, for every $v \in H^1_\nu(\Omega^\varepsilon_\infty)$

$$\begin{aligned}&\left|\frac{1}{t_p - t_{0p}}\langle G(t) - G(t_0), v\rangle_{(H^1_\nu)^* \times H^1_\nu} - \langle G_{\delta_p}(t_0), v\rangle_{(H^1_\nu)^*\times H^1_\nu}\right| \\ &\le \|v\|_{1,\nu}\left[\int_{\mathbb{R}^n}\left|\frac{e^{it\cdot x} - e^{it_0\cdot x}}{t_p - t_{0p}} - ix_p e^{it_0\cdot x}\right|^2 e^{-2\nu\|x\|}dx\right]^{1/2} \\ &\le C(\nu,n)|t_p - t_{0p}|\|v\|_{1,\nu}.\end{aligned}$$

This implies that the limit

$$\lim_{\substack{t_p \to t_{0p} \\ t_q = t_{0q},\,\forall q\ne p}} \frac{1}{t_p - t_{0p}}\langle G(t) - G(t_0), v\rangle_{(H^1_\nu(\Omega^\varepsilon_\infty))^*\times H^1_\nu(\Omega^\varepsilon_\infty)} \tag{32}$$

exists and is equal to $\langle G_{\delta_p}(t_0), v\rangle_{(H^1_\nu(\Omega^\varepsilon_\infty))^*\times H^1_\nu(\Omega^\varepsilon_\infty)}$, where $\delta_p = (\delta_{pq})_{q=1,\dots,n}$.

Let us take $v \in H^1_\nu(\Omega^\varepsilon_\infty)$ and estimate $|\langle G_k(t), v\rangle_{(H^1_\nu(\Omega^\varepsilon_\infty))^*\times H^1_\nu(\Omega^\varepsilon_\infty)}|$:

$$\begin{aligned}|\langle G_k(t), v\rangle_{(H^1_\nu)^*\times H^1_\nu}| &= \left|\int_{\mathbb{R}^n}\left(\prod_{p=1}^n (ix_p)^{k_p}\right)e^{it\cdot x}v(x)\,dx\right| \\ &\le \|v\|_{1,\nu}\prod_{p=1}^n\left(\int_{\mathbb{R}}|x_p|^{2k_p}e^{2|\mathrm{Im}(t_p)|\,|x_p|}e^{-2\nu|x_p|/\sqrt{n}}\,dx_p\right)^{1/2} \\ &\le \|v\|_{1,\nu}\frac{\sqrt{(2k)!}}{(\nu/\sqrt{n})^{|k|}(\nu/(2\sqrt{n}))^{n/2}}.\end{aligned}$$

The estimate (30) follows now immediately by Stirling's formula. □

Now, for fixed $t \in \mathcal{D}_{\nu_n}$ and $k \in \mathbb{N}^n$, let $\psi^\varepsilon_k(t)$ be the weak solution in $H^1_{-\nu}(\Omega^\varepsilon_\infty)$ of the following variational problem

$$\psi^\varepsilon_k \in H^1_{-\nu}(\Omega^\varepsilon_\infty) : \Psi(\varepsilon)[\psi^\varepsilon_k(t), v] = \langle G_k(t), v\rangle_{(H^1_\nu)^*\times H^1_\nu}, \ \forall v \in H^1_\nu(\Omega^\varepsilon_\infty). \tag{33}$$

By Proposition 2 and Lemma 1, ψ^ε_k exists and is unique.

Theorem 3. *The mapping*

$$\mathcal{D}_{\nu_n} \ni t \to \psi\left(\frac{\cdot}{\varepsilon}, \varepsilon, t\right) \in H^1_{-\nu}(\Omega^\varepsilon_\infty) \tag{34}$$

is holomorphic in $\mathcal{D}_{\nu_n}$ with values in the Banach space $H^1_{-\nu}(\Omega^\varepsilon_\infty)$. Moreover, $\psi^\varepsilon_k(t)$ defined by (33) is given by

$$\psi^\varepsilon_k(t) = D^k_t \psi\left(\frac{\cdot}{\varepsilon}, \varepsilon, t\right) = \partial^{k_1}_{t_1} \dots \partial^{k_n}_{t_n} \psi\left(\frac{\cdot}{\varepsilon}, \varepsilon, t\right) \tag{35}$$

and the $\|\cdot\|_{1,-\nu}$-norm of $\psi^\varepsilon_k(t)$ is uniformly bounded with respect to $t \in \mathcal{D}_{\nu_n}$ by

$$\|\psi^\varepsilon_k(t)\|_{1,-\nu} \leq C_{\nu,n} \frac{2^{|k|} k!}{\gamma \nu^{|k|}} n^{|k|/2}. \tag{36}$$

Proof. The proof is similar to that of Lemma 1 and is based on the fact that

$$\Psi(\varepsilon)\left[\psi\left(\frac{\cdot}{\varepsilon}, \varepsilon, t\right), v\right] = \langle G(t), v\rangle_{(H^1_\nu(\Omega^\varepsilon_\infty))^* \times H^1_\nu(\Omega^\varepsilon_\infty)}, \quad \forall v \in H^1_\nu(\Omega^\varepsilon_\infty) \tag{37}$$

and on the properties of the bilinear form $\Psi(\varepsilon)[\cdot,\cdot]$ stated in Proposition 1. In order to prove the analyticity of the $H^1_{-\nu}(\Omega^\varepsilon_\infty)$-valued mapping $t \to \psi(\cdot/\varepsilon, \varepsilon, t)$ in the strip $\mathcal{D}_{\nu_n}$ it is enough to show that for every $v \in H^1_\nu(\Omega^\varepsilon_\infty)$ the $\mathbb{C}$-valued function

$$\mathcal{D}_{\nu_n} \ni t \to \Psi(\varepsilon)\left[\psi\left(\frac{\cdot}{\varepsilon}, \varepsilon, t\right), v\right] \in \mathbb{C} \tag{38}$$

is holomorphic. From the definition of $\psi^\varepsilon_k(t)$ in (33) and for any fixed $p \in \{1, 2, \dots, n\}$ we have for every $v \in H^1_\nu(\Omega^\varepsilon_\infty)$

$$\Psi(\varepsilon)\left[(D_{t_p}\psi)(\cdot/\varepsilon, \varepsilon, t_0), v\right] =$$

$$= \lim_{\substack{t_p \to t_{0p} \\ t_q = t_{0q}\ \forall q \neq p}} \frac{1}{t_p - t_{0p}} \Psi(\varepsilon)\left[\psi(\cdot/\varepsilon, \varepsilon, t) - \psi(\cdot/\varepsilon, \varepsilon, t_0), v\right]$$

$$= \lim_{\substack{t_p \to t_{0p} \\ t_q = t_{0q}\ \forall q \neq p}} \frac{1}{t_p - t_{0p}} \langle G(t) - G(t_0), v\rangle_{(H^1_\nu)^* \times H^1_\nu}$$

$$= \langle G_{\delta_p}(t_0), v\rangle_{(H^1_\nu(\Omega^\varepsilon_\infty))^* \times H^1_\nu(\Omega^\varepsilon_\infty)} = \Psi(\varepsilon)[\psi^\varepsilon_{\delta_p}(t_0), v].$$

Here we denoted again by $\delta_p = (\delta_{pq})_{p,q=1,\dots,n} \in \mathbb{N}^n$. So we have shown that $\psi^\varepsilon_{\delta_p}(t) = (D_{t_p}\psi)(\cdot/\varepsilon, \varepsilon, t)$. Analogously, it can be shown that $\psi^\varepsilon_k(t) = (D^k_t\psi)(\cdot/\varepsilon, \varepsilon, t)$ for all $k \in \mathbb{N}^n$.

To show (36), we now use that from (33), Proposition 2 and by Lemma 1 we have that

$$\|\psi^\varepsilon_k(t)\|_{H^1_{-\nu}(\Omega^\varepsilon_\infty)} = \left\| D^k_t \psi\left(\frac{\cdot}{\varepsilon}, \varepsilon, t\right) \right\|_{1,-\nu} \leq \frac{1}{\gamma}\|G_k(t)\|_{(H^1_\nu)^*} \leq C_{\nu,n} \frac{2^{|k|} k!}{\gamma \nu^{|k|}} n^{|k|/2},$$

uniformly with respect to $t \in \mathcal{D}_{\nu_n}$. □

The following result on analyticity of the unit-cell solution $\phi(\cdot,\varepsilon,t)$ with respect to t is a direct consequence of Theorem 3 and of the structure of the Fourier-Bochner integral kernel ψ: $\psi(x/\varepsilon,\varepsilon,t) = e^{it\cdot x}\phi(x/\varepsilon,\varepsilon,t)$.

Theorem 4. *For a given $\nu \in (0,\nu_0)$ the mapping*

$$\mathcal{D}_{\nu_n} \ni t \to \phi\left(\frac{\cdot}{\varepsilon},\varepsilon,t\right) \in H^1_{-2\nu}(\Omega^\varepsilon_\infty) \tag{39}$$

is a holomorphic function of the variable $t \in \mathcal{D}_{\nu_n}$ with values in the Banach space $H^1_{-2\nu}(\Omega^\varepsilon_\infty)$. Moreover,

$$\left\| D^k_t \phi\left(\frac{\cdot}{\varepsilon},\varepsilon,t\right)\right\|_{H^1_{-2\nu}(\Omega^\varepsilon_\infty)} \le C_{\nu,n,\gamma}(1+\|t\|)\frac{(k+\mathbf{1})!}{(\nu/2)^{|k|}}n^{|k|/2}, \quad \forall t \in \mathcal{D}_d, \tag{40}$$

where $\mathbf{1} := (1,1,\ldots,1)^\top \in \mathbb{R}^n$.

3.5 Elliptic systems

In this section we want to generalize the construction of the Fourier-Bochner integral kernel $\phi(y,\varepsilon,t)$ and the representation formula (17) of the solution $u^\varepsilon(x)$ to the case of second order strongly elliptic systems and to prove the inf-sup conditions in Proposition 1 for the sesquilinear form $\Psi(\varepsilon)$. Throughout this section (and only here) we denote vector functions by $\underline{u}$ and matrix functions by $\underline{\underline{\phi}}$. We consider the model problem

$$L^\varepsilon\left(\frac{x}{\varepsilon},\partial_x\right)\underline{u}^\varepsilon(x) = \underline{f}(x) \quad \text{in } \Omega^\varepsilon_\infty, \tag{41}$$

where now $\underline{f} \in L^2(\mathbb{R}^n)^d$, with $d \ge 2$, and the differential operator

$$L^\varepsilon\left(\frac{x}{\varepsilon},\partial_x\right) = -\sum_{p,q=1}^{n} \partial_{x_p}\underline{\underline{a}}^{pq}\left(\frac{x}{\varepsilon}\right)\partial_{x_q} + \underline{\underline{a}}^0\left(\frac{x}{\varepsilon}\right)$$

has real, $\varepsilon\widehat{Y}$-periodic matrix-valued coefficients $\underline{\underline{a}}^{pq}(x/\varepsilon)$, $\underline{\underline{a}}^0(x/\varepsilon)$ with

$$\underline{\underline{a}}^{pq}(y)\,,\ \underline{\underline{a}}^0(y) \in L^\infty_{\mathrm{per}}(\widehat{Q})^{d\times d}.$$

We assume that L^ε is formally self-adjoint, i.e.,

$$a^{pq}_{jk}(y) = a^{qp}_{kj}(y),\ a^0_{jk}(y) = a^0_{kj}(y) \quad \forall p,q = 1,\ldots,n,\ \forall j,k = 1,\ldots,d.$$

Further, we assume homogeneous Neumann boundary conditions on the boundary of the interior holes $\partial\Omega^\varepsilon_\infty$ of the periodic lattice $\Omega^\varepsilon_\infty$

$$\gamma^\varepsilon_1\underline{u}^\varepsilon(x) := \sum_{p,q=1}^{n} n_p\underline{\underline{a}}^{pq}\left(\frac{x}{\varepsilon}\right)\partial_{x_q}\underline{u}^\varepsilon(x) = 0 \text{ on } \partial\Omega^\varepsilon_\infty. \tag{42}$$

The variational formulation for (41), (42) reads: find $\underline{u}^\varepsilon \in H^1(\Omega^\varepsilon_\infty)^d$ such that

$$B^\varepsilon(\underline{u}^\varepsilon, \underline{v}) = (\underline{f}, \underline{v})_{L^2(\Omega^\varepsilon_\infty)}, \quad \forall \underline{v} \in H^1(\Omega^\varepsilon_\infty)^d,$$

where for $\underline{u}, \underline{v} \in H^1(\Omega^\varepsilon_\infty)^d$

$$B^\varepsilon(\underline{u}, \underline{v}) := \int_{\Omega^\varepsilon_\infty} \left\{ \sum_{p,q=1}^{n} \partial_{x_p} \underline{v}^\top(x) \underline{\underline{a}}^{pq}\left(\frac{x}{\varepsilon}\right) \partial_{x_q} \underline{u}(x) + \underline{v}^\top(x) \underline{\underline{a}}^0\left(\frac{x}{\varepsilon}\right) \underline{u}(x) \right\} dx.$$

Under strong ellipticity assumptions on the differential operator L^ε

$$B^\varepsilon(\underline{u}, \underline{u}) \geq \gamma_0 \|\underline{u}\|^2_{H^1(\Omega^\varepsilon_\infty)^d}, \quad \forall \underline{u} \in H^1(\Omega^\varepsilon_\infty)^d, \tag{43}$$

with $\gamma_0 > 0$ independent of ε, a unique weak solution $\underline{u}^\varepsilon \in H^1(\Omega^\varepsilon_\infty)^d$ exists and $\|\underline{u}^\varepsilon\|_{H^1(\Omega^\varepsilon_\infty)^d} \leq C\|\underline{f}\|_{L^2(\Omega^\varepsilon_\infty)^d}$, with $C > 0$ independent of ε.

Due to the ellipticity condition (43), there exists $\nu_0 = \nu_0(\gamma_0) > 0$ independent of ε, such that for all $\nu \in (0, \nu_0)$, the sesquilinear form $\Psi(\varepsilon)[\cdot,\cdot]$: $H^1_{-\nu}(\Omega^\varepsilon_\infty)^d \times H^1_\nu(\Omega^\varepsilon_\infty)^d \to \mathbb{C}$ given by

$$\Psi(\varepsilon)[\underline{u}, \underline{v}] := \int_{\Omega^\varepsilon_\infty} \sum_{p,q=1}^{n} \overline{\partial_{x_p} \underline{v}(x)}^\top \underline{\underline{a}}^{pq}\left(\frac{x}{\varepsilon}\right) \partial_{x_q} \underline{u}(x) + \overline{\underline{v}(x)}^\top \underline{\underline{a}}^0\left(\frac{x}{\varepsilon}\right) \underline{u}(x)\, dx$$

is continuous and 'coercive', i.e., the Babuška-Brezzi *inf-sup* condition

$$\inf_{\underline{u} \in H^1_{-\nu}(\Omega^\varepsilon_\infty)^d} \sup_{\underline{v} \in H^1_\nu(\Omega^\varepsilon_\infty)^d} \frac{|\Psi(\varepsilon)[\underline{u}, \underline{v}]|}{\|\underline{u}\|_{H^1_{-\nu}(\Omega^\varepsilon_\infty)} \|\underline{v}\|_{H^1_\nu(\Omega^\varepsilon_\infty)}} \geq \gamma \tag{44}$$

holds with a constant $\gamma > 0$ independent of ε and for every $0 \neq \underline{v} \in H^1_\nu(\Omega^\varepsilon_\infty)^d$

$$\sup_{\underline{u} \in H^1_{-\nu}(\Omega^\varepsilon_\infty)^d} |\Psi(\varepsilon)[\underline{u}, \underline{v}]| > 0. \tag{45}$$

The continuity of $\Psi(\varepsilon)$ is obvious. In order to prove (44), it is enough to show the existence of positive constants $\nu_0 > 0$, $\gamma > 0$ and $C > 0$ independent of ε such that

$$\forall \underline{u} \in C^\infty(\Omega^\varepsilon_\infty)^d,\ supp\, \underline{u} \subset\subset \mathbb{R}^n, \quad \exists \underline{v} = \underline{v_u} \in H^1_\nu(\Omega^\varepsilon_\infty)^d :$$

$$|\Psi(\varepsilon)[\underline{u}, \underline{v_u}]| \geq \gamma \|\underline{u}\|^2_{H^1_{-\nu}(\Omega^\varepsilon_\infty)} \text{ and } \|\underline{v_u}\|_{H^1_\nu(\Omega^\varepsilon_\infty)} \leq C\|\underline{u}\|_{H^1_{-\nu}(\Omega^\varepsilon_\infty)},$$

and $C > 0$ is independent of the support of $\underline{u}$. Let therefore $\underline{u} \in C^\infty(\Omega^\varepsilon_\infty)^d$, $supp\, \underline{u} \subset\subset \mathbb{R}^n$, be arbitrary, fixed. For $\nu > 0$ we define $\underline{v_u}(x) := \underline{u}(x) e^{-2\nu\|x\|}$. Obviously, $\underline{v_u} \in H^1_\nu(\Omega^\varepsilon_\infty)^d$ and

$$\|\underline{v_u}\|_{H^1_\nu(\Omega^\varepsilon_\infty)} \leq C(\|\underline{u}\|_{H^1_{-\nu}(\Omega^\varepsilon_\infty)} + \nu\|\underline{u}\|_{L^2_{-\nu}(\Omega^\varepsilon_\infty)})$$

holds with a constant $C > 0$ independent of ν and $\underline{u}$. A direct calculation shows further that

$$\Psi(\varepsilon)[\underline{u}, \underline{v_u}] = B^\varepsilon(\underline{u}e^{-\nu\|x\|}, \underline{u}e^{-\nu\|x\|})$$
$$- \nu^2 \int_{\Omega^\varepsilon_\infty} \sum_{p,q=1}^{n} \overline{(x_q/\|x\|)\underline{u}(x)}^\top \underline{\underline{a^{pq}}}\left(\frac{x}{\varepsilon}\right)(x_p/\|x\|)\underline{u}(x)e^{-2\nu\|x\|}dx.$$

Due to (43), there holds

$$\Psi(\varepsilon)[\underline{u}, \underline{v_u}] \geq \gamma_0\|\underline{u}e^{-\nu\|x\|}\|^2_{H^1(\Omega^\varepsilon_\infty)^d} - C'\nu^2\|\underline{u}\|^2_{L^2_{-\nu}(\Omega^\varepsilon_\infty)^d},$$

with $C' > 0$ depending on the L^∞-norm of the coefficients $\underline{\underline{a^{pq}}}$ but independent of ε and ν. Since

$$\|\underline{u}e^{-\nu\|x\|}\|^2_{H^1(\Omega^\varepsilon_\infty)^d} = \|\underline{u}\|^2_{H^1_{-\nu}(\Omega^\varepsilon_\infty)^d} + \nu^2\|\underline{u}\|^2_{L^2_{-\nu}(\Omega^\varepsilon_\infty)^d}$$
$$- 2\nu\sum_{p=1}^{n}\left((\partial_{x_p}\underline{u})^\top\underline{u}, x_p/\|x\|e^{-2\nu\|x\|}\right)_{L^2(\Omega^\varepsilon_\infty)}$$
$$\geq \frac{1}{2}\|\underline{u}\|^2_{H^1_{-\nu}(\Omega^\varepsilon_\infty)^d} - \nu^2\|\underline{u}\|^2_{L^2_{-\nu}(\Omega^\varepsilon_\infty)^d},$$

we obtain finally the lower bound

$$\Psi(\varepsilon)[\underline{u}, \underline{v_u}] \geq \frac{\gamma_0}{2}\|\underline{u}\|^2_{H^1_{-\nu}(\Omega^\varepsilon_\infty)^d} - \left(C' + \frac{\gamma_0}{2}\right)\nu^2\|\underline{u}\|^2_{L^2_{-\nu}(\Omega^\varepsilon_\infty)^d}.$$

Taking $\nu_0 := \sqrt{\gamma_0/(4C' + 2\gamma_0)}$ it follows that $\Psi(\varepsilon)[\underline{u}, \underline{v_u}] \geq \gamma\|\underline{u}\|^2_{H^1_{-\nu}(\Omega^\varepsilon_\infty)}$, with $\gamma = \gamma_0/4$.

The proof of (45) follows analogously. First, by a density argument we may restrict ourselves only to $0 \neq \underline{v} \in C^\infty(\Omega^\varepsilon_\infty)^d$, $supp\,\underline{v} \subset\subset \mathbb{R}^n$. Then, the assertion follows if we take $\underline{u} = \underline{u_v} := \underline{v}e^{2\nu\|x\|} \in H^1_{-\nu}(\Omega^\varepsilon_\infty)^d$ and proceed as in the proof of (44).

We are interested now in the representation of the solution $\underline{u^\varepsilon}$ as Fourier-Bochner integral with respect to $H^1_{-\nu}(\Omega^\varepsilon_\infty)^d$ and in particular in the form of the integral kernel in this representation. To this end, we associate to our problem (41) a system of unit-cell problems with right hand side $\underline{e_k}$, $k = 1, \ldots, d$, the canonical basis in $\mathbb{R}^d$: for each $k = 1, \ldots, n$, find $\underline{\phi_k(y, \varepsilon, t)} \in H^1_{\text{per}}(\widehat{Q})^d$ such that

$$\mathcal{L}(\varepsilon, t; y, \partial_y)\underline{\phi_k}(y, \varepsilon, t) := e^{-i\varepsilon y\cdot t}L^\varepsilon(y, \varepsilon^{-1}\partial_y)e^{i\varepsilon y\cdot t}\underline{\phi_k}(y, \varepsilon, t) = \underline{e_k} \quad \text{in } \widehat{Q}$$
$$\sum_{p,q=1}^{n} n_p\underline{\underline{a^{pq}}}(y)(i\varepsilon t_q + \partial_{y_q})\underline{\phi_k}(y, \varepsilon, t) = 0 \quad \text{on } \widehat{\Gamma}_N := \partial\widehat{Q}\backslash\partial\widehat{Y}.$$

(46)

These unit-cell solutions are related to the solutions of

$$\Psi(\varepsilon)[\underline{\psi}_k^\varepsilon, \underline{v}] = \langle \underline{e_k} e^{it\cdot x}, \underline{v}\rangle_{(H^1_\nu(\Omega^\varepsilon_\infty)^d)^* \times H^1_\nu(\Omega^\varepsilon_\infty)^d}, \quad \forall \underline{v} \in H^1_\nu(\Omega^\varepsilon_\infty)^d \tag{47}$$

through

$$\underline{\psi}_k^\varepsilon(x) = \underline{\psi_k}\left(\frac{x}{\varepsilon}, \varepsilon, t\right) = e^{it\cdot x} \underline{\phi_k}\left(\frac{x}{\varepsilon}, \varepsilon, t\right) \quad k = 1, \ldots, n.$$

For $\nu \in (0, \nu_0)$, since $e^{it\cdot x} \in (H^1_\nu(\Omega^\varepsilon_\infty))^*$, the conditions (44), (45) ensure that (47) admits a unique weak solution $\underline{\psi_k}(\cdot/\varepsilon, \varepsilon, t) \in H^1_{-\nu}(\Omega^\varepsilon_\infty)^d$ and $\|\underline{\psi_k}(\cdot/\varepsilon, \varepsilon, t)\|_{H^1_{-\nu}(\Omega^\varepsilon_\infty)^d} \leq 1/\gamma \|e^{it\cdot x}\|_{(H^1_\nu(\Omega^\varepsilon_\infty))^*}$. Hence, for $\nu \in (0, \nu_0)$, $\|\underline{\psi_k}(\cdot/\varepsilon, \varepsilon, t)\|_{H^1_{-\nu}(\Omega^\varepsilon_\infty)^d} \leq 1/(\gamma \nu^{n/2})$ holds uniformly in $t \in \mathbb{R}^n$ and $\varepsilon > 0$.

Assume now that $\underline{f}(x) \in L^2(\mathbb{R}^n)^d$ is defined everywhere. Then, the solution $\underline{u}^\varepsilon \in H^1(\Omega^\varepsilon_\infty)^d \subset H^1_{-\nu}(\Omega^\varepsilon_\infty)^d$ admits the following Fourier-Bochner integral representation

$$\underline{u}^\varepsilon(x) = \overset{(B)}{\int_{\mathbb{R}^n}} e^{it\cdot x} \underline{\underline{\phi}}\left(\frac{x}{\varepsilon}, \varepsilon, t\right) \underline{\hat{f}}(t) dt, \tag{48}$$

where the matrix-valued integral kernel $\underline{\underline{\phi}}(y, \varepsilon, t)$ is related to the unit-cell solutions $\underline{\phi_k}(y, \varepsilon, t)$ through $(\underline{\underline{\phi}}^\top \underline{e_i}) \cdot \underline{e_k} = \underline{\phi_k} \cdot \underline{e_i}$, for all $i, k = 1, \ldots, d$.

Remark 3. In the case of linear elasticity systems $n = d \in \{2, 3\}$ and

$$a^{pq}_{jk}(y) = a^{qp}_{kj}(y) = a^{jq}_{pk}(y), \; \underline{\underline{a}}^0 = 0 \quad \forall p, q, j, k \in \{1, \ldots, n\}.$$

If $\Omega^\varepsilon_\infty$ is a perforated, infinite lattice and the cavities' boundaries are fixed, i.e., homogeneous Dirichlet boundary conditions on $\partial\Omega^\varepsilon_\infty$ for the deformation field $\underline{u}^\varepsilon$ are prescribed, the previous analysis applies as well. The case when $\underline{\underline{a}}^0 = 0$ and Neumann boundary conditions on $\partial\Omega^\varepsilon_\infty$ are prescribed is however not covered by the previous theory. For the equations of linear elasticity in a perforated lattice $\Omega^\varepsilon_\infty$ with homogeneous Dirichlet boundary conditions on the cavities' boundaries the Lamé operator takes the form

$$-\mu\Delta - (\lambda + \mu)\text{grad div},$$

i.e., $a^{pq}_{jk} = \lambda\delta_{jp}\delta_{kq} + \mu\delta_{jk}\delta_{pq} + \mu\delta_{jq}\delta_{kp}$, where λ, μ are the Lamé constants and δ denotes the Kronecker symbol.

4 Approximation of u^ε

In the previous section we represented u^ε as a Fourier-Bochner integral in exponentially weighted Sobolev spaces $H^1_{-\nu}(\Omega^\varepsilon_\infty)$ and we analyzed the properties of the integral kernel introduced as the inverse Fourier symbol of the elliptic operator. In this two-scale symbol, the dependence on the parameters $\varepsilon > 0$ and $t \in \mathbb{R}^n$ is *analytic*, even if the unit-cell problem admits only very low elliptic regularity. This analyticity is exploited in the construction of finite-dimensional approximations of the solution as well as in proving robust error estimates for these approximations.

4.1 Preliminaries

Asymptotic expansion in ε about $\varepsilon = 0$ gives classical homogenizations (outer expansion). Here, we present a different approach. To achieve robust (in ε) approximations, we approximate the Fourier-Bochner integral (17) by a finite sum via truncation of a (generalized) Poisson summation formula. We use for this Sinc approximations based on the *Cardinal function*. For $h > 0$ and $k \in \mathbb{Z}^n$, we define $S(k, h)$ by

$$S(k,h)(x) = \prod_{p=1}^{n} \frac{\sin[\pi(x_p - k_p h)/h]}{\pi(x_p - k_p h)/h}. \tag{49}$$

We shall refer to $S(k, h)$ as the k'th *Sinc* function, with step size h, evaluated at $x \in \mathbb{R}^n$. It holds, see, e.g., [20], Theorem 1.10.1:

Lemma 2.

$$S(k,h)(x) = \frac{1}{(2\pi)^n} \int\limits_{[-\pi/h,\pi/h]^n} h^n e^{it\cdot(kh-x)} dt, \tag{50}$$

and

$$\int\limits_{\mathbb{R}^n} S(k,h)(x) S(l,h)(x) dx = \frac{h^{2n}}{(2\pi)^n} \int\limits_{[-\pi/h,\pi/h]^n} e^{it\cdot h(k-l)} dt = h^n \delta_{k-l}. \tag{51}$$

For a function $f \in L^\infty(\mathbb{R}^n)$ the Cardinal function of f [20] is defined by

$$C(f,h)(x) = \sum_{k\in\mathbb{Z}^n} f(kh) S(k,h)(x).$$

The Cardinal function plays the same role for Sinc methods as polynomials play for the more classical numerical methods, in that operations such as differentiation, integration on Sinc approximations of functions produce approximations of these operations on the functions themselves.

Definition 2. Assume that X is a Banach space and let $\mathcal{D}_d$ denote the strip region of width d given by

$$\mathcal{D}_d = \{z \in \mathbb{C}^n \; : \; |\mathrm{Im}\,(z)| \leq d\}.$$

Then we define

$$\mathbf{H}^p(\mathcal{D}_d; X) = \{\, g : \mathcal{D}_d \to X \,|\, g \text{ is analytic in } \mathcal{D}_d \text{ and } \mathbf{N}_p(g, \mathcal{D}_d; X) < \infty \,\},$$

where

$$\mathbf{N}_p(g, \mathcal{D}_d; X) = \begin{cases} \lim\limits_{\delta \to 0+} \left(\displaystyle\int\limits_{\partial \mathcal{D}_d(\delta)} \|g(z)\|_X^p |dz| \right)^{1/p} &, \quad \text{if } 1 \le p < \infty \\ \lim\limits_{\delta \to 0+} \sup\limits_{z \in \mathcal{D}_d(\delta)} \|g(z)\|_X &, \quad \text{if } p = \infty, \end{cases}$$

and for $0 < \delta < 1$, $\mathcal{D}_d(\delta)$ is defined by

$$\mathcal{D}_d(\delta) = \{ z \in \mathbb{C}^n \;:\; |\mathrm{Re}\,(z)| < 1/\delta,\, |\mathrm{Im}\,(z)| < d(1-\delta)\}.$$

Definition 3. We say, a function f fulfils the 'usual assumptions', if $f \in L^2(\mathbb{R}^n)$, and its Fourier transformation can be extended to a holomorphic function $\hat{f}$ in the strip $\mathcal{D}_d$, with $d = \nu_n$ as in Theorem 4, satisfying the following growth condition:

$$|\hat{f}(z)| \le C(f) e^{-\alpha \|z\|}, \quad \forall z \in \mathcal{D}_d, \tag{52}$$

for some positive constants $C(f), \alpha > 0$.

A typical example of functions that satisfy the 'usual assumptions' is that of functions arising through convolution with a Gauss kernel. Let $f \in L^2(\mathbb{R}^n)$ be compactly supported. Without loss of generality we may assume that $\operatorname{supp} f \subset \{x \;:\; \|x\| \le 1\}$. Let further $\delta > 0$ and denote by ρ_δ the following Dirac sequence

$$\rho_\delta(y) := \delta^{-n/2} e^{-\|y\|^2/\delta}, \quad y \in \mathbb{R}^n.$$

Define $f_\delta := f * \rho_\delta$, i.e.,

$$f_\delta(x) = \delta^{-n/2} \int\limits_{\mathbb{R}^n} f(y) e^{-\|x-y\|^2/\delta} dy, \quad x \in \mathbb{R}^n. \tag{53}$$

Then, f_δ defined by (53) is an analytic, exponentially decaying function that satisfies the 'usual assumptions'. Indeed, the Fourier transform of f_δ is given by the product of the Fourier transforms of f and ρ_δ

$$\hat{f}_\delta(\xi) = \hat{f}(\xi)\hat{\rho}_\delta(\xi) = e^{-\delta\|\xi\|^2/4} \hat{f}(\xi)$$

which is an analytic function that decays exponentially in a complex strip neighborhood of $\mathbb{R}^n$.

The following theorem holds:

Theorem 5. *Under the 'usual assumptions' on f the mapping*

$$\mathcal{D}_d \ni t \to g(t,\cdot) = g_\varepsilon(t,\cdot) := \frac{1}{(2\pi)^{n/2}} \hat{f}(t)\phi\left(\frac{\cdot}{\varepsilon},\varepsilon,t\right) \in H^1_{-2\nu}(\Omega^\varepsilon_\infty) \tag{54}$$

is in $\mathbf{H}^p(\mathcal{D}_d; H^1_{-2\nu}(\Omega^\varepsilon_\infty))$. Moreover, there exist $C(\gamma,\nu,n)$ such that $g_\varepsilon(t,\cdot)$ satisfies the growth condition:

$$\|g_\varepsilon(t,\cdot)\|_{1,-2\nu} \le C(\gamma,\nu,n)C(f)\left(1+\frac{1}{\alpha}\right)e^{-\frac{\alpha}{2}\|t\|}, \quad \forall t \in \mathcal{D}_d, \tag{55}$$

where α and $C(f)$ are as in (52).

Proof. Strictly speaking, g in (54) depends on ε. However, all estimates which follow will be robust with respect to ε and we therefore do not write the dependence on ε explicitly.
From the 'usual assumptions' on f and from (40) it follows that there exists a positive constant $C = C(\gamma,\nu,n) > 0$ such that

$$\|g(t,\cdot)\|_{1,-2\nu} \le C(\gamma,\nu,n)C(f)(1+\|t\|)e^{-\alpha\|t\|}, \qquad \forall t \in \mathcal{D}_d. \tag{56}$$

It follows therefore that

$$\|g(t,\cdot)\|_{1,-2\nu} \le C(\gamma,\nu,n)C(f)\left(1+\frac{1}{\alpha}\right)e^{-\frac{\alpha}{2}\|t\|}, \qquad \forall t \in \mathcal{D}_d. \tag{57}$$

For $1 \le p < \infty$ we have that

$$\begin{aligned}
\mathbf{N}_p(g,\mathcal{D}_d; H^1_{-2\nu}(\Omega^\varepsilon_\infty)) &= \left(\int_{\partial\mathcal{D}_d} \|g(z,\cdot)\|^p_{1,-2\nu}|dz|\right)^{1/p} \\
&\le C(\gamma,\nu)C(f)\left(1+\frac{1}{\alpha}\right)\left(\int_{\partial\mathcal{D}_d} e^{-\frac{p\alpha}{2}\|z\|}|dz|\right)^{1/p} \\
&\le C(\gamma,\nu)C(f)\left(1+\frac{1}{\alpha}\right)\Bigg[n\left(\int_{|\operatorname{Im} z|=d} e^{-p\alpha|z|/(2\sqrt{n})}|dz|\right)\times \\
&\qquad \times\left(\int_{|\operatorname{Im} z|<d} e^{-p\alpha|z|/(2\sqrt{n})}dz\right)^{n-1}\Bigg]^{1/p} \\
&\le C(\gamma,\nu)C(f)\left(1+\frac{1}{\alpha}\right)\left(\frac{8n\sqrt{2n}}{\alpha p}\right)^{1/p} e^{-\alpha d/(2\sqrt{2n})}\times \\
&\qquad \times\left(\left(\frac{4\sqrt{2n}}{p\alpha}\right)^2\left(1-e^{-p\alpha d/(2\sqrt{2n})}\right)\right)^{(n-1)/p}.
\end{aligned}$$

The case $p = \infty$ can be treated analogously. □

4.2 Definition of the approximation of u^ε

Let $L \geq 1$ and assume in what follows that $h \leq \pi/(2L)$. We define

$$\mathbb{Z}^n_{(N)} := \{k \in \mathbb{Z}^n \ : \ |k_p| \leq N, \forall p = 1, \cdots, n\}$$

and we introduce the Cardinal function $C(g,h)(z,x)$ of g as a $H^1_{-2\nu}(\Omega^\varepsilon_\infty)$-valued functional in $\mathcal{D}_d$

$$C(g,h)(z,x) := \sum_{k \in \mathbb{Z}^n} g(kh,x) S(k,h)(z)$$

and its truncated series

$$C_N(g,h)(z,x) := \sum_{k \in \mathbb{Z}^n_{(N)}} g(kh,x) S(k,h)(z).$$

Set

$$E(f,h)(z,x) := g(z,x) - C(g,h)(z,x)$$

$$E_N(f,h)(z,x) := g(z,x) - C_N(g,h)(z,x)$$

in $H^1_{-2\nu}(\Omega^\varepsilon_\infty)$. Define $\delta(f,h)(\cdot)$, $\delta_N(f,h)(\cdot)$ as generalized inverse Fourier transforms with integral kernels $E(f,h)(t,x)$, $E_N(f,h)(t,x)$, respectively, which are given by

$$\delta(f,h)(x) = \lim_{\delta \to 0+} \int^{(B)}_{\mathbb{R}^n} \left(\prod_{p=1}^n e^{-\delta |t_p|} \right) e^{it \cdot x} E(f,h)(t,x) \, dt, \tag{58}$$

$$\delta_N(f,h)(x) = \lim_{\delta \to 0+} \int^{(B)}_{\mathbb{R}^n} \left(\prod_{p=1}^n e^{-\delta |t_p|} \right) e^{it \cdot x} E_N(f,h)(t,x) \, dt. \tag{59}$$

We will show that the above definitions make sense and that the limits in (58) and (59) are well defined as Bochner integrals of $H^1_{-2\nu}((-L,L)^n \cap \Omega^\varepsilon_\infty)$, respectively $H^0(\Omega^\varepsilon_\infty)$-valued functions.

By the properties of the *Sinc* functions $S(k,h)(\cdot)$ in Lemma 2 we will see that

$$\delta_N(f,h)(x) = \lim_{\delta \to 0+} \int^{(B)}_{\mathbb{R}^n} \left(\prod_{p=1}^n e^{-\delta |t_p|} \right) e^{it \cdot x} \left\{ g(t,x) - \sum_{k \in \mathbb{Z}^n_{(N)}} g(kh,x) S(k,h)(t) \right\} dt \tag{60}$$

$$= \begin{cases} \displaystyle\int^{(B)}_{\mathbb{R}^n} e^{it \cdot x} g(t,x) \, dt - h^n \sum_{k \in \mathbb{Z}^n_{(N)}} g(kh,x) e^{ikh \cdot x} & , \quad \text{if } x \in \left[-\frac{\pi}{h}, \frac{\pi}{h} \right]^n, \\ \displaystyle\int_{\mathbb{R}^n} e^{it \cdot x} g(t,x) \, dt & , \quad \text{else} \end{cases}$$

in $H^0_{-2\nu}(\Omega^\varepsilon_\infty) \cap H^1_{-2\nu}((-L,L)^n \cap \Omega^\varepsilon_\infty)$. Therefore, we define approximations $u^\varepsilon_{N,h}$ of u^ε by

$$\begin{aligned} u^\varepsilon_{N,h}(x) &= \mathbf{1}_{[-\frac{\pi}{h},\frac{\pi}{h}]^n}(x)\frac{1}{(2\pi)^{n/2}}h^n \sum_{k\in\mathbb{Z}^n_{(N)}} \phi\left(\frac{x}{\varepsilon},\varepsilon,kh\right)\hat{f}(kh)e^{ikh\cdot x} \\ &= \mathbf{1}_{[-\frac{\pi}{h},\frac{\pi}{h}]^n}(x)\frac{1}{(2\pi)^{n/2}}h^n \sum_{k\in\mathbb{Z}^n_{(N)}} \psi\left(\frac{x}{\varepsilon},\varepsilon,kh\right)\hat{f}(kh). \end{aligned} \tag{61}$$

Then, the solution of (17) can be split as

$$u^\varepsilon(x) = u^\varepsilon_{N,h}(x) + \delta_N(f,h)(x).$$

Remark 4. Since we have real valued input data we have a real valued solution u^ε. The same holds for the finite approximation $u^\varepsilon_{N,h}$. Indeed, since

$$\hat{f}(-\xi) = \overline{\hat{f}(\xi)}, \quad \psi\left(\frac{\cdot}{\varepsilon},\varepsilon,-\xi\right) = \overline{\psi\left(\frac{\cdot}{\varepsilon},\varepsilon,\xi\right)},$$

it follows that

$$\begin{aligned} u^\varepsilon_{N,h}(x) &= \mathbf{1}_{[-\frac{\pi}{h},\frac{\pi}{h}]^n}(x)\frac{1}{(2\pi)^{n/2}}h^n\psi\left(\frac{x}{\varepsilon},\varepsilon,0\right)\hat{f}(0) + 2\,\mathbf{1}_{[-\frac{\pi}{h},\frac{\pi}{h}]^n}(x)\frac{1}{(2\pi)^{n/2}}h^n \\ &\times \sum_{k\in\mathbb{Z}^n_{(N)}\setminus\{0\},\,k_1\geq 0} \left\{\operatorname{Re}\psi\left(\frac{x}{\varepsilon},\varepsilon,kh\right)\operatorname{Re}\hat{f}(kh) - \operatorname{Im}\psi\left(\frac{x}{\varepsilon},\varepsilon,kh\right)\operatorname{Im}\hat{f}(kh)\right\}, \end{aligned}$$

which is real valued.

4.3 Exponential convergence

We will now show that (61) approximates u^ε in (17) at a robust exponential rate, i.e., independent of ε. Our main result on the trapezoidal approximation $u^\varepsilon_{N,h}(x)$ of the Fourier-Bochner integral (17) is

Theorem 6. *Under the assumptions in Theorem 4, the error $\delta_N(f,h)(\cdot) = u^\varepsilon(\cdot) - u^\varepsilon_{N,h}(\cdot)$, with $u^\varepsilon_{N,h}(\cdot)$ as in (61), decays exponentially with respect to N and uniformly with respect to ε in the $\|\cdot\|_{0,-2\nu}$, $\|\cdot\|_{H^1_{-2\nu}((-L,L)^n\cap\Omega^\varepsilon_\infty)}$-norms:*

$$\begin{aligned} &\|\delta_N(f,h)(\cdot)\|_{0,-2\nu} + \|\delta_N(f,h)(\cdot)\|_{H^1_{-2\nu}((-L,L)^n\cap\Omega^\varepsilon_\infty)} \\ &\quad\leq C(\gamma,\nu,n)C(f)\left(1+\frac{1}{\alpha}\right)^2\frac{1}{\alpha^n}e^{-(\pi\alpha dN/n)^{1/2}}. \end{aligned} \tag{62}$$

The constants $C(\gamma,\nu,n), C(f)$ are independent of ε, N, L.

We start with the following preliminary result:

Lemma 3. *Assume that f satisfies the 'usual assumptions'. Let $g(t,\cdot)$ be as in Definition 3 and let $z \in \mathcal{D}_d$ be arbitrary, fixed. Then $E(f,h)(z,\cdot) \in H^1_{-2\nu}(\Omega^\varepsilon_\infty)$ can be represented in the following form*

$$E(f,h)(z,\cdot) := g(z,\cdot) - \sum_{k\in\mathbb{Z}^n} S(k,h)(z)g(kh,\cdot)$$

$$= \sum_{p=1}^{n}\left\{ \sum_{k_1,\dots,k_{p-1}\in\mathbb{Z}} S((k_1,\dots,k_{p-1}),h)(z_1,\dots,z_{p-1})\frac{\sin(\pi z_p/h)}{2\pi i}\right.$$

$$\stackrel{(B)}{\int_{\mathbb{R}}} \frac{g((k_1h,\dots,k_{p-1}h,\tau - id^-,z_{p+1},\dots,z_n),\cdot)}{(\tau - z_p - id)\sin[\pi(\tau - id)/h]}$$

$$\left. - \frac{g((k_1h,\dots,k_{p-1}h,\tau + id^-,z_{p+1},\dots,z_n),\cdot)}{(\tau - z_p + id)\sin[\pi(\tau + id)/h]}\,d\tau\right\}.$$

Here the integrals have to be understood as Bochner-integrals of $H^1_{-2\nu}(\Omega^\varepsilon_\infty)$-valued functions, and the infinite sums as absolutely convergent series in the Banach space $H^1_{-2\nu}(\Omega^\varepsilon_\infty)$.

Proof. We just sketch the proof which is similar to that in the 1-d case given in [13]. Let $0 < \delta < d$, let m denote a positive integer and let $\mathcal{D}(m,\delta)$ denote the region

$$\mathcal{D}(m,\delta) = \left\{ z \in \mathbb{C} \,|\, |\mathrm{Re}\, z| < \left(m + \frac{1}{2}\right)h,\ |\mathrm{Im}\, z| < \delta \right\}.$$

Taking advantage of the exponential decay of the $H^1_{-2\nu}(\Omega^\varepsilon_\infty)$- norm of $g(z,\cdot)$ in $\mathcal{D}_d$, we deduce that $E(f,h)(t,\cdot)$ is the sum of the following absolutely convergent series in the Banach space $H^1_{-2\nu}(\Omega^\varepsilon_\infty)$:

$$\begin{aligned} E(f,h)(z,\cdot) &= g(z,\cdot) - \sum_{k\in\mathbb{Z}^n} S(k,h)(z)g(kh,\cdot) \\ &= \sum_{p=1}^{n}\left\{ \sum_{k_1,\dots,k_{p-1}\in\mathbb{Z}} S((k_1,\dots,k_{p-1}),h)(z_1,\dots,z_{p-1})\right. \\ &\times \Bigg[g((k_1h,\dots,k_{p-1}h,z_p,\dots,z_n),\cdot) \\ &\left. - \sum_{k_p\in\mathbb{Z}} S(k_p,h)(z_p)g((k_1h,\dots,k_{p-1}h,k_ph,z_{p+1},\dots,z_n),\cdot)\Bigg]\right\}. \end{aligned} \tag{63}$$

One can see that $E(f,h)(z,\cdot)$, for $m \to \infty$, is approximated by

$$g(z,\cdot) - \sum_{k\in\mathbb{Z}^n_{(m)}} S(k,h)(z)g(kh,\cdot) =$$

$$= \sum_{p=1}^{n} \Bigg\{ \sum_{(k_1,\ldots,k_{p-1})\in\mathbb{Z}_{(m)}^{p-1}} S((k_1,\ldots,k_{p-1}),h)(z_1,\ldots,z_{p-1})$$

$$\times \Bigg[g((k_1 h,\ldots,k_{p-1}h, z_p,\ldots,z_n),\cdot) \tag{64}$$

$$- \sum_{k_p\in\mathbb{Z}_{(m)}} S(k_p,h)(z_p) g((k_1 h,\ldots,k_{p-1}h,k_p h, z_{p+1},\ldots,z_n),\cdot)\Bigg]\Bigg\}.$$

Since $z \in \mathcal{D}_d$, one can choose $0 < \delta < d$ sufficiently close to d and $m \in \mathbb{N}$ sufficiently large such that $z_p \in \mathcal{D}(m,\delta)$, $\forall\, p = 1,\ldots,n$. Then, each of the terms in (64) can be represented as the following Bochner integral on the boundary $\partial\mathcal{D}(m,\delta)$

$$S((k_1,\ldots,k_{p-1}),h)(z_1,\ldots,z_{p-1})\Bigg\{ g((k_1 h,\ldots,k_{p-1}h,z_p,\ldots,z_n),\cdot)$$

$$- \sum_{k_p\in\mathbb{Z}_{(m)}} S(k_p,h)(z_p) g((k_1 h,\ldots,k_{p-1}h,k_p h,z_{p+1},\ldots,z_n),\cdot)\Bigg\} =$$

$$= S((k_1,\ldots,k_{p-1}),h)(z_1,\ldots,z_{p-1})$$

$$\left\{ \frac{\sin(\pi z_p/h)}{2\pi i} \overset{(B)}{\int\limits_{\partial\mathcal{D}(m,\delta)}} \frac{g((k_1 h,\ldots,k_{p-1}h,\zeta,z_{p+1},\ldots,z_n),\cdot)}{(\zeta - z_p)\sin(\pi\zeta/h)} d\zeta \right\}. \tag{65}$$

Taking now the limit $m \to \infty$ in both sides of (65) the lemma follows. □

Remark 5. We do not actually need the strong 'usual assumptions' on f to deduce the above integral representation for $E(f,h)(z,\cdot)$ for z in the strip $\mathcal{D}_d$. These assumptions on f just imply that the integrand g defined in (54) is in $\mathbf{H}^p(\mathcal{D}_d; H^1_{-2\nu}(\Omega^\varepsilon_\infty))$, for every $1 \le p \le \infty$, as shown in Theorem 5. For the proof of Lemma 3 it is sufficient to know that $g \in \mathbf{H}^p(\mathcal{D}_d; H^1_{-2\nu}(\Omega^\varepsilon_\infty))$, for some $1 \le p \le \infty$, and such a property on g holds under more general hypothesis on f than the 'usual assumptions', as e.g., $f \in H^s_{\text{comp}}(\mathbb{R}^n)$ for some $s > 1/2$. In this case $g \in \mathbf{H}^\infty(\mathcal{D}_d; H^1_{-2\nu}(\Omega^\varepsilon_\infty))$ and again the integral representation in Lemma 3 is valid.

Let us denote by f_1 and f_2 the functions given by

$$f_1(t,x) = \begin{cases} e^{xd+ixt} & , \quad \text{if } x < -\dfrac{\pi}{h} \\[2ex] \dfrac{i}{2} \dfrac{e^{\left(x-\frac{\pi}{h}\right)(d+it)}}{\sin[\pi(t-id)/h]} & , \quad \text{if } -\dfrac{\pi}{h} < x < \dfrac{\pi}{h} \\[2ex] 0 & , \quad \text{if } x > \dfrac{\pi}{h}, \end{cases} \tag{66}$$

and

$$f_2(t,x) = \begin{cases} 0 & , \quad \text{if } x < -\dfrac{\pi}{h} \\ -\dfrac{i}{2}\dfrac{e^{-(x+\frac{\pi}{h})(d-it)}}{\sin[\pi(t+id)/h]} & , \quad \text{if } -\dfrac{\pi}{h} < x < \dfrac{\pi}{h} \\ e^{-xd+ixt} & , \quad \text{if } x > \dfrac{\pi}{h}. \end{cases} \tag{67}$$

Lemma 4. *Let f satisfy the 'usual assumptions' in Definition 3 with some $\alpha, d > 0$ and let $L > 0$ be arbitrary. Define*

$$h = \left(\frac{\pi d}{\alpha N}\right)^{1/2} \tag{68}$$

and assume $N \geq 4dL^2/(\alpha\pi)$, i.e., $\pi/h \geq 2L$. Then, with $E(f,h)(t,\cdot)$ as in Lemma 3 we have the following representation of the remainder $\delta(f,h)(\cdot)$

$$\delta(f,h)(x) =$$

$$\sum_{p=1}^{n} \sum_{k_1,\ldots,k_{p-1}\in\mathbb{Z}} \overset{(B)}{\int_{\mathbb{R}^{n-p}}} \overset{(B)}{\int_{\mathbb{R}}} h^{p-1}\left(\prod_{q=1}^{p-1} e^{ik_q h x_q}\mathbf{1}_{[-\frac{\pi}{h},\frac{\pi}{h}]}(x_q)\right)\left(\prod_{q>p} e^{it_q x_q}\right)$$

$$\Big\{ g((k_1h,\ldots,k_{p-1}h,\tau - id^-, t_{p+1},\ldots,t_n),x) f_1(\tau,x_p)$$

$$+ g((k_1h,\ldots,k_{p-1}h,\tau + id^-, t_{p+1},\ldots,t_n),x) f_2(\tau,x_p)\Big\}\, d\tau\, dt_{p+1}\ldots dt_n,$$

where the integral kernels $f_1(\tau,\cdot)$ and $f_2(\tau,\cdot)$ are as in (66), (67). Moreover, there exists a constant $C = C(\gamma,\nu,n)C(f)\left(1+1/\alpha\right)^2(1/\alpha)^n > 0$, which depends on f, α, d, γ, but is independent of N, such that

$$\|\delta(f,h)(\cdot)\|_{0,-2\nu} + \|\delta(f,h)(\cdot)\|_{H^1_{-2\nu}((-L,L)^n\cap\Omega^{\varepsilon}_{\infty})} \leq Ce^{-(\pi d\alpha N)^{1/2}}$$

holds.

For the proof of this lemma we refer to the Appendix below.

Proof of Theorem 6. From the definitions (58) and (59) of $\delta(f,h)(\cdot)$ and $\delta_N(f,h)(\cdot)$ and the properties of the *Sinc* functions $S(k,h)(\cdot)$ in Lemma 2 it follows that

$$\delta_N(f,h)(x) =$$

$$= \delta(f,h)(x) + \sum_{k\in\mathbb{Z}^n\setminus\mathbb{Z}^n_{(N)}} \lim_{\delta\to 0+} \overset{(B)}{\int_{\mathbb{R}^n}} \left(\prod_{p=1}^{n} e^{-\delta|t_p|}\right) e^{it\cdot x} g(kh,x) S(k,h)(t)\, dt$$

$$= \delta(f,h)(\cdot) + \frac{1}{(2\pi)^{n/2}}\mathbf{1}_{[-\frac{\pi}{h},\frac{\pi}{h}]^n}(x) \sum_{k\in\mathbb{Z}^n\setminus\mathbb{Z}^n_{(N)}} h^n \hat{f}(kh)\psi\left(\frac{x}{\varepsilon},\varepsilon,kh\right), \tag{69}$$

in $H^0_{-2\nu}(\Omega^\varepsilon_\infty)$, respectively

$$\delta_N(f,h)(x) = \delta(f,h)(x) + \frac{1}{(2\pi)^{n/2}} \sum_{k\in\mathbb{Z}^n\setminus\mathbb{Z}^n_{(N)}} h^n \hat{f}(kh)\psi\left(\frac{x}{\varepsilon},\varepsilon,kh\right), \tag{70}$$

in $H^1_{-2\nu}((-L,L)^n \cap \Omega^\varepsilon_\infty)$. Since

$$\|\psi(\cdot/\varepsilon,\varepsilon,t)\|_{1,-2\nu} \le C(\gamma,\nu,n), \quad \forall t \in \mathcal{D}_d \subset \mathbb{C}^n,$$

for some positive constant $C = C(\gamma,\nu,n) > 0$, it follows that

$$\begin{aligned}\|\delta_N(f,h)(\cdot)\|_{0,-2\nu} &\le \|\delta(f,h)(\cdot)\|_{0,-2\nu} \\ &\quad + \frac{1}{(2\pi)^{n/2}} \sum_{k\in\mathbb{Z}^n\setminus\mathbb{Z}^n_{(N)}} h^n|\hat{f}(kh)| \left\|\psi\left(\frac{\cdot}{\varepsilon},\varepsilon,kh\right)\right\|_{0,-2\nu} \\ &\le \|\delta(f,h)(\cdot)\|_{0,-2\nu} + C(\gamma,\nu,n) \sum_{k\in\mathbb{Z}^n\setminus\mathbb{Z}^n_{(N)}} h^n|\hat{f}(kh)|\end{aligned}$$

and

$$\begin{aligned}\|\delta_N(f,h)(\cdot)\|_{H^1_{-2\nu}((-L,L)^n\cap\Omega^\varepsilon_\infty)} &\le \|\delta(f,h)(\cdot)\|_{H^1_{-2\nu}((-L,L)^n\cap\Omega^\varepsilon_\infty)} \\ &\quad + \frac{1}{(2\pi)^{n/2}} \sum_{k\in\mathbb{Z}^n\setminus\mathbb{Z}^n_{(N)}} h^n|\hat{f}(kh)| \left\|\psi\left(\frac{\cdot}{\varepsilon},\varepsilon,kh\right)\right\|_{1,-2\nu} \\ &\le \|\delta(f,h)(\cdot)\|_{H^1_{-2\nu}((-L,L)^n\cap\Omega^\varepsilon_\infty)} + C(\gamma,\nu,n) \sum_{k\in\mathbb{Z}^n\setminus\mathbb{Z}^n_{(N)}} h^n|\hat{f}(kh)|.\end{aligned}$$

Since $|\hat{f}(kh)| \le C(f)e^{-\alpha\|k\|h}$,

$$\begin{aligned}h^n \sum_{k\in\mathbb{Z}^n\setminus\mathbb{Z}^n_{(N)}} |\hat{f}(kh)| &\le n2^n\, C(f) h^n\, e^{-\alpha(N+1)h/\sqrt{n}} \tfrac{1}{(1-e^{-\alpha h/\sqrt{n}})^n} \\ &\le n2^n\, C(f) \frac{1}{(\alpha/\sqrt{n})^n} e^{-\alpha N h/\sqrt{n}},\end{aligned}$$

which implies with our choice of h that the sum in (69) satisfies the estimate (62). It remains to show that $\|\delta(f,h)(\cdot)\|_{0,-2\nu} + \|\delta(f,h)(\cdot)\|_{H^1_{-2\nu}((-L,L)^n\cap\Omega^\varepsilon_\infty)}$ satisfies (62), and this is just the statement of Lemma 4. □

As a corollary, the following approximation result holds

Corollary 2. *Let us assume that f satisfies the 'usual assumptions' and that*

$$h = \left(\frac{\pi d}{\alpha N}\right)^{1/2}, \quad N \ge \frac{4dL^2}{\alpha\pi}. \tag{71}$$

Let

$$\mathcal{V}^N_\varepsilon := \text{Span}\left\{\text{Re}\,\psi\left(\frac{\cdot}{\varepsilon},\varepsilon,kh\right),\ \text{Im}\,\psi\left(\frac{\cdot}{\varepsilon},\varepsilon,kh\right)\ :\ k \in \mathbb{Z}^n_{(N)}\right\}. \tag{72}$$

Then

$$\inf_{v\in\mathcal{V}^N_\varepsilon} \|u^\varepsilon - v\|_{H^1_{-2\nu}((-L,L)^n\cap\Omega^\varepsilon_\infty)} + \inf_{v\in\mathcal{W}^N_\varepsilon} \|u^\varepsilon - v\|_{H^0_{-2\nu}(\Omega^\varepsilon_\infty)} \\ \leq C(\gamma,\nu,n)C(f)\left(1+\frac{1}{\alpha}\right)^2 \frac{1}{\alpha^n} e^{-(\pi\alpha dN/n)^{1/2}}, \tag{73}$$

where $C(f)$ and $\alpha = \alpha(f)$ are those from Definition 3, in particular independent of ε, N.

4.4 Spectral convergence

The usual assumptions on the data f entail essentially boundedness of all derivatives. In this section we assume that $f(x) \in H^s(\mathbb{R}^n)$ in (10) has finite regularity ($s > n/2$) and is compactly supported, i.e., $f \in H^s_{\text{comp}}(\mathbb{R}^n)$. We will show that for any $\varepsilon > 0$ the solution $u^\varepsilon(x)$ of (17) can be approximated by

$$u^\varepsilon_N \in \mathcal{V}^N_\varepsilon = \text{Span}\,\{\text{Re}\,\psi(\cdot/\varepsilon,\varepsilon,kh),\ \text{Im}\,\psi(\cdot/\varepsilon,\varepsilon,kh)\}_{k\in\mathbb{Z}^n_{(N)}}$$

with respect to $\|\cdot\|_{1,-\nu}$ at an algebraic convergence rate independent of ε.

Theorem 7. *Assume that f in (10) is $H^s(\mathbb{R}^n)$ with $s > n/2$ and $\text{supp}\, f \subset \Omega_M := (-M,M)^n$, with $M > 0$. Let $d := \min\{1/M, \nu_n\}$, $\nu_n = 1/(2\sqrt{n})$, and $N \geq 1$. Then, for any $\varepsilon > 0$ and for all N*

$$\inf_{v\in\mathcal{W}^N_\varepsilon} \|u^\varepsilon - v\|_{1,-\nu} \leq C_{\nu,s,n} N^{-(s-n/2)/2} \|f\|_{H^s(\mathbb{R}^n)}, \tag{74}$$

where $C_{\nu,s,n} > 0$ is independent of ε, N, and M.

Proof. By a density argument it can be assumed that $f \in C^\infty_0(\Omega_M)$. It is known then that the Fourier transform $\hat{f}$ of f can be continued analytically in $\mathbb{C}^n$ and $\hat{f}$ is uniformly bounded in a strip of width $1/M$. Therefore, the integrand in the Bochner integral representation (17) of u^ε is analytic in a strip with values in the Banach space $H^1_{-\nu}(\Omega^\varepsilon_\infty)$

$$g(t,\cdot) = \frac{1}{(2\pi)^{n/2}} \hat{f}(t)\psi\left(\frac{\cdot}{\varepsilon},\varepsilon,t\right) \in \mathcal{A}\left(\mathcal{D}_{\min\{1/M,\nu_n\}}; H^1_{-\nu}(\Omega^\varepsilon_\infty)\right).$$

Define $d = \min\{1/M, \nu_n\}$, and let $h = \sqrt{d/N}$. Let us split the solution u^ε again as

$$u^\varepsilon(x) = \sum_{k\in\mathbb{Z}^n_{(N)}} \frac{1}{(2\pi)^{n/2}} h^n \hat{f}(kh)\psi\left(\frac{x}{\varepsilon},\varepsilon,kh\right) \\ + \sum_{k\in\mathbb{Z}^n\setminus\mathbb{Z}^n_{(N)}} \frac{1}{(2\pi)^{n/2}} h^n \hat{f}(kh)\psi\left(\frac{x}{\varepsilon},\varepsilon,kh\right) \\ + \int_{\mathbb{R}^n}^{(B)} g(t,x)\,dt - h^n \sum_{k\in\mathbb{Z}^n} g(kh,x). \tag{75}$$

We define u_N^ε as the first sum in the right hand side of (75). It has been shown in Theorem 3 that $\psi(x/\varepsilon, \varepsilon, t)$ is analytic in the strip $\mathcal{D}_{\nu_n}$ with values in $H^1_{-\nu}(\Omega^\varepsilon_\infty)$ and the $\|\cdot\|_{1,-\nu}$-norm of $\psi(\cdot/\varepsilon, \varepsilon, t)$ is uniformly bounded in $\mathcal{D}_{\nu_n}$. Moreover, the norm $\|\psi(x/\varepsilon, \varepsilon, t)\|_{L^\infty(\mathcal{D}_{\nu_n}; H^1_{-\nu}(\Omega^\varepsilon_\infty))}$ is bounded uniformly with respect to ε. Hence, the second sum in (75) satisfies the estimate in (74)

$$
\begin{aligned}
&\left\| \sum_{k\in\mathbb{Z}^n\setminus\mathbb{Z}^n_{(N)}} \frac{1}{(2\pi)^{n/2}} h^n \hat{f}(kh)\psi\left(\frac{x}{\varepsilon}, \varepsilon, kh\right) \right\|_{1,-\nu} \leq \\
&\leq C_{\nu,n} \left(\sum_{k\in\mathbb{Z}^n\setminus\mathbb{Z}^n_{(N)}} h^n |\hat{f}(kh)|^2 (1+\|kh\|^2)^s \right)^{1/2} \left(\sum_{k\in\mathbb{Z}^n\setminus\mathbb{Z}^n_{(N)}} h^n \|kh\|^{-2s} \right)^{1/2} \\
&\leq C_{\nu,n}\, \|f\|_{H^s(\mathbb{R}^n)} h^{(n-2s)/2} \left(n2^n n^s \sum_{\substack{k_1\geq N \\ k_j\geq 0,\ \forall j\geq 2}} (k_1+\ldots+k_n)^{-2s} \right)^{1/2} \\
&= C_{\nu,n,s} \|f\|_{H^s(\mathbb{R}^n)} h^{(n-2s)/2} \left(\frac{N^{-2s+n}}{(2s-1)\cdot\ldots\cdot(2s-n)} \right)^{1/2} \\
&\leq C_{\nu,n,s} \|f\|_{H^s(\mathbb{R}^n)} \left(\sqrt{d}N\right)^{-(s-n/2)} .
\end{aligned}
$$

It remains to find a similar bound for the remainder in (75). Let us introduce the following notations: for each $k \in \mathbb{Z}^n$, $p = 1, \ldots, n$, we denote by $k_{(p)} := (k_1, \ldots, k_{p-1})$ and $k^c_{(p)} := (k_{p+1}, \ldots, k_n)$. Analogously, for $t \in \mathbb{R}^n$, $t_{(p)} := (t_1, \ldots, t_{p-1})$, $t^c_{(p)} := (t_{p+1}, \ldots, t_n)$. Further, we use the notation $\mathcal{F}_{(p)} f$ to denote the partial Fourier transform $\mathcal{F}_{x_{(p)} \to t_{(p)}} f$ of f with respect to $x_{(p)} \in \mathbb{R}^{p-1}$, respectively $\mathcal{F}^c_{(p)} f$ for the partial Fourier transform $\mathcal{F}_{x_{(p)^c} \to t^c_{(p)}} f$ of f in $x^c_{(p)} \in \mathbb{R}^{n-p}$. Let

$$
\delta(g) := \int_{\mathbb{R}^n}^{(B)} g(t,x)\, dt - \sum_{k\in\mathbb{Z}^n} h^n\, g(kh, x).
$$

We can write $\delta(g)$ as

$$
\delta(g) = \int_{\mathbb{R}^n}^{(B)} \left\{ g(t,x) - \sum_{k\in\mathbb{Z}^n} g(kh,x) S(k,h)(t) \right\} dt. \tag{76}
$$

The integrand in (76) is well defined as an $H^1_{-\nu}(\Omega^\varepsilon_\infty)$-valued function and it can be writen in the following form

$$
g(t,\cdot) - \sum_{k\in\mathbb{Z}^n} g(kh,\cdot) S(k,h)(t) =
$$

$$= \sum_{p=1}^{n} \sum_{k_{(p)} \in \mathbb{Z}^{p-1}} S(k_{(p)}, h)(t_{(p)}) \tag{77}$$

$$\times \left(g((k_{(p)}h, t_p, t^c_{(p)}), \cdot) - \sum_{k_p \in \mathbb{Z}} S(k_p, h)(t_p) g((k_{(p)}h, k_p h, t^c_{(p)}), \cdot) \right),$$

the series in (77) being absolutely convergent with respect to the Banach space topology of $H^1_{-\nu}(\Omega^\varepsilon_\infty)$. Indeed, taking the $H^1_{-\nu}(\Omega^\varepsilon_\infty)$-norm of each term in (77) we can bound the resulting series by

$$C_{\nu,n} \sum_{p=1}^{n+1} h^{-(p-1)} \left(\sum_{k_{(p)} \in \mathbb{Z}^{p-1}} h^{p-1} (1 + \|k_{(p)}h\|^2)^{-s} \right)^{1/2}$$

$$\times \left(\sum_{k_{(p)} \in \mathbb{Z}^{p-1}} h^{p-1} (1 + \|k_{(p)}h\|^2)^{s} |\hat{f}(k_{(p)}h, t_p, t^c_{(p)})|^2 \right)^{1/2}$$

$$\leq C_{\nu,n,s} \sum_{p=1}^{n+1} M^{(n-p+1)/2} h^{-(p-1)} \|f\|_{H^s(\mathbb{R}^n)}.$$

Further, for each fixed $k_{(p)} \in \mathbb{Z}^{p-1}$ and $t^c_{(p)} \in \mathbb{R}^{n-p}$ (see Remark 5),

$$g((k_{(p)}h, t_p, t^c_{(p)}), \cdot) - \sum_{k_p \in \mathbb{Z}} S(k_p, h)(t_p) g((k_{(p)}h, k_p h, t^c_{(p)}), \cdot)$$

$$= \frac{\sin(\pi t_p/h)}{2\pi i} \int_{\mathbb{R}}^{(B)} \frac{g((k_{(p)}h, \tau - id^-, t^c_{(p)}), \cdot)}{(\tau - t_p - id) \sin[\pi(\tau - id)/h]}$$

$$- \frac{g((k_{(p)}h, \tau + id^-, t^c_{(p)}), \cdot)}{(\tau - t_p + id) \sin[\pi(\tau + id)/h]} \, d\tau.$$

Summing up, the integrand in (76) can be written now in the following form

$$g(t, \cdot) - \sum_{k \in \mathbb{Z}^n} g(kh, \cdot) S(k, h)(t)$$

$$= \sum_{p=1}^{n} \left\{ \sum_{k_{(p)} \in \mathbb{Z}^{p-1}} S(k_{(p)}, h)(t_{(p)}) \frac{\sin(\pi t_p/h)}{2\pi i} \int_{\mathbb{R}}^{(B)} \frac{g((k_{(p)}h, \tau - id^-, t^c_{(p)}), \cdot)}{(\tau - t_p - id) \sin[\pi(\tau - id)/h]} \right.$$

$$\left. - \frac{g((k_{(p)}h, \tau + id^-, t^c_{(p)}), \cdot)}{(\tau - t_p + id) \sin[\pi(\tau + id)/h]} \, d\tau \right\}. \tag{78}$$

Substituting (78) in (76), changing the order of integration and integrating with respect to $t_1, \ldots t_{p-1}$ first, we obtain an equivalent integral representation for $\delta(g)$

$$\delta(g) = \sum_{p=1}^{n} \sum_{k_1,\dots,k_{p-1}\in\mathbb{Z}} \int_{\mathbb{R}^{n-p}}^{(B)} \int_{\mathbb{R}}^{(B)} h^{p-1}$$
$$\times\left\{ \frac{i}{2} \frac{e^{-\pi/h(d+i\tau)}}{\sin[\pi(\tau - id)/h]} g((k_1 h, \dots, k_{p-1}h, \tau - id^-, t_{p+1}, \dots, t_n)\cdot) \right. \quad (79)$$
$$\left. - \frac{i}{2} \frac{e^{-\pi/h(d-i\tau)}}{\sin[\pi(\tau + id)/h]} g((k_1 h, \dots, k_{p-1}h, \tau + id^-, t_{p+1}, \dots, t_n), \cdot) \right\}$$
$$\times\, d\tau dt_{p+1} \dots dt_n.$$

The estimate for the $\|\cdot\|_{1,-\nu}$-norm of the remainder $\delta(g)$ follows now by systematically exploiting the H^s regularity of f $(s > n/2)$

$$\|\delta(g)\|_{1,-\nu} \le C_{\nu,n} e^{-\pi d/h} \sum_{p=1}^{n} \left(\sum_{k_{(p)}\in\mathbb{Z}^{p-1}} h^{p-1}(1 + \|k_{(p)}h\|^2)^{-s} \right)^{1/2}$$
$$\times \left\{ \sum_{k_{(p)}\in\mathbb{Z}^{p-1}} h^{p-1}(1 + \|k_{(p)}h\|^2)^s \left(\int_{\mathbb{R}^{n-p}\times\mathbb{R}}^{(B)} |\hat{f}(k_{(p)}h, \tau \pm id^-, t^c_{(p)})| d\tau dt^c_{(p)} \right)^2 \right\}^{1/2}$$
$$\le C_{\nu,n} e^{-\pi d/h} \sum_{p=1}^{n} \left(\int_{\mathbb{R}^{p-1}} (1 + \|t_{(p)}\|^2)^{-s} dt_{(p)} \right)^{1/2}$$
$$\times \left\{ \int_{\mathbb{R}^{p-1}} (1 + \|t_{(p)}\|^2)^s \left(\int_{\mathbb{R}^{n-p}\times\mathbb{R}} |\hat{f}(t_{(p)}, \tau \pm id^-, t^c_{(p)})| d\tau dt^c_{(p)} \right)^2 \right\}^{1/2}. \quad (80)$$

At this point we use the following observation: assuming that one has $\phi \in H^r_{\text{comp}}(-M, M)$, $d < 1/M$ and $r > 1/2$, then for arbitrary $\tau \in \mathbb{R}$, $\hat{\phi}(\tau \pm id) = (\widehat{e^{\pm xd}\phi})(\tau)$. Hence,

$$\int_{\mathbb{R}} |\hat{\phi}(\tau \pm id)| d\tau \le \left(\int_{\mathbb{R}} (1 + |\tau \pm id|^2)^{-r} d\tau \right)^{1/2}$$
$$\times \left(\int_{\mathbb{R}} (1 + |\tau \pm id|^2)^r |\widehat{e^{\pm xd}\phi}(\tau)| d\tau \right)^{1/2}$$
$$\le C_{r,d} \frac{1}{(r - 1/2)^{1/2}} \|e^{\pm xd}\phi\|_{H^r(\mathbb{R})}$$
$$\le C_{r,d} \frac{1}{(r - 1/2)^{1/2}} \|\phi\|_{H^r(\mathbb{R})}.$$

Therefore, for arbitrary fixed $t_{(p)} \in \mathbb{R}^{p-1}$ and $t^c_{(p)} \in \mathbb{R}^{n-p}$ it holds

$$\int_{\mathbb{R}} |\hat{f}(t_{(p)}, \tau \pm id, t^c_{(p)})| d\tau =$$

$$= \int_{\mathbb{R}} (1 + |\tau \pm id|)^s (1 + |\tau \pm id|)^{-s} |\hat{f}(t_{(p)}, \tau \pm id, t^c_{(p)})| d\tau$$

$$\leq C \frac{1}{(s - 1/2)^{1/2}} \left\| \mathcal{F}_{(p)} \mathcal{F}^c_{(p)} \left(e^{\pm x_p d} f(t_{(p)}, x_p, t^c_{(p)}) \right) \right\|_{H^s(\mathbb{R})} . \quad (81)$$

Finally, the estimate for $||\delta(g)||_{1,-\nu}$ follows from (80), (81) and by Parseval's relation

$$||\delta(g)||_{1,-\nu} \leq C_{\nu,n} e^{-\pi\sqrt{dN}} \sum_{p=1}^{n} \frac{1}{(s - p/2)}$$

$$\times \left[\int_{\mathbb{R}^{n-1}} (1 + ||t_{(p)}||^2)^s (1 + ||t^c_{(p)}||^2)^s \left\| \mathcal{F}_{(p)} \mathcal{F}^c_{(p)} \left(e^{\pm x_p d} f(t_{(p)}, x_p, t^c_{(p)}) \right) \right\|^2_{H^s(\mathbb{R})} \right]^{\frac{1}{2}}$$

$$\leq C_{\nu,s,n} e^{-\pi\sqrt{dN}} ||f||_{H^s(\mathbb{R}^n)}.$$

□

5 Approximation on the bounded domain

In Section 4 we derived finite dimensional approximation spaces $\mathcal{V}^N_\varepsilon$ (72) on the unbounded domain, that have robust (with respect to ε) approximation properties. It was shown that the rates of convergence are independent of the regularity of the coefficients and of the unit-cell problem (Proposition 6, Corollary 2). We base a generalized FEM in Ω^ε on the spaces $\mathcal{V}^N_\varepsilon$ from the unbounded domain. However, the spaces $\mathcal{V}^N_\varepsilon$ are not $H^1_{\Gamma^\varepsilon_D}(\Omega^\varepsilon)$-conforming, i.e., $\mathcal{V}^N_\varepsilon \not\subset H^1_{\Gamma^\varepsilon_D}(\Omega^\varepsilon)$ in general. Thus, $\mathcal{V}^N_\varepsilon$ cannot be used as FE-spaces directly and have to be modified. Our approach for enforcing $H^1_{\Gamma^\varepsilon_D}(\Omega^\varepsilon)$-conformity is to use a standard FEM in a neighborhood of the boundary and the spaces $\mathcal{V}^N_\varepsilon$ in the interior of the domain.

5.1 Density of the approximation spaces

A minimal requirement for the generalized FEM (7) is that $\cup_N V_N$ be dense (in the $H^1(\Omega^\varepsilon)$-norm) in the set of all solutions. Since we will base our spaces V_N on the spaces $\mathcal{V}^N_\varepsilon$ in (72), we first ascertain in Proposition 3 that $\cup_N \mathcal{V}^N_\varepsilon$ is dense in the set of solutions of (1), (2) with boundary conditions as in (5).

Proposition 3. *Let $f \in L^2(\Omega)$ and let $u^\varepsilon \in H^1_{\Gamma^\varepsilon_D}(\Omega^\varepsilon)$ be the solution of (1), (5). Then, for all bounded Lipschitz domains $\Omega_0 \subset\subset \Omega$, with $\Omega_0^\varepsilon := \Omega_0 \cap \Omega^\varepsilon_\infty$*

$$\lim_{N\to\infty} \inf_{v \in \bigcup_{\mu=1}^N \mathcal{V}_\varepsilon^\mu} \|u^\varepsilon - v\|_{H^1(\Omega_0^\varepsilon)} = 0,$$

i.e., $\bigcup_{N>0} \mathcal{V}_\varepsilon^N|_\Omega$ is dense with respect to the $H^1(\Omega_0^\varepsilon)$-topology in the space of solutions to (6) corresponding to right-hand sides $f \in L^2(\Omega)$.

The proof of this result for the case when the domain has no holes, i.e., $\Omega^\varepsilon = \Omega$ can be found in [10]. In the general case the proof follows analogously.

5.2 Conforming approximation spaces via scale resolution at the boundary

The approximation spaces $\mathcal{V}_\varepsilon^N$ defined in (72) are, when restricted to Ω^ε, in general not $H^1_{\Gamma^\varepsilon_D}(\Omega^\varepsilon)$-conforming. We propose scale resolution at the boundary with standard FEM and use $\mathcal{V}_\varepsilon^N$ in the interior only. In Proposition 4, we show that such composite FE-spaces V_N are still dense in the set of all $H^1_{\Gamma^\varepsilon_D}(\Omega^\varepsilon)$ solutions. Enforcement of the essential boundary conditions is only one reason for using scale resolution near the boundary. For problems posed on the unbounded domain the solution $u^\varepsilon(x)$ has a two-scale behavior and these scales are separated, but the situation is different for problems on bounded domains, since near the boundary, scale separation occurs only for special situations. For example, if the boundary is flat and aligned with the periodicity as in the case of an infinite, perforated layer, then two-scale separation occurs in the tangential direction at the boundary. In general, however, scale separation does not take place at curved or nonaligned parts of the boundary (see, e.g., [16]).

Different approaches to enforcing Dirichlet boundary conditions are possible. In the asymptotic homogenization approach ($\varepsilon \to 0$) one constructs so-called *correctors*. The explicit construction of so-called "boundary correctors" (which compensate for boundary conditions violated by outer asymptotic expansions, but which are, generally, not of boundary layer type) is tedious and limited to special cases, in which some geometry simplifications are assumed, as in the case when the periodic structure is aligned with the boundaries. For the general case (e.g., when the boundaries are curved and intersect the cavities of the microscopic periodic lattice) only little is known about the detailed structure of the boundary correctors (see also [16]). Such situations can be dealt with by scale resolution at the boundary. The geometry approximation comes as an additional argument for scale resolution at the boundary.

Let $\widetilde{V}_N \subset H^1(\Omega^\varepsilon)$ be any finite dimensional subspace (it need not satisfy the homogeneous Dirichlet boundary conditions) such that

$$\inf_{v \in \widetilde{V}_N} \|u_\varepsilon - v\|_{H^1(\Omega^\varepsilon_{(\varepsilon)})} \le \delta,$$

where $\Omega^\varepsilon_{(\varepsilon)} := \{x \in \Omega^\varepsilon \,|\, \mathrm{dist}(x, \partial\Omega) > \varepsilon\}$. Let $\chi \in H^1_0(\Omega) \cap W^{1,\infty}(\Omega)$ be a cut-off function such that

$$\chi \equiv 1 \text{ on } \Omega_{(2\varepsilon)} := \{x \in \Omega \,|\, \mathrm{dist}(x, \partial\Omega) > 2\varepsilon\},$$
$$\chi \equiv 0 \text{ on } \Omega\backslash\Omega_{(\varepsilon)} = \{x \in \Omega \,|\, \mathrm{dist}(x, \partial\Omega) \leq \varepsilon\}.$$

Assume further that $V_{0,\Gamma^\varepsilon_D}(\Omega^\varepsilon\backslash\Omega_{(2\varepsilon)}) \subset H^1_{0,\Gamma^\varepsilon_D}(\Omega^\varepsilon\backslash\Omega_{(2\varepsilon)})$ is a finite dimensional subspace of $H^1_{0,\Gamma^\varepsilon_D}(\Omega^\varepsilon\backslash\Omega_{(2\varepsilon)}) := \{v \in H^1(\Omega^\varepsilon\backslash\Omega_{(2\varepsilon)}) \;:\; v|_{\Gamma^\varepsilon_D} = 0\}$. Then we define the $H^1_{\Gamma^\varepsilon_D}(\Omega)$-conforming FE-space V_N as

$$V_N := \chi\widetilde{V}_N + (1-\chi)V_{0,\Gamma^\varepsilon_D}(\Omega^\varepsilon\backslash\Omega_{(2\varepsilon)}).$$

Proposition 4. *Assume that the cut-off function χ introduced above satisfies:*

$$\|\chi\|_{W^{1,\infty}(\Omega)} \leq C_\chi\varepsilon^{-1}.$$

Then, for each $s \in [0, 1/2)$ there exist $C(s) = C(s, \Omega, C_\chi) > 0$ and $C > 0$ such that there holds

$$\begin{aligned}\inf_{v\in V_N} \|u^\varepsilon - v\|_{H^1(\Omega^\varepsilon)} &\leq C(s)\varepsilon^{-(1-s)}\delta \\ &+ C \inf_{w\in V_{0,\Gamma^\varepsilon_D}(\Omega^\varepsilon\backslash\Omega_{(2\varepsilon)})} \Big(\|\nabla(u^\varepsilon - w)\|_{L^2(\Omega^\varepsilon\backslash\Omega_{(2\varepsilon)})} \\ &\qquad + \varepsilon^{-1}\|(u^\varepsilon - w)\|_{L^2(\Omega^\varepsilon\backslash\Omega_{(2\varepsilon)})}\Big).\end{aligned} \tag{82}$$

The proof of this result for domains without holes ($\widehat{Q} = (-\pi, \pi)^n$) can be found in [10], in the general case the result can be proved analogously.

Remark 6. Let $V_{0,\Gamma^\varepsilon_D}(\Omega^\varepsilon\backslash\Omega_{(2\varepsilon)}) = S^{1,1}_{0,\Gamma^\varepsilon_D}(\Omega^\varepsilon\backslash\Omega_{(2\varepsilon)}, \mathcal{T}_h)$ be the space of piecewise linears with respect to a quasiuniform triangulation $\mathcal{T}_h$ of $\Omega^\varepsilon\backslash\Omega_{(2\varepsilon)}$ that are zero on the boundary Γ^ε_D. Note that $\cup_{h>0} S^{1,1}_{0,\Gamma^\varepsilon_D}(\Omega^\varepsilon\backslash\Omega_{(2\varepsilon)}, \mathcal{T}_h)$ is dense in $H^1_{0,\Gamma^\varepsilon_D}(\Omega^\varepsilon\backslash\Omega_{(2\varepsilon)})$ so that the infimum in the right hand side of the error estimate (82) tends to zero as $h \to 0$. If u^ε is piecewise regular then the size of this infimum can be quantified. Assuming that $u^\varepsilon(x)|_K \in H^2(K)$ for all $K \in \mathcal{T}_h$ we have the standard estimate [6] for the piecewise linear interpolant $u^\varepsilon_{\mathrm{Int}}$ of u^ε on $\mathcal{T}_h$

$$\begin{aligned}h\|\nabla(u^\varepsilon - u^\varepsilon_{\mathrm{Int}})\|_{L^2(\Omega^\varepsilon\backslash\Omega_{(2\varepsilon)})} + \|u^\varepsilon - u^\varepsilon_{\mathrm{Int}}\|_{L^2(\Omega^\varepsilon\backslash\Omega_{(2\varepsilon)})} \leq \\ Ch^2 \Big\{\textstyle\sum_{K\in\mathcal{T}_h} |u^\varepsilon|^2_{H^2(K)}\Big\}^{1/2}.\end{aligned}$$

For smooth coefficients $a(y)$, $a_0(y)$ we have $\Big\{\sum_{K\in\mathcal{T}_h} |u^\varepsilon|^2_{H^2(K)}\Big\}^{1/2} \leq C\varepsilon^{-1}$. Hence, we obtain with $w = u^\varepsilon_{\mathrm{Int}}$ in (82) the following error estimate with respect to the subspace $V_N = \chi\widetilde{V}_N + (1-\chi)S^{1,1}_{0,\Gamma^\varepsilon_D}(\Omega^\varepsilon\backslash\Omega_{(2\varepsilon)}, \mathcal{T}_h)$

$$\inf_{v\in V_N} \|u_\varepsilon - v\|_{H^1(\Omega^\varepsilon)} \leq C_1(s)\varepsilon^{-(1-s)}\delta + C_2 h/\varepsilon,$$

with C_i, $i = 1, 2$ independent of h and ε. Note that here $h < \varepsilon$. The space V_N is essentially one in which *scale resolution* is achieved at the boundary and in which the functions from $\widetilde{V}_N$ are used in the interior.

Appendix. Proof of Lemma 4

The present section is devoted to the proof of the estimates concerning the remainder $\delta(f,h)(\cdot)$ stated in Lemma 4 of Section 4.3. The proof of this theorem in the one dimensional setting was given in [13]. We start from the Fourier-Bochner integral representation

$$u^{\varepsilon}(x) = \frac{1}{(2\pi)^{n/2}} \overset{(B)}{\int_{\mathbb{R}^n}} \hat{f}(t) e^{it\cdot x} \phi\left(\frac{x}{\varepsilon}, \varepsilon, t\right) dt = \overset{(B)}{\int_{\mathbb{R}^n}} e^{it\cdot x} g(t,x)\, dt,$$

with integral kernel $g(t,\cdot)$

$$g(t,\cdot) = \frac{1}{(2\pi)^{n/2}} \hat{f}(t) \phi\left(\frac{\cdot}{\varepsilon}, \varepsilon, t\right) \in \mathcal{A}(\mathcal{D}_d; H^1_{-2\nu}(\Omega^{\varepsilon}_{\infty}))$$

being analytic, $H^1_{-2\nu}(\Omega^{\varepsilon}_{\infty})$-valued functional in the strip region $\mathcal{D}_{d=\nu_n}$, $\nu_n = \nu/(2\sqrt{n})$, $\nu \in (0, \nu_0)$. By Theorem 5, the kernel satisfies the growth condition

$$\|g(t,\cdot)\|_{1,-2\nu} \leq C(\gamma,\nu,n) C(f) \left(1 + \frac{1}{\alpha}\right) e^{-\frac{\alpha}{2}\|t\|}, \quad \forall\, t \in \mathcal{D}_d.$$

Recall some definitions from Section 3.2

$$\mathbb{Z}^n_{(N)} := \{k \in \mathbb{Z}^n \,:\, |k_p| \leq N,\, \forall\, p = 1, \ldots, n\}, \tag{83}$$

$$u^{\varepsilon}_{N,h}(x) := \mathbf{1}_{[-\frac{\pi}{h},\frac{\pi}{h}]^n}(x) \frac{1}{(2\pi)^{n/2}} h^n \sum_{k\in\mathbb{Z}^n_{(N)}} \hat{f}(kh) \psi\left(\frac{x}{\varepsilon}, \varepsilon, kh\right) \tag{84}$$

$$\delta_N(f,h)(\cdot) := u^{\varepsilon}(\cdot) - u^{\varepsilon}_{N,h}(\cdot). \tag{85}$$

Assume that $L \geq 1$ is given and let $h = (\pi d/\alpha N)^{1/2}$ satisfy

$$\frac{\pi}{h} \geq 2L,$$

i.e., $N \geq 4dL^2/(\alpha\pi)$. We will split the remainder $\delta_N(f,h)(x)$ as

$$\begin{aligned} \delta_N(f,h)(x) = {} & \lim_{\delta\to 0+} \overset{(B)}{\int_{\mathbb{R}^n}} \left(\prod_{p=1}^{n} e^{-\delta|t_p|}\right) e^{it\cdot x} E(f,h)(t,x)\, dt \\ & + \sum_{k\in\mathbb{Z}^n\setminus\mathbb{Z}^n_{(N)}} \mathbf{1}_{[-\frac{\pi}{h},\frac{\pi}{h}]^n}(x) h^n e^{ikh\cdot x} g(kh,x). \end{aligned} \tag{86}$$

The $\|\cdot\|_{0,-2\nu}$, respectively $\|\cdot\|_{H^1_{-2\nu}((-L,L)^n\cap\Omega^\varepsilon_\infty)}$ norm of the second term in $\delta_N(f,h)(\cdot)$ can be estimated as follows

$$\begin{aligned}
&\Big\| \sum_{k\in\mathbb{Z}^n\setminus\mathbb{Z}^n_{(N)}} \mathbf{1}_{[-\frac{\pi}{h},\frac{\pi}{h}]^n}(x) h^n e^{ikh\cdot x} g(kh,\cdot)\Big\|_{0,-2\nu;H^1_{-2\nu}((-L,L)^n\cap\Omega^\varepsilon_\infty)} \\
&\le C_{\gamma,\nu,n} C(f) h^n \sum_{k\in\mathbb{Z}^n\setminus\mathbb{Z}^n_{(N)}} e^{-\alpha\|k\|h} \le C_{\gamma,\nu,n} C(f)\, n h^n \frac{e^{-\alpha(N+1)h/\sqrt{n}}}{(1-e^{-\alpha h/\sqrt{n}})^n} \\
&\le C_{\gamma,\nu,n} \frac{C(f)}{\alpha^n} e^{-\sqrt{\pi d\alpha N/n}}. \qquad (87)
\end{aligned}$$

It remains to estimate the integral term in (86)

$$\delta(f,h)(x) := \lim_{\delta\to 0+} \overset{(B)}{\int_{\mathbb{R}^n}} \left(\prod_{p=1}^n e^{-\delta|t_p|}\right) e^{it\cdot x} E(f,h)(t,x)\,dt. \qquad (88)$$

We have derived in Lemma 3 the following Bochner-integral representation for the integral kernel $E(f,h)(t,\cdot)\in H^1_{-2\nu}(\Omega^\varepsilon_\infty)$ in (88)

$$\begin{aligned}
E(f,h)(t,\cdot) &:= g(t,\cdot) - \sum_{k\in\mathbb{Z}^n} S(k,h)(t) g(kh,\cdot) \\
&= \sum_{p=1}^n \Big\{ \sum_{k_1,\dots,k_{p-1}\in\mathbb{Z}} S((k_1,\dots,k_{p-1}),h)(t_1,\dots,t_{p-1}) \frac{\sin(\pi t_p/h)}{2\pi i} \qquad (89) \\
&\overset{(B)}{\int_{\mathbb{R}}} \Big[\frac{g((k_1h,\dots,k_{p-1}h,\tau-id^-,t_{p+1},\dots,t_n),\cdot)}{(\tau-t_p-id)\sin[\pi(\tau-id)/h]} \qquad (90) \\
&- \frac{g((k_1h,\dots,k_{p-1}h,\tau+id^-,t_{p+1},\dots,t_n),\cdot)}{(\tau-t_p+id)\sin[\pi(\tau+id)/h]} \Big] d\tau \Big\}.
\end{aligned}$$

Taking advantage of the exponential decay of the $H^1_{-2\nu}(\Omega^\varepsilon_\infty)$- norm of $g(z,\cdot)$ in $\mathcal{D}_d$ we easily deduce that $\|E(f,h)(t,\cdot)\|_{1,-2\nu}$ is uniformly bounded with respect to $t\in\mathbb{R}^n$ by a constant depending on n,d but independent of t. We can thus change the order of integration and commute the sum in (89) with the Bochner integrals. Recalling the short-hand notation: for $t\in\mathbb{R}^n$, $t_{(p)} := (t_1,\dots,t_{p-1})$, $t^c_{(p)} := (t_{p+1},\dots,t_n)$, we obtain that

$$\begin{aligned}
\delta(f,h)(x) &= \lim_{\delta\to 0+} \overset{(B)}{\int_{\mathbb{R}^n}} \left(\prod_{p=1}^n e^{-\delta|t_p|}\right) e^{it\cdot x} E(f,h)(t,x)\,dt \\
&= \lim_{\delta\to 0+} \sum_{p=1}^n \sum_{k_{(p)}\in\mathbb{Z}^{p-1}} \overset{(B)}{\int_{\mathbb{R}^n}} \overset{(B)}{\int_{\mathbb{R}}} S(k_{(p)},h)(t_{(p)}) \frac{\sin(\pi t_p/h)}{2\pi i}
\end{aligned}$$

$$\times\left(\prod_{q=1}^{n} e^{-\delta|t_q|}\right) e^{it\cdot x}\left\{\frac{g((k_{(p)}h,\tau-id^-,t^c_{(p)}),x)}{(\tau-t_p-id)\sin[\pi(\tau-id)/h]}\right.$$

$$\left.-\frac{g((k_{(p)}h,\tau+id^-,t^c_{(p)}),x)}{(\tau-t_p+id)\sin[\pi(\tau+id)/h]}\right\} d\tau\, dt$$

$$=\lim_{\delta\to 0+}\sum_{p=1}^{n}\sum_{k_{(p)}\in\mathbb{Z}^{p-1}}\int_{\mathbb{R}^{n-1}}^{(B)}\int_{\mathbb{R}}^{(B)} S(k_{(p)},h)(t_{(p)})\frac{1}{2\pi i}\left(\prod_{q\neq p} e^{-\delta|t_q|e^{it_qx_q}}\right)$$

$$\left\{\frac{g((k_{(p)}h,\tau-id^-,t^c_{(p)}),x)}{\sin[\pi(\tau-id)/h]}F_\delta(\tau,x_p)\right.$$

$$\left.-\frac{g((k_{(p)}h,\tau+id^-,t^c_{(p)}),x)}{\sin[\pi(\tau+id)/h]}\tilde{F}_\delta(\tau,x_p)\right\} d\tau\, dt_{(p)}dt^c_{(\nu)}.$$

Here,

$$F_\delta(\tau,x) := \int_{\mathbb{R}} \sin(\pi t/h)\,\frac{e^{it\cdot x}e^{-\delta|t|}}{\tau-t-id}\,dt \tag{91}$$

$$\tilde{F}_\delta(\tau,x) := \int_{\mathbb{R}} \sin(\pi t/h)\,\frac{e^{it\cdot x}e^{-\delta|t|}}{\tau-t+id}\,dt. \tag{92}$$

Integration with respect to $t_1,\ldots,t_{p-1}$ corresponds to the Fourier transform of the Sinc-functions $S((k_1,\ldots,k_{p-1}),h)(\cdot)$ evaluated at $(x_1,\ldots,x_{p-1})$. We get therefore that

$\delta(f,h)(x) =$

$$\sum_{p=1}^{n}\sum_{k_{(p)}\in\mathbb{Z}^{p-1}}\int_{\mathbb{R}^{n-p}}^{(B)}\int_{\mathbb{R}}^{(B)} h^{p-1}\left(\prod_{q=1}^{p-1} e^{ik_qhx_q}\mathbf{1}_{[-\frac{\pi}{h},\frac{\pi}{h}]}(x_q)\right)\left(\prod_{q>p} e^{it_qx_q}\right)$$

$$\left\{g((k_1h,\ldots,k_{p-1}h,\tau-id^-,t_{p+1},\ldots,t_n),x)f_1(\tau,x_p)\right.$$

$$\left.+\,g((k_1h,\ldots,k_{p-1}h,\tau+id^-,t_{p+1},\ldots,t_n),x)f_2(\tau,x_p)\right\} d\tau\, dt^c_{(p)},$$

where $f_1(\tau,\cdot)$ and $f_2(\tau,\cdot)$ are as in (66), (67). Taking now the $\|\cdot\|_{0,-2\nu}$ norm in both sides we get

$$\|\delta(f,h)(\cdot)\|_{0,-2\nu}$$

$$\leq\sum_{p=1}^{n}\sum_{k_{(p)}\in\mathbb{Z}^{p-1}}\int_{\mathbb{R}^{n-p+1}} h^{p-1}\left\{\|g((k_{(p)}h,\tau-id^-,t^c_{(p)}),\cdot)\|_{0,-2\nu}\|f_1(\tau,\cdot)\|_{L^\infty}\right.$$

$$\left.+\,\|g((k_{(p)}h,\tau+id^-,t^c_{(p)}),\cdot)\|_{0,-2\nu}\|f_2(\tau,\cdot)\|_{L^\infty}\right\} d\tau\, dt^c_{(p)}.$$

Since the $\|\cdot\|_{1,-2\nu}$ norm of $g(z,\cdot)$ decays exponentially in the strip $\mathcal{D}_d$, as stated in (57), and the L^∞ norms of $f_1(\tau,\cdot)$ respectively $f_2(\tau,\cdot)$ are bounded by

$$\|f_1(t,\cdot)\|_{L^\infty(\mathbb{R})} + \|f_2(t,\cdot)\|_{L^\infty(\mathbb{R})} \le \frac{1}{\sinh(\pi d/h)}, \tag{93}$$

respectively

$$\|f_1(t,\cdot)\|_{W^{1,\infty}(-L,L)} + \|f_2(t,\cdot)\|_{W^{1,\infty}(-L,L)} \le C(d)(1+|t|)\frac{1}{\sinh(\pi d/h)}, \tag{94}$$

we get

$$\begin{aligned}\|\delta(f,h)(\cdot)\|_{0,-2\nu} &\le C_{\nu,\gamma,n} e^{-\sqrt{\pi d\alpha N}} \left(1+\frac{1}{\alpha}\right)\frac{1}{\alpha^{n-p+1}} \times \\ &\quad \times \sum_{p=1}^{n} \sum_{k_1,\ldots,k_{p-1}\in\mathbb{Z}} h^{p-1}\left(\prod_{q=1}^{p-1} e^{-\alpha/(2\sqrt{n})|k_q|h}\right) \\ &\le C_{\nu,\gamma,n}\left(1+\frac{1}{\alpha}\right)\frac{1}{\alpha^n} e^{-\sqrt{\pi d\alpha N}}.\end{aligned}$$

A similar argument is used to get estimates for the $\|\cdot\|_{H^1_{-2\nu}((-L,L)^n\cap\Omega^\varepsilon_\infty)}$ norm, using the bounds for the $W^{1,\infty}$ norms of $f_1(\tau,\cdot)$ respectively $f_2(\tau,\cdot)$ from (93, 94)

$$\begin{aligned}&\|\delta(f,h)(\cdot)\|_{H^1_{-2\nu}((-L,L)^n\cap\Omega^\varepsilon_\infty)} \\ &\le C_{\nu,n}\sum_{p=1}^{n}\sum_{k_{(p)}\in\mathbb{Z}^{p-1}}\int_{\mathbb{R}^{n-p}}\int_{\mathbb{R}} h^{p-1}\Big\{\Big(1+|\tau|+\sum_{q=1}^{p-1}|k_q|h+\sum_{q>p}|t_q|\Big) \\ &\Big(\|g((k_{(p)}h,\tau-id^-,t^c_{(p)}),\cdot)\|_{1,-2\nu}\|f_1(\tau,\cdot)\|_{L^\infty} \\ &+\|g((k_{(p)}h,\tau+id^-,t^c_{(p)}),\cdot)\|_{1,-2\nu}\|f_2(\tau,\cdot)\|_{L^\infty}\Big)\Big\}\,d\tau\,dt^c_{(p)} \\ &\le C_{\nu,\gamma,n}e^{-\sqrt{\pi d\alpha N}}\left(1+\frac{1}{\alpha}\right)^2\frac{1}{\alpha^{n-p+1}} \\ &\quad\times\sum_{p=1}^{n}\sum_{k_{(p)}\in\mathbb{Z}^{p-1}} h^{p-1}\left(\prod_{q=1}^{p-1}e^{-\alpha/(2\sqrt{n})|k_q|h}\right) \\ &\le C_{\nu,\gamma,n}\left(1+\frac{1}{\alpha}\right)^2\frac{1}{\alpha^n}e^{-\sqrt{\pi d\alpha N}}.\end{aligned}$$

The proof of Lemma 4 is now complete.

References

1. Babuška I., Aziz A.K. (1973) Survey lectures on the mathematical foundation of the finite element method. The Mathematical Foundations of the Finite Element Method with Applications to Partial Differential Equations, A. K. Aziz, ed., Academic Press, New York, 5–359
2. Babuška I., Caloz G., Osborn J. (1994) Special finite element methods for a class of second order elliptic problems with rough coefficients. SIAM J. Numer. Anal., **31** No. **4** 945–981
3. Bakhvalov N., Panasenko G. (1989) Homogenization: Averaging Process in Periodic Media. Kluwer Publ., Dordrecht
4. Bensoussan A., Lions J.L., Papanicolau G. (1978) Asymptotic Analysis for Periodic Structures. North Holland, Amsterdam
5. Brenner S.C., L.R. Scott L.R. (1994) The Mathematical Theory of Finite Element Methods. Texts in Applied Mathematics **15**, Springer Verlag, New York
6. Ciarlet P.G. (1976) The Finite Element Method for Elliptic Problems. North-Holland Publishing Company
7. Cioranescu D., J. Saint Jean Paulin J (1999) Homogenization of Reticulated Structures. Springer Applied Mathematical Sciences
8. Duarte C.A., Babuška I., Oden J.T. (2000) Generalized Finite Element Methods for Three-Dimensional Structural Mechanics Problems. Computers and Structures, **77** 215–232
9. Hou T.Y., Wu X.H., Cai Z. (1999) Convergence of a multiscale finite element method for elliptic problems with rapidly oscillating coefficients. Math. Comp. **68**, 913–943
10. Matache A.-M. (2000) Spectral and p-Finite Elements for Problems with Microstructure. Doctoral Dissertation, ETH Zürich
11. Matache A.-M., Schwab C. (2000) Homogenization via p-FEM for Problems with Microstructure. Applied Numerical Mathematics, vol **33**, Issue 1–4
12. Matache A.-M., Schwab C. Two-Scale FEM for Homogenization Problems. in preparation
13. Matache A.-M., Babuška I., Schwab C. (2000) Generalized p-FEM in Homogenization. Numerische Mathematik, **86** Issue 2, 319–375
14. Morgan R.C., Babuška I. (1991) An approach for constructing families of homogenized equations for periodic media. I: An integral representation and its consequences. SIAM J. Math. Anal. Vol. 22, No. 1, 1–15
15. Morgan R.C., Babuška I. (1991) An approach for constructing families of homogenized equations for periodic media. II: Properties of the kernel. SIAM J. Math. Anal. Vol. 22, No. 1, 16–33
16. Neuss-Radu M. (1999) Boundary Layers in the Homogenization of Elliptic Problems. Doctoral Dissertation, Heidelberg University
17. Oleinik O.A., Shamaev A.S., Yosifian G.A. (1992) Mathematical Problems in Elasticity and Homogenization. North-Holland
18. Schwab C. (1998) p- and hp- Finite Element Methods. Oxford University Press
19. Schwab C., Matache A.-M. (2000) High order generalized FEM for lattice materials. In: P. Neittaanmäki, T. Tiihonen and P. Tarvainen (Eds.) Numerical Mathematics and Advanced Applications, 3rd European Conference, World Scientific Publ.

20. Stenger F. (1993) Numerical Methods Based on Sinc and Analytic Functions. Springer Verlag

The Failure of Uniform Exponential Decay for Boundary Layers

Maria Neuss-Radu

Institut für Angewandte Mathematik, Universität Heidelberg, INF 294, D-69120 Heidelberg, Germany

Abstract. In the homogenization of elliptic boundary value problems with periodically oscillating coefficients, boundary layers are used to describe the boundary behavior of the solution. In the classical case, when the domain is the half space $\Omega = \{x \in \mathbb{R}^n, x_n > 0\}$, the boundary layers are defined on the semi-infinite strip $]0,1[^{n-1}\times]0,\infty[$, and their energies decrease exponentially with respect to the second variable. In [11] we have shown that the property of uniform exponential decay of the boundary layers does not hold in general.

In our contribution, we improve the result of [11] by showing that for general domains the optimal decay of the energy of the boundary layers is polynomial of degree -1.

1 Setting of the problem

Let $\Omega \subset \mathbb{R}^n$, $\Omega \neq \mathbb{R}^n$ be an unbounded domain with Lipschitz-boundary. We consider the function space $\dot{H}_0^1(\Omega)$ to be defined as follows. Let $C^\infty(\bar{\Omega})$ be the space of all infinitely differentiable functions on Ω such that all derivatives have continuous extensions to $\partial\Omega$. We define the space $\dot{H}^1(\Omega)$ as the closure of

$$\{u \in C^\infty(\bar{\Omega})/\mathbb{R} : \|\nabla u\|_{L_2(\Omega)} \leq C\}$$

in the norm

$$\|u\|_{\dot{H}^1(\Omega)} := \|\nabla u\|_{L_2(\Omega)} \,.$$

The function space $\dot{H}_0^1(\Omega)$ is now defined as the subspace of $\dot{H}^1(\Omega)$ consisting of functions with vanishing trace on $\partial\Omega$. We denote by $\dot{H}_0^1(\Omega)'$ the dual space of $\dot{H}_0^1(\Omega)$, and by $< \cdot, \cdot >$ the duality product between $\dot{H}_0^1(\Omega)$ and its dual.

We are interested in the behavior of the solution to the following boundary value problem:

Given $f \in \dot{H}_0^1(\Omega)'$, *find* $u^\varepsilon \in \dot{H}_0^1(\Omega)$, *such that*

$$\int_\Omega A_{ij}\left(\frac{x}{\varepsilon}\right)\partial_j u^\varepsilon(x)\partial_i\varphi(x)\,dx = <f,\varphi> , \ \forall\varphi \in \dot{H}_0^1(\Omega) \qquad (1)$$

Here, the coefficients A_{ij} satisfy:

(i) A_{ij} are periodic with period 1; i.e., A_{ij} are defined on the unit cube $Y = [0,1]^n$ and extended by periodicity over the whole space $\mathbb{R}^n$

(ii) $\mathcal{A}_{ij} = \mathcal{A}_{ji}$
(iii) $\mathcal{A}_{ij}$ are bounded and strongly elliptic; i.e., there exist $0 < \lambda \le \Lambda$ such that:

$$\lambda \left|\eta\right|^2 \le \mathcal{A}_{ij}(y)\eta_i\eta_j \le \Lambda \left|\eta\right|^2 \ , \ \forall \eta \in \mathbb{R}^n, \forall y \in Y$$

For small values of the parameter $\varepsilon > 0$, the method of homogenization leads to the macroscopic approximation

$$u_\varepsilon^1 = u^0 + \varepsilon \partial_{x_k} u^0 w^k \left(\frac{\cdot}{\varepsilon}\right) \tag{2}$$

for the microscopic solution u^ε of (1) , see for example [6] or [13]. This approximation is given by using the solution u^0 of the homogenized problem:

Find $u^0 \in \dot{H}_0^1(\Omega)$, *such that*

$$\int_\Omega \mathcal{A}_{ij}^0 \partial_j u^0(x) \partial_i \varphi(x)\, dx = < f, \varphi > \, , \ \forall \varphi \in \dot{H}_0^1(\Omega) \tag{3}$$

with the homogenized coefficients $\mathcal{A}_{ik}^0$ given by:

$$\mathcal{A}_{ik}^0 := \langle \mathcal{A}_{ij}(\delta_{jk} + \partial_{y_j} w^k)\rangle_Y = \langle (\delta_{il} + \partial_{y_l} w^i)\mathcal{A}_{lj}(\delta_{jk} + \partial_{y_j} w^k)\rangle_Y , \tag{4}$$

and the periodic solutions w^k, $k = 1, \ldots, n$ to the cell problems:

$$\textit{Find } w^k \in H_{per}^1(Y): \quad \partial_{y_i}\left(\mathcal{A}_{ij}(y)(\partial_{y_j} w^k(y) + \delta_{jk})\right) = 0 \, , \ \forall y \in Y. \tag{5}$$

$H_{per}^1(Y)$ is the completion of the space $C_{per}^\infty(Y)$ of infinitely differentiable, periodic functions on the cell $Y = [0,1]^n$ with respect to the H^1-norm. The symbol $\langle . \rangle_Y$ denotes the mean value operator over Y.

The approximation u_ε^1 turns out to be optimal in the interior of the domain. That is, if the domain is $\mathbb{R}^n$, we have the estimate

$$\left\|\nabla(u^\varepsilon - u_\varepsilon^1)\right\|_{L_2(\mathbb{R}^n)} \le C\varepsilon$$

see [3], and if $\Omega \ne \mathbb{R}^n$, then for each open subdomain $\Omega' \subset\subset \Omega$, we have

$$\left\|\nabla(u^\varepsilon - u_\varepsilon^1)\right\|_{L_2(\Omega')} \le C\varepsilon$$

with the constant C depending on dist$(\Omega', \partial\Omega)$, see [1]. However, a study of the accuracy of the approximation u_ε^1 up to the boundary, leads to error estimates which are only of order $\varepsilon^{\frac{1}{2}}$, see [9]. This is due to the fact that the periodic structure of the approximation u_ε^1 is not consistent with the homogeneous Dirichlet boundary conditions for the solution u^ε. In order to catch the boundary effects on the solution u^ε, we have to improve the approximation u_ε^1 by adding a boundary corrector. Such correctors are given by means of boundary layers.

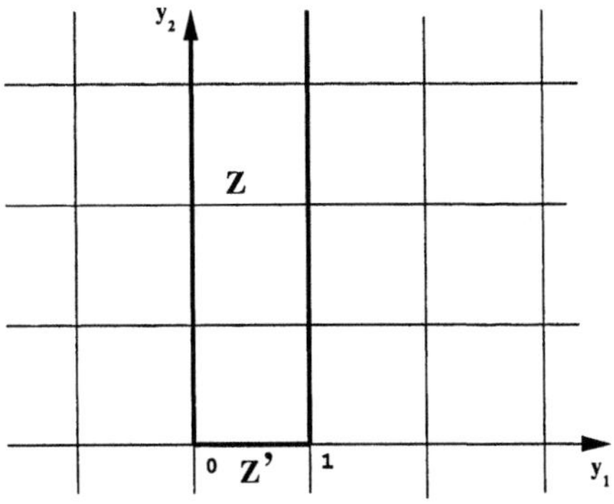

Fig. 1. The boundary layer cell Z

2 Boundary layers for the half space

The classical geometry for which boundary layers have been constructed is the half space $\Omega = \{x \in \mathbb{R}^n , x_n > 0\}$. In this case, the boundary layers are defined as follows:

Definition 1. For the coefficient matrix $\mathcal{A}_{ij}$ with the properties (i), (ii), (iii) from Section 1, the boundary layers are the solutions $w^{bl,k}$, $k = 1, \ldots , n$ of the problems:

Find $w^{bl,k}$ such that:

$$\begin{aligned} &-\partial_{y_i}\left(\mathcal{A}_{ij}(y)\partial_{y_j} w^{bl,k}(y)\right) = 0\,,\ y \in Z\,,\\ &w^{bl,k}(y',0) = -w^k(y',0)\,,\ y' \in Z'\,,\\ &\forall\, y_n = t : y' \to w^{bl,k}(y',t) \text{ is periodic}\,,\\ &\nabla w^{bl,k} \text{ decreases exponentially as } y_n \to \infty\,. \end{aligned} \tag{6}$$

Here $Z' =]0,1[^{n-1}$ and $Z = Z' \times]0,\infty[$ is called the boundary layer cell (see Fig. 1).

The condition on the decay can be formulated more precisely by introducing the following weighted spaces, see [10]. For $\gamma > 0$, set

$$W_\gamma(Z) = \{w \in L_2(]0,R[, H^1_{per}(Z')), \forall R < \infty,\, e^{\gamma y_n}\nabla w \in L_2(Z)\}$$

and

$$W^0_\gamma(Z) = \{w \in W_\gamma(Z),\, w(y',0) = 0\}.$$

They are Hilbert spaces for the scalar product given by:

$$< v, w >_{W_\gamma(Z)} = \int_Z e^{2\gamma y_n} \nabla v \nabla w \, dy.$$

This scalar product then induces the norm

$$\|w\|_{W_\gamma(Z)} = \|e^{\gamma y_n} \nabla w\|_{L_2(Z)} \,.$$

Results on the existence, uniqueness and decay for the boundary layers for the half space have been given for example by [4,5,7,8,10,11,14]. We use here the following theorem from [10] in a slightly modified form.

Theorem 1. *Let the coefficient matrix $\mathcal{A} \in (L_\infty(Z))^{n\times n}$ be bounded and strongly elliptic with constants $0 < \lambda \le \Lambda$, i.e.:*

$$\lambda\,|\eta| \le \mathcal{A}_{ij}(y)\eta_i\eta_j \le \Lambda\,|\eta| \;,\; \forall y \in Z \,,\; \forall \eta \in \mathbb{R}^n .$$

We consider the following problem:

$$\begin{aligned}
&-\partial_{y_i}\left(\mathcal{A}_{ij}(y)\partial_{y_j} w(y)\right) = f(y)\,,\; y \in Z\,,\\
&w(y',0) = 0\,,\; y' \in Z'\,,\\
&\forall\, y_n = t : y' \to w(y',t) \;\; \textit{is periodic.}
\end{aligned} \tag{7}$$

Assume that there exists $\gamma_0 > 0$ such that $e^{\gamma_0 y_n} f \in L_2(Z)$, and denote by ρ_1 the smallest nonzero eigenvalue of the Laplace-operator on Z' for periodic boundary conditions. Then for

$$0 < \gamma < min\left\{\gamma_0, \frac{\lambda\sqrt{\rho_1}}{2\Lambda}\right\}$$

there exists a unique solution $w \in W_\gamma(Z)$ to the problem (7).

Proof. See [10], Theorem 10.1.

From Theorem 1, the existence, uniqueness and exponential decay of the boundary layers $w^{bl,k}$ immediately follow. However, if we want to extend these results to more general domains we get into difficulties. For example, if we consider the situation of half spaces Ω_ϕ with $\partial\Omega_\phi$ having a rational slope, then for certain angles ϕ the period of the boundary layer cell gets very large and implicitly the corresponding eigenvalue $\rho_1^\phi \to 0$. Thus, no uniform exponential decay is obtained. If $\partial\Omega_\phi$ has an irrational slope, the definition of the boundary layers has to be extended such that boundary layer cells with infinite period are included. This is also the case if we consider boundary layers for general smooth domains, see [12]. For these generalized boundary layers, the application of Theorem 1 is not possible.

What can be said about the rate of decay of the boundary layers for general domains? From Lemma 16 in [2] it follows that for Hölder continuous coefficients $\mathcal{A}_{ij}$, the gradients of the boundary layers are bounded from above by a constant times the inverse to the distance to the boundary. On the other hand, in [11] we have shown that we cannot get uniform exponential decay of the boundary layers for general domains.

In the next section, we will improve the result of [11] by giving an explicit example where the decay of the boundary layer gradient is not better than polynomial of degree -1.

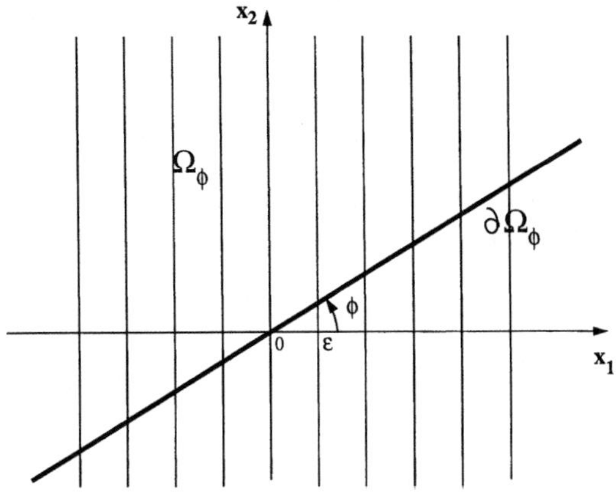

Fig. 2. Layered media

3 A special layered medium problem

Let Ω_ϕ be a half plane in $\mathbb{R}^2$, inclined at an angle ϕ to the x_1-axis (see Fig. 2):

$$\Omega_\phi = \{x \in \mathbb{R}^2,\ x \cdot n_\phi > 0\},$$

with the inner unit normal $n_\phi = (-\sin\phi, \cos\phi)$. The coefficients $\mathcal{A}_{ij}(\frac{x_1}{\varepsilon})$, $\varepsilon > 0$ of the bilinear form are given by

$$\mathcal{A}_{ij}(y_1) = \frac{1}{1 + \frac{1}{2}\sin(2\pi y_1)}\delta_{ij}. \tag{8}$$

This means that the domain Ω_ϕ consists of homogeneous and isotropic layers of width ε periodically distributed in x_1-direction. In this special framework periodicity parallel to the boundary is preserved for any value of ϕ, such that we can use Definition 1 to define boundary layers $w^{k,\phi}$ for every value of ϕ. Furthermore, the cell-solutions are

$$\begin{aligned} w^1(y_1) &= -\frac{1}{4\pi}\cos(2\pi y_1) \\ w^2 &= 0, \end{aligned} \tag{9}$$

so we will have only one boundary layer $w^{1,\phi}$ corresponding to the cell-solution w^1:

Find $w^{1,\phi}$ such that:

$$\begin{aligned} &-\partial_{y_i}\left(\mathcal{A}_{ij}(y_1)\partial_{y_j} w^{1,\phi}(y)\right) = 0\,,\ y \in Z_\phi\,, \\ &w^{1,\phi}(y_1, y_2) = -w^1(y_1)\,,\ y = (y_1, y_2) \in Z'_\phi\,, \\ &w^{1,\phi}\ \textit{is periodic in the direction}\,(\cos\phi, \sin\phi)\,, \\ &\nabla w^{1,\phi}\ \textit{decreases exponentially in the direction}\,(-\sin\phi, \cos\phi) \end{aligned} \tag{10}$$

where Z_ϕ, the boundary layer cell, is the semi-infinite strip inclined at an angle ϕ to the x_1-axis of width $\delta = \frac{1}{\cos\phi}$:

$$Z_\phi = \{y \in \mathbb{R}^2 \,,\; y \cdot n_\phi > 0 \,,\; 0 < y \cdot t_\phi < \delta\}.$$

Here, $t_\phi = (\cos\phi, \sin\phi)$ is the tangential unit vector to the boundary $\partial\Omega_\phi$, and Z'_ϕ is the part of the boundary of Z_ϕ given by:

$$Z'_\phi = \{y \in \mathbb{R}^2 \,,\; y \cdot n_\phi = 0 \,,\; 0 \le y \cdot t_\phi \le \delta\}.$$

To reduce this problem to the one covered by Theorem 1, we make the following change of coordinates:

$$\begin{cases} y_1 = \cos\phi\,\xi_1 - \sin\phi\,\xi_2 \\ y_2 = \sin\phi\,\xi_1 + \cos\phi\,\xi_2 \end{cases} \tag{11}$$

In the new coordinates (ξ_1, ξ_2), the boundary layer cell is given by $\tilde{Z}_\phi =]0,\delta[\times]0,\infty[$, and the cell-solution w^1 becomes $\tilde{w}^1(\xi_1,\xi_2) = w^1(y_1(\xi_1,\xi_2))$. The boundary layer

$$\tilde{w}^{1,\phi}(\xi_1,\xi_2) = w^{1,\phi}(y_1(\xi_1,\xi_2), y_2(\xi_1,\xi_2))$$

is the solution of the problem:

Find $\tilde{w}^{1,\phi}$ *such that* $\tilde{w}^{1,\phi} + \hat{\tilde{w}}^1 \in W^0_\gamma(\tilde{Z}_\phi)$ *and*

$$\int_{\tilde{Z}_\phi} \tilde{A}_{ij}(\xi_1,\xi_2)\partial_j \tilde{w}^{1,\phi} \partial_i \varphi \, d\xi = 0 \,,\; \forall \varphi \in W^0_\gamma(\tilde{Z}_\phi), \tag{12}$$

where $\hat{\tilde{w}}^1 \in W^0_\gamma(\tilde{Z}_\phi)$ is an extension of the boundary values of $\tilde{w}^1$ into $\tilde{Z}_\phi$, and $\tilde{A}_{ij}$ are the coefficients A_{ij} in the new coordinates (ξ_1,ξ_2).

Concerning the decay of the boundary layers $\tilde{w}^{1,\phi}$ we can prove the following result:

Theorem 2. *There exist no* $\alpha > 1$ *and* $K < \infty$ *such that, for all* $0 < \phi < \frac{\pi}{2}$, *the boundary layers* $\tilde{w}^{1,\phi}$ *satisfy*

$$\left|\nabla \tilde{w}^{1,\phi}(\xi)\right| \le K \xi_2^{-\alpha}, \quad \forall \xi \in \tilde{Z}_\phi.$$

Proof. For the proof we consider the following auxiliary problems with constant coefficients:

Find v^ϕ such that:

$$\begin{aligned} &-\partial_i\left(\tilde{A}^0_{ij}\partial_j v^\phi(\xi)\right) = 0, \quad \xi \in \tilde{Z}_\phi =]0,\delta[\times]0,\infty[, \\ &v^\phi(\xi_1,0) = -\tilde{w}^1(\xi_1,0), \quad \xi_1 \in]0,\delta[, \\ &v^\phi \in W_\gamma(\tilde{Z}_\phi) \end{aligned}$$

where $\tilde{A}^0_{ij}$ are the homogenized coefficients with respect to the coordinates (ξ_1, ξ_2).

We will proceed as follows: we first estimate the energy of the boundary layers by the energy of the solutions to the auxiliary problems. Then, using the knowledge about the latter, we prove the theorem by an indirect argument.

Let us consider the energy of the boundary layers $\tilde{w}^{1,\phi}$ on the domain $\tilde{Z}^k_\phi$ given by:

$$\tilde{Z}^k_\phi := \bigcup_{(\mu_1,\mu_2)\in\mathbb{Z}^2} \left\{\tilde{Y}_\mu\,,\ \tilde{Y}_\mu \subset \{(\xi_1,\xi_2)\in \tilde{Z}_\phi\,,\ k\delta < \xi_2 < (k+1)\delta\}\right\}$$

where $\tilde{Y}_\mu$ is the translation of the unit cell $\tilde{Y}$ by the vector (μ_1, μ_2). As usual the lateral boundaries of $\tilde{Z}_\phi$ are identified by periodicity. If we set

$$\tilde{a}(u,v) = \int_{\tilde{Z}^k_\phi} \tilde{A}_{ij}(\xi_1,\xi_2)\partial_{\xi_j} u \partial_{\xi_i} v\, d\xi$$

then the following estimate holds:

$$\tilde{a}(\tilde{w}^{1,\phi},\tilde{w}^{1,\phi}) \geq \frac{1}{8}\int_{\tilde{Z}^k_\phi} \tilde{A}^0_{ij}\partial_j v^\phi \partial_i v^\phi\, d\xi - \\ -C\left\{\left\|\nabla^2 v^\phi\right\|^2_{L_2(\tilde{Z}_\phi)} + \frac{1}{\delta^2}\left\|\nabla v^\phi\right\|^2_{L_2(\tilde{Z}_\phi)}\right\}. \tag{13}$$

The proof of this estimate can be found in [11], Theorem 4.1. The main idea is to split the bilinear form $\tilde{a}(\tilde{w}^{1,\phi},\tilde{w}^{1,\phi})$ by using the auxiliary term $v^\phi + \partial_k v^\phi \tilde{w}^k$. The two terms which are obtained are estimated by using the formula of the homogenized coefficients and the estimates between the initial and homogenized solution for $\varepsilon = 1$.

Now we use the fact that for angles $\phi \to \frac{\pi}{2}$ the boundary values:

$$v^\phi(\xi_1, 0) = -\frac{1}{4\pi}\cos(2\pi(\cos\phi\, \xi_1))$$

have long wave oscillations. Because of the constant coefficients this leads to a slow decay of the solutions v^ϕ. This property of the v^ϕ together with estimate (13) lead to the failure of the uniform exponential decay of the boundary layers $\tilde{w}^{1,\phi}$. The argument is the following. For our special coefficients (8) the estimate (13) yields:

$$\left\|\nabla\tilde{w}^{1,\phi}\right\|^2_{L_2(\tilde{Z}^k_\phi)} \geq Ce^{-4\pi\kappa(k+1)} + O(1/\delta)$$

where $\kappa = \sqrt{\frac{\tilde{A}^0_{11}}{\tilde{A}^0_{22}}}$. Now, let us assume that there exist $\alpha > 1$, independent of δ, and $K > 0$, such that for all $0 < \phi < \frac{\pi}{2}$ we have $\left|\nabla\tilde{w}^{1,\phi}(\xi_1,\xi_2)\right| \leq K\xi_2^{-\alpha}$. This implies

$$\left\|\nabla\tilde{w}^{1,\phi}\right\|^2_{L_2(\tilde{Z}^k_\phi)} \leq K^2 k^{-2\alpha}\delta^{2(1-\alpha)}$$

Taking $k = 2$, for δ sufficiently large, we obtain:

$$e^{-6\kappa\pi} \leq CK2^{-\alpha}\delta^{1-\alpha}$$

But this cannot hold for arbitrarily large $\delta > 0$ with a fixed constant K.

References

1. Allaire, G., Amar, M. (1999) Boundary Layer Tails in Periodic Homogenization. ESAIM: Control Optim. Calc. Var. **4**, 209–243
2. Avellaneda, M., Lin, F.-H. (1987) Compactness Methods in the Theory of Homogenization. Comm. Pure Appl. Math. **40**, 803–847
3. Babuska, I. (1976) Solution of Interface Problems by Homogenization I. SIAM Journal of Mathematical Analysis. **7**, 603–634
4. Babuska, I. (1976) Solution of Interface Problems by Homogenization II. SIAM Journal of Mathematical Analysis. **7**, 635–645
5. Bensoussan, A., Lions, J. L., Papanicolaou, G. (1976) Boundary Layer Analysis in Homogenization of Diffusion Equations with Dirichlet Conditions on the Half Space. Proc. of Intern. Symp. SDE Kyoto. 21–40
6. Bensoussan, A., Lions, J. L., Papanicolaou, G. (1978) Asymptotic Analysis for Periodic Structures. North Holland
7. Bakhvalov, N., Panasenko, N. G. (1989) Homogenization: Averaging Processes in Periodic Media. Kluwer, Dordrecht
8. Dumontet, H. (1986) Study of a Boundary Layer Problem in Elastic Composite Materials. Mathematical Modelling and Numerical Analysis. **20**, 265–286
9. Jikov, V. V., Kozlov, S. M., Oleinik, O. A. (1994) Homogenization of Differential Operators and Integral Functionals. Springer, Berlin Heidelberg New York
10. Lions, J. L. (1981) Some Methods in Mathematical Analysis of Systems and their Control. Gordon and Breach, New York
11. Neuss-Radu, M. (2000) A result on the decay of the boundary layers in the homogenization theory. Asymptotic Analysis. **23**, 313–328
12. Neuss-Radu, M. The boundary behavior of a composite material. To appear in Mathematical Modelling and Numerical Analysis
13. Sanchez-Palencia, E. (1980) Non-Homogenous Media and Vibration Theory. Lecture Notes in Physics **127**. Springer, Berlin Heidelberg New York
14. Sanchez-Huber, J., Sanchez-Palencia, E. (1993) Exercises sur les méthodes asymptotiques et l'homogénéisation. Masson, Paris

Homogenization of Random Nonstationary Convection-Diffusion Problem

Marina Kleptsyna[1] and Andrey Piatnitski[2]

[1] Institute for Information Transmission Problems
19, Bol'shoi Karetnii, Moscow, 101447, Russia
[2] Narvik Institute of Technology, HiN, Department of Mathematics,
Postbox 385, 8505 Narvik, Norway
and
P.N.Lebedev Physical Institute RAS
Leninski prospect 53, Moscow, 117924, Russia

Abstract. We deal with nonstationary convection-diffusion problem in a medium with a periodic microstructure whose characteristics are stationary ergodic rapidly oscillating in time random functions. Under the assumption that the coefficients of the corresponding parabolic equation possess certain mixing properties, we show that, in appropriate moving coordinates, the law of the solutions of the original problem converges weakly in the energy functional space to a solution of a stochastic partial differential equation. This limit stochastic PDE is well-posed and thus the limit measure is uniquely defined.

Contents

1 Introduction

We consider a model convection-diffusion problem in a medium with periodic microstructure whose characteristics are random rapidly changing in time functions. For simplicity we suppose that all the characteristics depend on time via a stationary ergodic process ξ_s, $0 \le s < \infty$. The Cauchy problem for the corresponding parabolic operator takes the form

$$\frac{\partial}{\partial t}u^\varepsilon - \frac{\partial}{\partial x_i}a^{ij}\Big(\frac{x}{\varepsilon}, \xi_{\frac{t}{\varepsilon^2}}\Big)\frac{\partial}{\partial x_j}u^\varepsilon - \frac{1}{\varepsilon}b^i\Big(\frac{x}{\varepsilon}, \xi_{\frac{t}{\varepsilon^2}}\Big)\frac{\partial}{\partial x_i}u^\varepsilon = 0,$$
$$(x,t) \in R^n \times (0,T), \qquad (1)$$
$$u^\varepsilon(x,0) = u_0(x),$$

and we study the limit behaviour of u^ε, as $\varepsilon \downarrow 0$.

Previously, divergence form parabolic operators of this type with a diffusion process $\xi_\cdot$ were considered in [1]. It was shown that these operators admit homogenization and that the limit operator has constant nonrandom coefficients.

The picture changes essentially in the presence of large zero-order term. As was shown in [2], in this case the limit dynamics might remain random, and the following homogenization result holds: the law of the solutions of the original problem converges weakly, in a proper functional space, to the solution of a limit martingale problem.

The basic ideas and methods of homogenization theory can be found in [3] and [4]. Averaging problems for elliptic and stationary parabolic operators with lower order terms have been considered in [5–7]. The special case of incompressible convection terms has been widely studied in [8–12].

In [13] a particular case of nonstationary parabolic operators with periodic coefficients was investigated.

Due to the structure of the equation studied in this work, the asymptotics of u^ε might involve a large convection term of order $1/\varepsilon$. In this connection the averaging results for problem (1) will be obtained in moving coordinates $x' = x - \bar{b}t/\varepsilon$ with a constant vector $\bar{b}$. This change of variables will allow us to avoid the drift of order $1/\varepsilon$ in the limit dynamics.

We show that the family of measures generated by the solutions of (1), converges weakly, as $\varepsilon \to 0$, to a solution of a limit functional martingale problem. The latter problem involves a finite dimensional white noise with the coefficients of the form $\Lambda \nabla u$, where $\Lambda = \{\Lambda^{ij}\}$ is a constant matrix; the homogenized drift operator is a second order elliptic operator with constant coefficients. The matrix Λ is of special interest for, in the case of a diffusive behaviour of ξ_s, this matrix cannot be expressed in terms of the solutions of auxiliary local "cell" problems and requires more delicate techniques.

In the paper we introduce various correctors being usually solutions of auxiliary (random) PDE problems, prove a priori estimates and combine this technique with some ideas adopted from [14,15]. It should be noted that some of the correctors constructed here are not standard, they depend not only on the value of the process ξ_s at the current time but on the whole "future" of the process. This makes a certain difference with the approaches developed in [2,1,15], where the diffusive behaviour of the correctors was essentially used.

In Section 2 we pose the problem and provide all the conditions on the coefficients and on the process ξ_s.

The third section is aimed at proving the tightness results for the corresponding families of measures.

The last section deals with the passage to the limit; we construct the coefficients of the limit martingale problem and then, using the uniqueness of a solution, derive weak convergence of the distributions of u^ε.

2 Setting of the problem

We study the asymptotic behaviour of solutions of the following Cauchy problem

$$\frac{\partial}{\partial t}u^\varepsilon(x,t) - \frac{\partial}{\partial x_i}a^{ij}\left(\frac{x}{\varepsilon},\xi_{\frac{t}{\varepsilon^2}}\right)\frac{\partial}{\partial x_j}u^\varepsilon(x,t) - \frac{1}{\varepsilon}b^i\left(\frac{x}{\varepsilon},\xi_{\frac{t}{\varepsilon^2}}\right)\frac{\partial}{\partial x_i}u^\varepsilon(x,t) = 0,$$
$$u^\varepsilon(x,0) = u_0(x), \tag{2}$$

as $\varepsilon \downarrow 0$; here ξ_s is an ergodic stationary process with values in R^d, it is defined on a probability space $(\Omega, \mathcal{F}, \mathbf{P})$, the notation $\mathbf{E}$ will be used for the expectation.

Further assumptions on ξ_t will be made in terms of strong mixing coefficient (see, for instance, [16], §9.2). For the reader's convenience we recall here its definition.

Denote by $\mathcal{F}_{\le t}$ and $\mathcal{F}_{\ge t}$, respectively, the σ-algebras $\sigma(\xi_s,\ s \le t)$ and $\sigma(\xi_s,\ s \ge t)$. The function $\rho(s),\ s \ge 0$, with

$$\rho(s) = \sup_{\substack{\eta_1 \in L^2(\Omega,\mathcal{F}_{\le 0})\\ \eta_2 \in L^2(\Omega,\mathcal{F}_{\ge s})}} \frac{|\mathrm{cov}(\eta_1,\eta_2|}{\sqrt{\mathbf{E}((\eta_1-\mathbf{E}\eta_1)^2)\,\mathbf{E}((\eta_2-\mathbf{E}\eta_2)^2)}},$$

is said to be a maximum correlation coefficient of $\{\xi_\cdot\}$.

In what follows we always suppose that all the coefficients $a^{ij}(z,y)$ and $b^i(z,y)$ are periodic in z variable, and identify the space of periodic functions on R^n with the space of functions defined on the n-dimensional torus T^n.

Other assumptions are as follows:

A1 The functions $a^{ij}(z,y)$ and $b^i(z,y)$ are uniformly bounded as well as their first order derivatives in z and y:

$$|a^{ij}(z,y)| + |\nabla_z a^{ij}(z,y)| + |\nabla_y a^{ij}(z,y)| \le c,$$

$$|b^i(z,y)| + |\nabla_z b^i(z,y)| + |\nabla_y b^i(z,y)| \le c;$$

here and afterwards c stands for a generic positive constant.

A2 The matrix $a^{ij}(z,y)$ is symmetric and uniformly positive definite: there is $c > 0$ such that

$$a^{ij}\zeta_i\zeta_j \ge c|\zeta|^2, \quad \zeta \in R^n,$$

for all $(z,y) \in T^n \times R^d$.

A3 The following relation holds

$$\int_0^\infty \rho(s)\,ds < \infty.$$

Under assumptions A1–A3, for any initial condition $u_0 \in L^2(R^n)$ and any $\varepsilon > 0$, almost surely problem (2) has a unique solution $u^\varepsilon \in L^2(0,T;H^1(R^n)) \cup C(0,T;L^2(R^n))$; for each $\varepsilon > 0$ this solution generates a Radon probability measure on the space

$$V = L^2_w(0,T;H^1(R^n)) \cup C(0,T;L^2_w(R^n))$$

equipped with the Borel σ-algebra; symbol w stands for the weak topology. We denote this measure defined as the law of u^ε in V, by Q^ε.

3 Tightness

In this section we establish the tightness results for the family of distributions Q^ε defined above. To this end we obtain a number of a priori estimates for the solution u^ε of problem (2) and then derive the tightness in a standard way by means of Prokhorov criterium (see [14,15]).

In general, the family Q^ε is not tight. In order to recover a tight family, we introduce moving coordinates $(x', t) = (x - \bar{b}/\varepsilon, t)$ with a constant vector $\bar{b}$, consider $u^\varepsilon(x', t)$ and show that, under proper choice of $\bar{b}$, the family of distributions obtained is tight.

Proposition 1. *Uniformly in $\varepsilon > 0$ the following estimate holds*

$$\sup_{t \le T} \left(\|u^\varepsilon(t)\|^2 + \int_0^t \|\nabla u^\varepsilon(s)\|^2 ds \right) \le c. \tag{3}$$

There exists a constant vector $\bar{b}$ such that for any $\varphi(x) \in C_0^\infty(R^n)$ and any $\gamma > 0$ the relation holds

$$\lim_{\nu \to 0} \sup_{\varepsilon > 0} \mathbf{P} \left\{ \sup_{|t-s|<\nu} \left| (u^\varepsilon(t), \varphi(\cdot + \frac{\bar{b}}{\varepsilon} t)) - (u^\varepsilon(s), \varphi(\cdot + \frac{\bar{b}}{\varepsilon} s)) \right| \ge \gamma \right\} = 0. \tag{4}$$

Proof. Consider an auxiliary problem

$$\frac{\partial}{\partial s} p_\varepsilon(z,s) + A^* p_\varepsilon(z,s) = 0, \qquad (z,t) \in T^n \times (-\infty, (T+1)/\varepsilon^2), \tag{5}$$

$$p_\varepsilon|_{s=(T+1)/\varepsilon^2)} = 1;$$

with

$$A^* = \frac{\partial}{\partial z_i} a^{ij}(z, \xi_s) \frac{\partial}{\partial z_j} - \frac{\partial}{\partial z_i} (b^i(z, \xi_s) \cdot). \tag{6}$$

By the maximum principle (see [19]) the solution p_ε is strictly positive; moreover, considering the structure of the equation and integrating by parts over a set $T^n \times (s, (T+1)/\varepsilon^2)$ we get

$$\int_{T^n} p_\varepsilon(z,s) dz = 1$$

for any $s \le (T+1)/\varepsilon^2$. From the Harnack inequality (see [17]) it follows that

$$0 < c_1 \le p_\varepsilon(z,s) \le c_2 < \infty, \tag{7}$$

where c_1 and c_2 depend neither on ε nor on a trajectory of ξ_s.

Now, if we multiply (2) by $p_\varepsilon(x/\varepsilon, s/\varepsilon^2)u^\varepsilon(x,s)$ and integrate it over $R^n \times (0,t)$, then after simple transformations we find

$$\frac{1}{2}\int_{R^n}(u^\varepsilon(x,t))^2 p_\varepsilon(\frac{x}{\varepsilon},\frac{t}{\varepsilon^2})dx - \frac{1}{2}\int_{R^n}(u^\varepsilon(x,0))^2 p_\varepsilon(\frac{x}{\varepsilon},0)dx +$$

$$+\int_0^t\int_{R^n} a^{ij}(\frac{x}{\varepsilon},\xi_{s/\varepsilon^2})p_\varepsilon(\frac{x}{\varepsilon},\frac{s}{\varepsilon^2})\frac{\partial}{\partial x_i}u^\varepsilon(x,s)\frac{\partial}{\partial x_j}u^\varepsilon(x,s)dxds$$

$$-\frac{1}{2}\int_0^t\int_{R^n}(u^\varepsilon)^2\Big[\frac{\partial}{\partial t}p_\varepsilon(\frac{x}{\varepsilon},\frac{s}{\varepsilon^2})dxds + \frac{\partial}{\partial x_i}(a^{ij}(\frac{x}{\varepsilon},\xi_{s/\varepsilon^2})\frac{\partial}{\partial x_j}p_\varepsilon(\frac{x}{\varepsilon},\frac{s}{\varepsilon^2}))$$

$$-\frac{1}{\varepsilon}\frac{\partial}{\partial x_i}(b^i(\frac{x}{\varepsilon},\xi_{s/\varepsilon^2})p_\varepsilon(\frac{x}{\varepsilon},\frac{s}{\varepsilon^2}))\Big]\,dxds = 0$$

According to (5) the last integral here is equal to 0. Combining this with (7) we derive (3).

Now we proceed to (4). The technique we use relies on the following statement.

Lemma 1. *Let the initial condition u_0 in (2) belong to C_0^∞. Then there is a nonrandom function $\kappa_1(\varepsilon)$ vanishing as $\varepsilon \downarrow 0$, such that*

$$\sup_{(x,t)\in R^n\times(0,T)} \operatorname*{osc}_{Q_x^\varepsilon} u^\varepsilon \le \kappa_1(\varepsilon), \tag{8}$$

where we denote $Q_x^\varepsilon = x + (-\varepsilon/2, \varepsilon/2)^n$.

Proof. Let us consider an auxiliary periodic Cauchy problem:

$$\frac{\partial}{\partial s}v(z,s) - Av(z,s) = 0, \quad z \in T^n,\ s > s_0,$$

$$v|_{s=s_0} = v_0$$

with

$$A = \frac{\partial}{\partial z_i}a^{ij}(z,\xi_s)\frac{\partial}{\partial z_j} + b^i(z,\xi_s)\frac{\partial}{\partial z_i}, \tag{9}$$

Its solution $v(z,s)$ converges at an exponential rate, as $s - s_0 \to \infty$, to a constant. In particular,

$$\operatorname*{osc}_{T^n}\ v(\cdot, s_0 + \varepsilon^{-1/2}) \le c' \exp(-c\varepsilon^{-1/2})|v_0|_{L^\infty} \tag{10}$$

with c' and c independent of s_0 and of a trajectory $\xi_\cdot$. Indeed, under assumptions **A1** and **A2** above, the solution $v(z,s)$ satisfies uniformly in s_0 and $\{\xi_\cdot\}$ the Harnack inequality (see, for instance, [17]) which, in turn, implies the bound required.

The operator

$$A^\varepsilon = \frac{\partial}{\partial x_i} a^{ij}\Big(\frac{x}{\varepsilon}, \xi_{\frac{t}{\varepsilon^2}}\Big)\frac{\partial}{\partial x_j} + \frac{1}{\varepsilon} b^i\Big(\frac{x}{\varepsilon}, \xi_{\frac{t}{\varepsilon^2}}\Big)\frac{\partial}{\partial x_i} \tag{11}$$

commutes with any shift operator of the form $S_k u(x) = u(x+\varepsilon k)$, $k \in Z^n$. Therefore, by the maximum principle,

$$|u^\varepsilon(x+\varepsilon k, t) - u^\varepsilon(x,t)| \le M\varepsilon|k|, \qquad k \in Z^n, \tag{12}$$

where $M = \max\limits_x |\nabla u_0(x)|$.

Next we fix arbitrary $t_0 \ge 0$ and $x_0 \in R^n$, restrict $u^\varepsilon(x,t_0)$ onto $Q^\varepsilon_{x_0}$ and denote the periodic in x extension of this function, with period ε in all the coordinate directions, by $\tilde v^\varepsilon_0$. By virtue of (12) we have

$$|\tilde v^\varepsilon_0(x) - u^\varepsilon(x,t_0)| \le M\varepsilon^{1/4}, \tag{13}$$

for all x such that $|x - x_0| \le \varepsilon^{1/4}$.

Let $p(t,t',x,x')$ be a fundamental solution of problem (2). According to [18], under assumptions **A1-A2** for any x, x' such that $|x-x'| > \varepsilon^{1/4}$, the following estimate holds

$$p(t_0, t_0+\varepsilon^{3/2}, x, x') \le c' \exp(-c|x-x'|^2/\varepsilon^{3/2}); \tag{14}$$

where c and c' only depend on the constants in **A1-A2**. Thus, integrating (14) over $\{x : |x-x'| > \varepsilon^{1/4}\}$, we get

$$\int\limits_{|x-x'|>\varepsilon^{1/4}} p(t_0, t_0+\varepsilon^{3/2}, x, x')dx \le c\varepsilon \tag{15}$$

Denoting by $\tilde v^\varepsilon(x,t)$ a solution of the equation $\partial_t \tilde v^\varepsilon - A^\varepsilon \tilde v^\varepsilon = 0$ with the initial condition $\tilde v^\varepsilon|_{t=t_0} = \tilde v_0(x)$, the estimate (10) reads

$$\operatorname*{osc}_{R^n} \tilde v^\varepsilon(\cdot, t_0+\varepsilon^{3/2}) \le c' \exp(-c\varepsilon^{-1/2})|\tilde v_0|_{L^\infty}.$$

On the other hand, for any $x \in Q^\varepsilon_{x_0}$, from (15) and (13) it follows

$$|u^\varepsilon(x, t+\varepsilon^{3/2}) - \tilde v^\varepsilon(x+\varepsilon^{3/2})| \le c\varepsilon|u_0|_{L^\infty} + cM\varepsilon^{1/4}.$$

The last two inequalities imply the desired statement for all $t \ge \varepsilon^{3/2}$. For small t this statement is a trivial consequence of (15). This completes the proof of Lemma.

Next, we consider a sequence of problems

$$\begin{gathered}\frac{\partial}{\partial s}p^N(z,s) + A^*p^N = 0, \quad (z,s) \in T^n \times (-\infty, N), \\ p^N(z,N) = 1;\end{gathered} \tag{16}$$

with A^* given by (6). By the same arguments as above one can show that the functions p^N satisfy the bounds (7) uniformly in N and s, $-\infty < s \leq N$.

Lemma 2. *The sequence p^N converges, as $N \to \infty$, to a stationary ergodic process taking on values in $C(T^n)$ and satisfying the equation*

$$\frac{\partial}{\partial s}p + A^*p = 0, \quad \int\limits_{T^n} p(z,s)dz = 1, \qquad s \in (-\infty, +\infty). \tag{17}$$

Moreover, there are nonrandom constants $c > 0$ and $c_1 > 0$ such that

$$\max_{\substack{z \in T^n \\ k \leq s \leq k+1}} |p^N(z,s) - p(z,s)| \leq c_1 \exp(-c(N-k)). \tag{18}$$

Proof. To obtain the convergence and the estimate (18), consider the following problem

$$\frac{\partial}{\partial s}q + A^*q = 0, \quad q|_{s=N} = q_0, \tag{19}$$

with $q_0 \in L^2(T^n)$ subject to

$$\int\limits_{T^n} q_0(z)dz = 0. \tag{20}$$

It suffices to show that a solution to the latter problem decays exponentially as $(N-s) \to \infty$:

$$|q(z,s)| \leq c_1 \|q_0\|_{L^2(T^n)} \exp(-c(N-s)). \tag{21}$$

Indeed, it is obvious that for any $N > 0$ and $k > 0$ the difference $(p^{N+k} - p^N)$ obeys the equation

$$\frac{\partial}{\partial s}(p^{N+k} - p^N) + A^*(p^{N+k} - p^N) = 0, \quad s < N,$$

and, also, that the relations

$$\int\limits_{T^n} (p^{N+k}(z,s) - p^N(z,s))dz = 0, \qquad \|(p^{N+k} - p^N)\|_{L^\infty} \leq c,$$

hold for any $s \leq N$. Therefore, by (21) we get

$$\|(p^{N+k} - p^N)\|_{C(T^n \times [k,k+1])} \leq c_1 \exp(-c(N-k))$$

By the Cauchy criterion p^N converges to a limit continuous function $p(z,s)$ and (18) holds. Then, passing to the limit in (16), we obtain (17).

To establish (21) consider the adjoint problem

$$\begin{gathered} \frac{\partial}{\partial s}\nu(z,s) - A\nu(z,s) = 0, \quad (z,s) \in T^n \times (s_0, +\infty), \\ \nu(z,s_0) = \varphi(z), \end{gathered} \tag{22}$$

with an arbitrary $L^2(T^n)$-function φ, and A given by (9). Multiplying this equation by $q(z,s)$, taking the integral over $T^n \times [s_0, N]$ and integrating by parts, we have

$$\int_{T^n} \nu(z,N) q_0(z) dz = \int_{T^n} \varphi(z) q(z,s_0) dz. \tag{23}$$

By the Nash inequality (see [19])

$$\max_{z \in T^n} |\nu(z, s_0+1)| \le c \|\varphi\|_{L^2}. \tag{24}$$

Denote by $\nu^+(z,s_1)$ and $\nu^-(z,s_1)$ the positive and negative parts of $\nu(z,s_1)$ respectively:

$$\nu^+(z,s_1) = \max(\nu(z,s_1),0); \qquad \nu^-(z,s_1) = -\min(\nu(z,s_1),0).$$

Subtracting if necessary an appropriate constant we may assume that

$$\max_{z\in T^n} \nu(z,s_1) = -\min_{z \in T^n} \nu(z,s_1)$$

and hence $\|\nu^+(\cdot,s_1)\|_{L^\infty} = \|\nu^-(\cdot,s_1)\|_{L^\infty}$. Then for a solution to the problem

$$\begin{gathered} \frac{\partial}{\partial s}\nu^1(z,s) + A\nu^1(z,s) = 0, \quad (z,s) \in T^n \times (s_1, +\infty), \\ \nu^1(z,s_1) = \nu^+(z,s_1) \end{gathered}$$

the Harnack inequality implies

$$\max_{z\in T^n} \nu^1(z,s_1+1) \le c_2 \min_{z\in T^n} \nu^1(z,s_1+1),$$

where c_2 only depends on the constants from **A1** and **A2**. Combining this bound with evident estimate

$$0 \le \nu^1 \le \|\nu^+(\cdot,s_1)\|_{L^\infty},$$

we find

$$0 < c_3 \le \nu^1(z,s_1+1) \le \|\nu^+(\cdot,s_1)\|_{L^\infty}.$$

Similarly, ν^2 defined as a solution to the problem

$$\frac{\partial}{\partial s}\nu^2(z,s) + A\nu^2(z,s) = 0, \quad \nu^2(z,s_1) = \nu^-(z,s_1),$$

can be estimated as follows

$$0 < c_3 \leq \nu^2(z, s_1 + 1) \leq \|\nu^-(\cdot, s_1)\|_{L^\infty}.$$

From the last two inequalities it follows that there is a constant $c_4 > 0$ such that for all $s > s_0 + 1$

$$\operatorname*{osc}_{T^n} \nu(\cdot, s+1) \leq (1 - c_4) \operatorname*{osc}_{T^n} \nu(\cdot, s). \tag{25}$$

Considering (24) and (25) we derive

$$\operatorname*{osc}_{T^n} \nu(\cdot, s) \leq c_1 \exp(-c(s - s_0)) \|\varphi\|_{L^2(T^n)}$$

for all $s \geq s_0 + 1$. Finally, by (23), (20) and the latter bound we get

$$\left| \int_{T^n} \varphi(z) q(z, s_0) dz \right| = \left| \int_{T^n} \nu(z, N) q_0(z) dz \right| \leq$$

$$c_1 \exp(-c(N - s_0)) \|q_0\|_{L^2(T^n)} \|\varphi\|_{L^2(T^n)},$$

for any $\varphi \in L^2(T^n)$. Therefore,

$$\|q(\cdot, s_0)\|_{L^2(T^n)} \leq c_1 \exp(-c(N - s_0)) \|q_0\|_{L^2(T^n)}$$

and

$$|q(z, s_0)| \leq c_1 \exp(-c(N - s_0)) \|q_0\|_{L^2(T^n)};$$

the latter bound follows from the preceding one by virtue of the Nash estimate.

Next, we define functions $\tilde{p}^N(z, s)$ to be equal to $p^{N+s}(z, s)$ for each $s \in R$, where p^N are solutions to (16). Clearly, for each $N > 0$, $\tilde{p}^N$ is a stationary ergodic process with values in $C(T^n)$. As was proved above, $\tilde{p}^N(z, s)$ converges at the exponential rate, as $N \to \infty$, to $p(z, s)$. Consequently, $p(\cdot, s)$ is also stationary ergodic process. This completes the proof of Lemma.

Let us now introduce the following constant vector

$$\bar{b}^i = \mathbf{E} \int_{T^n} \left(\frac{\partial}{\partial z_j} a^{ij}(z, \xi_s) + b^i(z, \xi_s)\right) p(z, s) dz, \tag{26}$$

and stationary random process with values in R^n

$$\eta^i(s) = \int_{T^n} \left(\frac{\partial}{\partial z_j} a^{ij}(z, \xi_s) + b^i(z, \xi_s) - \bar{b}^i\right) p(z, s) dz. \tag{27}$$

All the processes involved in the definition of $\bar{b}$ are ergodic stationary thus by the Birkhoff theorem we have

$$\bar{b}^i = \lim_{T\to\infty} \frac{1}{T} \int_0^T \int_{T^n} (\frac{\partial}{\partial z_j} a^{ij}(z,\xi_s) + b^i(z,\xi_s))p(z,s)dzds \tag{28}$$

for a.a. $\omega \in \Omega$.

Lemma 3. *The process $\eta(s)$ satisfies functional Central Limit Theorem (invariance principle) with the covariance matrix given by*

$$\{\sigma^2\}^{ij} = \int_0^\infty \mathbf{E}(\eta^i(s)\eta^j(0) + \eta^i(0)\eta^j(s))ds.$$

That is

$$\varepsilon \int_0^{\cdot/\varepsilon^2} \eta(s)ds \xrightarrow[\varepsilon\to 0]{\mathcal{L}} \sigma w.,$$

in $(C[0,T])^n$, where $w.$ is a standard n-dimensional Brownian motion.

Proof. First we are going to show that there are constants $c > 0$ and $c_1 > 0$ such that

$$||\mathbf{E}\{\eta_0|\mathcal{F}_{\geq T}\}||_{L^2(\Omega)} \leq c_1(\exp(-cT) + \rho(T/2)) \tag{29}$$

for all $T > 0$. To this end we represent the function $p(z,s)$, $0 \leq s \leq T/2$, as a sum $p(z,s) = p^1(z,s) + p^2(z,s)$, where both p^1 and p^2 satisfy the equation $\frac{\partial}{\partial s}p^i(z,s) + A^*p^i(z,s) = 0$, $s < T/2$, with the initial conditions $p^1|_{s=T/2} = 1$ and $p^2|_{s=T/2} = p(z,T/2) - 1$ respectively. Then, $\eta(0) = \eta^1(0) + \eta^2(0)$ with

$$\eta^{m,i}(0) = \int_{T^n} ((\frac{\partial}{\partial z_j} a^{ij}(z,\xi_0) + b^i(z,\xi_0))p^m(z,0)dz, \quad m = 1,\, 2.$$

By the definition, $p^1(0)$ is measurable with respect to $\mathcal{F}_{\leq T/2}$, so is $\eta^1(0)$. Taking into account the mixing condition **A3**, we obtain

$$||\mathbf{E}\{\eta^1(0)|\mathcal{F}_{\geq T}||_{L^2(\Omega)} \leq \rho(T/2)||\eta^1(0)||_{L^2(\Omega)} \leq C\rho(T/2) \tag{30}$$

It follows from (18) that

$$|p^2(z,0)| = |p(z,0) - p^{T/2}(z,0)| \leq c_1 \exp(-cT/2),$$

and, therefore,

$$||\mathbf{E}\{\eta^2(0)|\mathcal{F}_{\geq T}||_{L^2(\Omega)} \leq ||\eta^2(0)||_{L^2(\Omega)} \leq C\exp(-cT/2).$$

Finally, (29) is a consequence of the latter bound and (30).

According to [16], Chapter 9, §2, Theorem 1, the relation (29) and Assumption **A3** ensure the function CLT for the process $\eta(-s)$, which, in turn, implies the function CLT for $\eta(s)$. Indeed, by virtue of Prokhorov's theorem (see [20]), the compactness in $(C([0,T])^n$ of the family $\left\{\varepsilon\int_0^{t/\varepsilon^2}\eta(s)ds\right\}$ is a consequence of the compactness of $\left\{\varepsilon\int_0^{t/\varepsilon^2}\eta(-s)ds\right\}$ and of the stationarity. The fact that the convergence of finite dimensional distributions of $\left\{\varepsilon\int_0^{t/\varepsilon^2}\eta(-s)ds\right\}$ implies the corresponding convergence of finite dimensional distributions of $\left\{\varepsilon\int_0^{t/\varepsilon^2}\eta(s)ds\right\}$ is trivial.

Next, we consider an auxiliary problem

$$\frac{\partial}{\partial s}\psi^i(z,s)+A^*\psi^i(z,s)=-\frac{\partial}{\partial z_j}(a^{ij}(z,\xi_s)p(z,s))-a^{ij}(z,\xi_s)\frac{\partial}{\partial z_j}p(z,s)+$$
$$+b^i(z,\xi_s)p(z,s)-\bar{b}^ip(z,s)-\eta^i(s)p(z,s),\quad (z,s)\in T^n\times(-\infty,+\infty). \tag{31}$$

with A^* given by (6).

Lemma 4. *Problem (31) has a stationary ergodic solution. Under the normalization*

$$\int\limits_{T^n}\psi(z,s)dz=0$$

the solution is unique and ergodic.

Proof. Denote the right hand side in (31) by $F(z,s)$ and consider a sequence of problems

$$\frac{\partial}{\partial s}\psi^N(z,s)+A^*\psi^N(z,s)=\mathbf{1}_{\{N-1<s\le N\}}F(z,s),\quad (z,s)\in T^n\times(-\infty,N)$$

$$\psi^N|_{s=N}=0.$$

Taking into account the fact that $\int_{T^n}F(z,s)dz\equiv 0$, one can easily verify that

$$\int\limits_{T^n}\psi^N(z,s)dz=0$$

for all s, where we put $\psi^N=0$ for $s>N$. By the estimate (21) we get

$$\|\psi^N\|_{L^\infty(T^n\times(k,k+1))}\le c_1\exp(-c(N-k))$$

for all $k\le N$. Summing up in N, $-\infty<N<+\infty$, we obtain a stationary solution $\psi=\sum\limits_{N=-\infty}^{+\infty}\psi^N$ to problem (31). Moreover,

$$\|\psi\|_{L^\infty(T^n\times(-\infty,+\infty)}\le C \tag{32}$$

for some nonrandom constant C. The uniqueness and ergodicity can now be proved in the same way as in Lemma 2.

To complete the proof of the compactness of $\{u^\varepsilon\}$ we consider, for an arbitrary test function $\varphi \in C_0^\infty(R^n)$, the expression

$$(\tilde{u}^\varepsilon, \tilde{p}^\varepsilon\varphi) + \varepsilon(\tilde{u}^\varepsilon, \tilde{\psi}^\varepsilon \nabla_x\varphi),$$

where

$$\tilde{u}^\varepsilon(x,t) = u^\varepsilon(x - \varepsilon^{-1}\bar{b}t - \varepsilon^{-1}\int_0^t \eta(s/\varepsilon^2)ds, t), \tag{33}$$

$$\tilde{p}^\varepsilon(x,t) = p(x/\varepsilon - \varepsilon^{-2}\bar{b}t - \varepsilon^{-2}\int_0^t \eta(s/\varepsilon^2)ds, t/\varepsilon^2)$$

and

$$\tilde{\psi}^\varepsilon(x,t) = \psi(x/\varepsilon - \varepsilon^{-2}\bar{b}t - \varepsilon^{-2}\int_0^t \eta(s/\varepsilon^2)ds, t/\varepsilon^2).$$

Considering the above equations (17) and (31), and using the notation

$$\tilde{\varphi} = \varphi(x + \varepsilon^{-1}\bar{b}t + \varepsilon^{-1}\int_0^t \eta(s/\varepsilon^2)ds),$$

and, for a generic function r,

$$\nabla_x^\varepsilon r(x) = \nabla_z r(z)|_{z=x/\varepsilon}, \qquad \partial_t^\varepsilon r(t) = \frac{\partial}{\partial s} r(s)|_{s=t/\varepsilon^2},$$

we obtain after integration by parts and simple rearrangements

$$\frac{d}{dt}[(\tilde{u}^\varepsilon, \tilde{p}^\varepsilon\varphi) + \varepsilon(\tilde{u}^\varepsilon, \tilde{\psi}^\varepsilon\nabla_x\varphi)] = \frac{d}{dt}[(u^\varepsilon, p^\varepsilon\tilde{\varphi}) + \varepsilon(u^\varepsilon, \psi^\varepsilon\nabla_x\tilde{\varphi})] =$$

$$= (A^\varepsilon u^\varepsilon, p^\varepsilon\tilde{\varphi}) + (u^\varepsilon, \tilde{\varphi}\frac{\partial}{\partial t}p^\varepsilon) + \varepsilon^{-1}(\bar{b} + \eta(t/\varepsilon^2)) \cdot (u^\varepsilon, p^\varepsilon\nabla_x\tilde{\varphi}) +$$

$$+\varepsilon(A^\varepsilon u^\varepsilon, \psi^\varepsilon\nabla_x\tilde{\varphi}) + \varepsilon(u^\varepsilon, \nabla_x\tilde{\varphi}\frac{\partial}{\partial t}\psi^\varepsilon) + (\bar{b} + \eta(t/\varepsilon^2)) \cdot (u^\varepsilon, \psi^\varepsilon\nabla_x\nabla_x\tilde{\varphi}) = \tag{34}$$

$$= (u^\varepsilon, p^\varepsilon a^\varepsilon\nabla_x\nabla_x\tilde{\varphi}) + (u^\varepsilon\nabla_x\nabla_x\tilde{\varphi}, [\nabla_x^\varepsilon(a\psi) + a^\varepsilon\nabla_x^\varepsilon\psi - b^\varepsilon\psi^\varepsilon + \bar{b}\psi^\varepsilon + \eta^\varepsilon\psi^\varepsilon]) +$$

$$+\varepsilon(u^\varepsilon\psi^\varepsilon, a^\varepsilon\nabla_x\nabla_x\nabla_x\tilde{\varphi}).$$

In view of (3) this implies the estimate

$$|(\tilde{u}^\varepsilon(t), \tilde{p}^\varepsilon(t)\varphi) - (\tilde{u}^\varepsilon(s), \tilde{p}^\varepsilon(s)\varphi)| \le c|t-s| \, ||\varphi||_{C^3} \tag{35}$$

Approximating, if necessary, $u_0 \in L^2(R^n)$ by a sequence $u_0^k \in C_0^\infty(R^n)$ and quoting (3), we may assume that $u_0 \in C_0^\infty(R^n)$.

Then, by (35) and Lemma 1 we get

$$|(\tilde{u}^\varepsilon(t), \varphi) - (\tilde{u}^\varepsilon(s), \varphi)| \le c|t-s| \, ||\varphi||_{C^3} + c\kappa(\varepsilon)||\varphi||_{C^1}. \tag{36}$$

If we denote

$$\hat{u}^\varepsilon(x,t) = u^\varepsilon(x - \varepsilon^{-1}\bar{b}t, t) \tag{37}$$

then

$$|(\hat{u}^\varepsilon(t),\varphi)-(\hat{u}^\varepsilon(s),\varphi)| = |(\tilde{u}^\varepsilon(t),\varphi(\cdot-\varepsilon^{-1}\int_0^t \eta(\tau/\varepsilon^2)d\tau))-$$

$$-(\tilde{u}^\varepsilon(s),\varphi(\cdot-\varepsilon^{-1}\int_0^s \eta(\tau/\varepsilon^2)d\tau))| \le$$

$$\le |(\tilde{u}^\varepsilon(t),\varphi(\cdot-\varepsilon^{-1}\int_0^s \eta(\tau/\varepsilon^2)d\tau))-(\tilde{u}^\varepsilon(s),\varphi(\cdot-\varepsilon^{-1}\int_0^s \eta(\tau/\varepsilon^2)d\tau))|+ \quad (38)$$

$$+|(\tilde{u}^\varepsilon(t),\varphi(\cdot-\varepsilon^{-1}\int_0^t \eta(\tau/\varepsilon^2)d\tau)-\varphi(\cdot-\varepsilon^{-1}\int_0^s \eta(\tau/\varepsilon^2)d\tau))| \le$$

$$\le c|t-s|\,\|\varphi\|_{C^3}+c\kappa(\varepsilon)\|\varphi\|_{C^1}+c\|\varphi\|_{C^1}\left|\varepsilon^{-1}\int_s^t \eta(\tau/\varepsilon^2)d\tau\right|.$$

For any $\varphi \in C_0^\infty(R^n)$ the first two terms on the r.h.s. of the latter formula go to zero, as $\varepsilon \to 0$ and $|t-s| \to 0$, uniformly in t, s and $\omega \in \Omega$. By Lemma 3 the integral $\varepsilon^{-1}\int_0^t \eta(\tau/\varepsilon^2)d\tau$ satisfies the function CLT in $(C[0,T])^n$ for any finite interval $[0,T]$. Thus, applying twice the Prokhorov criterion of compactness, we derive the compactness of the family $(\tilde{u}^\varepsilon,\varphi)$ in $C(0,T)$.

4 Passage to the limit

The aim of this section is to show that the family $\{\tilde{u}^\varepsilon\}$ converges in probability and a.s. towards the unique solution of Cauchy problem for a limit deterministic parabolic operator with constant coefficients. First we justify the convergence in the space V defined above. Then we will prove the convergence in the space with a stronger topology.

Since the compactness of $\{\tilde{u}^\varepsilon\}$ has been proved, in order to characterize the limit distribution it suffices to pass to the limit in the expressions of the form $(\tilde{u}^\varepsilon,\varphi)$, with $\varphi \in C_0^\infty$.

By (34) and Lemma 1 we have

$$(\tilde{u}^\varepsilon(t),\varphi)-(u_0,\varphi) = (\tilde{u}^\varepsilon(t),p^\varepsilon(t)\varphi)+\varepsilon(\tilde{u}^\varepsilon(t),\psi^\varepsilon(t)\nabla_x\varphi)-$$

$$-(u_0,p^\varepsilon(0)\varphi)-\varepsilon(u_0,\psi^\varepsilon(0)\nabla_x\varphi)+O(\varepsilon)=$$

$$=\int_0^t (u^\varepsilon(t),a^\varepsilon p^\varepsilon(t)\nabla_x\nabla_x\tilde{\varphi})dt+\int_0^t (u^\varepsilon(t),a^\varepsilon\nabla_x^\varepsilon\psi^\varepsilon(t)\nabla_x\nabla_x\tilde{\varphi})dt+ \quad (39)$$

$$+\int_0^t (u^\varepsilon(t),\nabla_x^\varepsilon(a^\varepsilon\psi^\varepsilon(t))\nabla_x\nabla_x\tilde{\varphi})dt-\int_0^t (u^\varepsilon(t),b^\varepsilon\psi^\varepsilon(t)\nabla_x\nabla_x\tilde{\varphi})dt$$

$$+\int_0^t (u^\varepsilon(t), \bar{b}\psi^\varepsilon(t)\nabla_x\nabla_x\tilde{\varphi})dt + \int_0^t (u^\varepsilon(t), \eta^\varepsilon(t)\psi^\varepsilon(t)\nabla_x\nabla_x\tilde{\varphi})dt + O(\varepsilon).$$

The following statement allows to pass to the limit in the above formula.

Lemma 5. *Let $\zeta(z,s)$ be a stationary ergodic process with values in $L^2(T^n)$ $(C(T^n))$, such that*

$$||\zeta||_{L^2(T^n\times(0,1))} \le C$$

for a nonrandom constant C. Then for any C_0^∞-function φ the relation holds true a.s.:

$$\lim_{\varepsilon\to 0}\left|\int_0^t (u^\varepsilon(t), \zeta^\varepsilon(t)\tilde{\varphi})dt - \overline{<\zeta>}\int_0^t (u^\varepsilon(t), \tilde{\varphi})dt\right| = 0 \tag{40}$$

where

$$\overline{<\zeta>} = \mathbf{E}\int_{T^n} \zeta(z,s)dz.$$

Proof. Without loss of generality one can assume $\overline{<\zeta>} = 0$. Denote $\mu(s) = \int_{T^n}\zeta(z,s)dz$. Then, by Lemma 1

$$\sup_{t\le T}\left|\int_0^t [(u^\varepsilon(\tau), \zeta^\varepsilon(\tau)\tilde{\varphi}) - \mu(\frac{\tau}{\varepsilon^2})(u^\varepsilon(\tau), \tilde{\varphi})]d\tau\right| \le C\kappa(\varepsilon)$$

with nonrandom C. Thus, it is sufficient to show that

$$\lim_{\varepsilon\to 0}\sup_{t\le T}\left|\int_0^t \mu(\frac{\tau}{\varepsilon^2})(u^\varepsilon(\tau), \tilde{\varphi})d\tau\right| = 0 \tag{41}$$

It follows from (36) that for any $\varphi \in C_0^\infty(R^n)$ there is a compact subset $K \subset C(0,T)$ such that $(\tilde{u}^\varepsilon(\cdot), \varphi) \in K$ a.s. The rest of the proof is standard, we consider a finite δ-net consisting of step functions and use the Birkhoff theorem to obtain the limit relation

$$\lim_{\varepsilon\to 0}\left|\int_0^t \mu(\frac{\tau}{\varepsilon^2})(u^\varepsilon(\tau), \tilde{\varphi})d\tau\right| = 0$$

valid for all $t \le T$. It remains to combine this relation with the estimate

$$\left|\int_s^t \mu(\frac{\tau}{\varepsilon^2})(u^\varepsilon(\tau), \tilde{\varphi})d\tau\right| \le c||\mu(\frac{\cdot}{\varepsilon^2})||_{L^2(0,T)}|t-s|^{1/2},$$

and (41) follows.

Denote by $\bar{a}^{ij}$ the "homogenized" matrix:

$$\bar{a}^{ij} = \mathbf{E} \int_{T^n} [a^{ij}(z,\xi_s)p(z,s) + a^{ik}(z,\xi_s)\frac{\partial}{\partial z_k}\psi^j(z,s) - b^i(z,\xi_s)\psi^j(z,s)]dz; \quad (42)$$

this definition is natural since

$$\int_{T^n} \left(\frac{\partial}{\partial z_k}(a^{ik}(z,\xi_s)\psi^j(z,s)) - \bar{b}^i\psi^j(z,s) - \eta^i(s)\psi^j(z,s) \right) dz = 0.$$

By Lemma 5, considering (39) and the last relation, we obtain

$$\lim_{\varepsilon\to 0} \sup_{t\le T} \left| (\tilde{u}^\varepsilon(t),\varphi) \quad (u_0,\varphi) - \int_0^t (\tilde{u}^\varepsilon(s), \bar{a}\nabla_x\nabla_x\varphi)ds \right| = 0$$

a.s. for any $\varphi \in C_0^\infty$. Thus, for a typical realization of ξ. any accumulation point of $\{\tilde{u}^\varepsilon\}$ in V satisfies the equation

$$\frac{\partial}{\partial t}u^0 - \bar{A}u^0 = 0, \quad u^0|_{t=0} = u_0, \quad (43)$$

where $\bar{A} = \bar{a}^{ij}\dfrac{\partial}{\partial x_i}\dfrac{\partial}{\partial x_j}$. Due to the uniqueness of solution of the latter problem, the family $\{\tilde{u}^\varepsilon\}$ converges a.s. in V, as $\varepsilon \to 0$, towards u^0.

In fact, the above convergence can be improved. Namely, we are going to show that under a proper choice of stationary ergodic process $\chi(z,s)$ with values in $C(T^n)$, the function

$$\begin{aligned} v^\varepsilon(x,t) \equiv u^\varepsilon(x,t) - u^0\Big(x + \bar{b}\tfrac{t}{\varepsilon} + \varepsilon^{-1}\int_0^t \eta(\tfrac{\tau}{\varepsilon^2})d\tau, t\Big) - \\ -\varepsilon\nabla_x u^0\Big(x + \bar{b}\tfrac{t}{\varepsilon} + \varepsilon^{-1}\int_0^t \eta(\tfrac{\tau}{\varepsilon^2})d\tau, t\Big)\chi(\tfrac{x}{\varepsilon}, \tfrac{t}{\varepsilon^2}) \end{aligned} \quad (44)$$

converges to 0 in $L^\infty(R^n \times [0,T]) \cap L^2(0,T;H^1(R^n))$-norm. To this end we substitute the expression (44) in the original problem (1), collect the terms of like powers of ε and equate the obtained expressions to 0. The first equation related to ε^{-1}, reads

$$\left(\frac{\partial}{\partial s}\chi(z,s) - A\chi(z,s)\right)\Big|_{z=\frac{x}{\varepsilon},\, s=\frac{t}{\varepsilon^2}} \nabla_x u^0(x,t) =$$

$$= \left[+\nabla_z a(z,\xi_s) + b(z,\xi_s) - \bar{b} - \eta(s)\right]\Big|_{z=\frac{x}{\varepsilon},\, s=\frac{t}{\varepsilon^2}} \nabla_x u^0(x,t)$$

Hence, we want $\chi(z,s)$ to satisfy the equation

$$\frac{\partial}{\partial s}\chi^j(z,s) - A\chi^j(z,s) = \nabla_{z_i} a^{ij}(z,\xi_s) + b^j(z,\xi_s) - \bar{b}^j - \eta^j(s) \quad (45)$$

Let us study the properties of this equation in detail. Multiplying it by $p(z,s)$ and integrating by parts over a cylinder $T^n \times [s_1, s_2]$, it is easy to see that

$$\int_{T^n} \chi(z,s)p(z,s)dz = const;$$

the definitions of $p(z,s)$, $\bar{b}$ and $\eta(s)$ have also been used here.

Then, arguing like in the proof of Lemma 4 one can show that the equation (45) has unique up to an additive constant stationary solution $\chi(z,s)$; moreover, $\chi(\cdot,s)$ is an ergodic process with values in $C(T^n)$. For definiteness we set $\int_{T^n} \chi(z,s)p(z,s)dz = 0$.

Under the above choice of χ we have

$$(\frac{\partial}{\partial t} - A^\varepsilon)v^\varepsilon = \left(-\frac{\partial}{\partial t} + \breve{a}^{ij}\frac{\partial}{\partial x_i}\frac{\partial}{\partial x_j}\right)u^0 + \Big(a^\varepsilon + \nabla_x^\varepsilon(a^\varepsilon\chi^\varepsilon) +$$

$$a^\varepsilon\nabla_x^\varepsilon\chi^\varepsilon + b^\varepsilon\chi^\varepsilon - \bar{b}\chi^\varepsilon - \eta(\frac{t}{\varepsilon^2})\chi^\varepsilon - \breve{a}\Big)\nabla_x\nabla_x u^0 + O(\varepsilon), \tag{46}$$

where $O(\varepsilon)$ is vanishing in the uniform metric uniformly in $\omega \in \Omega$, and

$$\breve{a}^{ij} = \mathbf{E}\int_{T^n} \big(a^{ij}(z,\xi_s) + \nabla_{z_k}(a^{ik}(z,\xi_s)\chi^j(z,s)) + a^{ik}(z,\xi_s)\nabla_{z_k}\chi^j(z,s) +$$

$$+ b^i(z,\xi_s)\chi^j(z,s)\big)\,p(z,s)dz$$

For brevity we will also use the following notation

$$\breve{a}^\varepsilon_{\mathrm{aux}}(x,t) = \{\breve{a}^{\varepsilon,ij}_{\mathrm{aux}}(x,t)\} = a(\frac{x}{\varepsilon},\xi_{t/\varepsilon^2}) + \nabla_x^\varepsilon(a(\frac{x}{\varepsilon},\xi_{t/\varepsilon^2})\chi(\frac{x}{\varepsilon},\frac{t}{\varepsilon^2})) +$$

$$+a(\frac{x}{\varepsilon},\xi_{t/\varepsilon^2})\nabla_x^\varepsilon\chi(\frac{x}{\varepsilon},\frac{t}{\varepsilon^2}) + b(\frac{x}{\varepsilon},\xi_{t/\varepsilon^2})\chi(\frac{x}{\varepsilon},\frac{t}{\varepsilon^2}) - (\bar{b} + \eta(\frac{t}{\varepsilon^2}))\chi(\frac{x}{\varepsilon},\frac{t}{\varepsilon^2})$$

and

$$< \breve{a}_{\mathrm{aux}} > (s) = \int_{T^n} \big(a^{ij}(z,\xi_s) + \nabla_{z_k}(a^{ik}(z,\xi_s)\chi^j(z,s)) +$$

$$+a^{ik}(z,\xi_s)\nabla_{z_k}\chi^j(z,s) + b^i(z,\xi_s)\chi^j(z,s)\big)\,p(z,s)dz$$

It will be clear later on that $\breve{a} = \bar{a}$. Thus, the first term on the r.h.s. of (46) vanishes. Now, to achieve an energy estimate, we multiply (46) by $p^\varepsilon v^\varepsilon$ and integrate over $R^n \times (0,T)$. After simple transformations we get

$$\int_{R^n} (v^\varepsilon(x,t))^2 p(\frac{x}{\varepsilon},\frac{t}{\varepsilon^2})dx + \int_0^t\int_{R^n} p(\frac{x}{\varepsilon},\frac{s}{\varepsilon^2})a^{ij}(\frac{x}{\varepsilon},\xi_{\frac{s}{\varepsilon^2}})\frac{\partial}{\partial x_i}v^\varepsilon(x,s)\frac{\partial}{\partial x_j}v^\varepsilon(x,s)dxds$$

$$= \varepsilon^2\int_{R^n} p(\frac{x}{\varepsilon},0)(\chi^i(\frac{x}{\varepsilon},0)\frac{\partial}{\partial x_i}u_0(x))^2dx +$$

$$+\int_0^t\int_{R^n}[\breve{a}_{\rm aux}^{\varepsilon,ij}(x,s)-\breve{a}^{ij}]p(\frac{x}{\varepsilon},\frac{s}{\varepsilon^2})\frac{\partial}{\partial x_i}\frac{\partial}{\partial x_j}u^0(x,s)v^\varepsilon(x,s)dxds=$$

$$=\varepsilon^2\int_{R^n}p(\frac{x}{\varepsilon},0)(\chi^i(\frac{x}{\varepsilon},0)\frac{\partial}{\partial x_i}u_0(x))^2dx+$$

$$+\int_0^t\int_{R^n}[\breve{a}_{\rm aux}^{\varepsilon,ij}(x,s)-<\breve{a}_{\rm aux}>^{ij}(\frac{s}{\varepsilon^2})]p(\frac{x}{\varepsilon},\frac{s}{\varepsilon^2})\frac{\partial}{\partial x_i}\frac{\partial}{\partial x_j}u^0(x,s)v^\varepsilon(x,s)dxds+$$

$$+\int_0^t\int_{R^n}[<\breve{a}_{\rm aux}>^{ij}(\frac{s}{\varepsilon^2})-\breve{a}^{ij}]p(\frac{x}{\varepsilon},\frac{s}{\varepsilon^2})\frac{\partial}{\partial x_i}\frac{\partial}{\partial x_j}u^0(x,s)v^\varepsilon(x,s)dxds.$$

The second integral on the r.h.s. can be estimated as follows:

$$\left|\int_0^t\int_{R^n}[\breve{a}_{\rm aux}^{\varepsilon,ij}(x,s)-<\breve{a}_{\rm aux}>^{ij}(\frac{s}{\varepsilon^2})]p(\frac{x}{\varepsilon},\frac{s}{\varepsilon^2})\frac{\partial}{\partial x_i}\frac{\partial}{\partial x_j}u^0(x,s)v^\varepsilon(x,s)dxds\right|=$$

$$=\varepsilon\left|\int_0^t\int_{R^n}\Psi^{ij,k}(\frac{x}{\varepsilon},\frac{s}{\varepsilon^2})\nabla_{x_k}(\frac{\partial}{\partial x_i}\frac{\partial}{\partial x_j}u^0(x,s)v^\varepsilon(x,s))dxds\right|\le c\varepsilon$$

with nonrandom c; here $\Psi^{ij,k}(z,s)$ stands for the functions that satisfy the relation

$$\mathrm{div}\Psi^{ij}(z,s)=(\breve{a}_{\rm aux}^{\varepsilon,ij}(z,s)-<\breve{a}_{\rm aux}>^{ij})p(z,s)$$

We proceed by estimating the last term. Clearly,

$$v^\varepsilon(x,t)=u^\varepsilon(x,t)-u^0(x+\bar{b}\frac{t}{\varepsilon}+\varepsilon^{-1}\int_0^t\eta(\frac{\tau}{\varepsilon^2})d\tau,t)+O(\varepsilon)$$

uniformly in x, t and ω. Denote $\tilde{u}^0(x,t)=u^0(x+\bar{b}\frac{t}{\varepsilon}+\varepsilon^{-1}\int_0^t\eta(\frac{\tau}{\varepsilon^2})d\tau,t)$. Then, by (36), the family

$$(u^\varepsilon(t)-\tilde{u}^0(t),\nabla_x\nabla_x\tilde{u}^0(t))=(\tilde{u}^\varepsilon(t)-u^0(t),\nabla_x\nabla_x u^0(t))$$

is compact in $(C[0,T])^{d^2}$ uniformly in ω, and the Birkhoff theorem implies the a.s. relation

$$\lim_{\varepsilon\to0}\sup_{t\le T}\left|\int_0^t\int_{R^n}[<\breve{a}_{\rm aux}>^{ij}(\frac{s}{\varepsilon^2})-\breve{a}^{ij}]p(\frac{x}{\varepsilon},\frac{s}{\varepsilon^2})\frac{\partial}{\partial x_i}\frac{\partial}{\partial x_j}\tilde{u}^0(x,s)v^\varepsilon(x,s)dxds\right|=0$$

We arrive at the following statement.

Lemma 6. *The difference* $(u^\varepsilon - \tilde{u}^0)$ *converges a.s.,as* $\varepsilon \to 0$, *towards* 0 *in* $L^\infty(0,T;L^2(R^n))$*-norm. So does* $(\tilde{u}^\varepsilon - u^0)$.

Now we proceed with the main results. Denote by $\hat{Q}^\varepsilon$ the distribution of $\hat{u}^\varepsilon(x,t) = u^\varepsilon(x - \varepsilon^{-1}\bar{b}t, t)$ in V with $\bar{b}$ defined by (26), and let $\{\bar{a}^{ij}\}$ be the matrix introduced in (42).

Theorem 1. *Let* $\{u^\varepsilon\}$ *be a family of solutions to problem (1), and assume* $u_0 \in L^2(R^n)$. *Then the distributions* $\hat{Q}^\varepsilon$ *converge weakly in* V *to the unique solution of the following stochastic partial differential equation*

$$du(t) = \left(\bar{a}^{ij} + \frac{1}{2}(\Lambda^2)^{ij}\right)\frac{\partial}{\partial x_i}\frac{\partial}{\partial x_j}u(t)dt + \Lambda\nabla_x u(t)dw_t \tag{47}$$

$$u|_{t=0} = u_0,$$

where

$$(\Lambda^2)^{ij} = \int_0^\infty \mathbf{E}(\eta^i(0)\eta^j(s) + \eta^j(0)\eta^i(s))ds$$

and w_s *is a standard* n*-dimensional Wiener process.*

Proof. It is easy to verify by virtue of Ito's formula that the function $u^0(x - \Lambda w_t, t)$ solves problem (47); according to [21] this problem is well-posed and, in particular, has unique solution. We then represent $u^\varepsilon(x - \varepsilon^{-1}\bar{b}t, t)$ as

$$u^\varepsilon(x - \varepsilon^{-1}\bar{b}t, t) = u^0\left(x + \varepsilon^{-1}\int_0^t \eta(\frac{s}{\varepsilon^2})ds\,,\,t\right) +$$

$$u^\varepsilon(x - \varepsilon^{-1}\bar{b}t, t) - u^0\left(x + \varepsilon^{-1}\int_0^t \eta(\frac{s}{\varepsilon^2})ds\,,\,t\right).$$

As was proved in Lemma 6, the $L^\infty(0,T;L^2(R^n))$-norm of the difference between the second and the third terms on the r.h.s. vanishes as $\varepsilon \to 0$. Thus for any $\varphi \in C_0^\infty$ the families $(u^\varepsilon(x - \varepsilon^{-1}\bar{b}t, t, \varphi)$ and $(u^0(x + \varepsilon^{-1}\int_0^t \eta(\frac{s}{\varepsilon^2}ds, t), \varphi)$ converge in law in $C(0,T)$ to the same limit distribution.

The mapping $F_\varphi : (C(0,T))^n \to C(0,T)$ defined by

$$F(\theta(\cdot)) = (u^0(\cdot + \theta(t), t), \varphi)$$

is continuous, therefore the convergence in law of $\{\varepsilon^{-1}\int_0^t \eta(s/\varepsilon^2)ds\}$ to Λw_t in $(C(0,T))^n$ implies the convergence in law of $(u^0(x + \varepsilon^{-1}\int_0^t \eta(\frac{s}{\varepsilon^2})ds, t), \varphi)$ towards $(u^0(x + \Lambda w_t, t), \varphi)$ and the theorem is proved.

The topology of V is quite poor. In fact, more strong convergence takes place.

Theorem 2. *The family* $\hat{u}^\varepsilon$ *converges in law in the functional space* $V_1 = L^\infty(0,T;L^2(R^n))$ *endowed with the topology of convergence in norm (strong topology).*

Proof. As was already proved, the difference

$$u^\varepsilon(x - \varepsilon^{-1}\bar{b}t, t) - u^0(x + \varepsilon^{-1} \int_0^t \eta(\frac{s}{\varepsilon^2})ds, t)$$

converges a.s. to zero in V^1, i.e. in $L^\infty(0,T; L^2(R^n))$-norm. According to the Prokhorov theorem and by Lemma 3, for any $\delta > 0$ there is a compact subset $K \in (C(0,T))^n$ such that

$$\sup_{\varepsilon>0} \mathbf{P} \left\{ \frac{1}{\varepsilon} \int_0^t \eta(\frac{s}{\varepsilon^2})ds \notin K \right\} < \delta.$$

The mapping $\Phi : (C(0,T))^n \longrightarrow V^1$ defined by

$$\Phi(\theta(\cdot)) = u^0(x + \theta(t), t),$$

is continuous. Therefore, for any $\delta > 0$ there is a compact subset K^1 of V^1 such that

$$\sup_{\varepsilon>0} \mathbf{P} \left\{ u^0(x + \frac{1}{\varepsilon} \int_0^t \eta(\frac{s}{\varepsilon^2})ds, t) \notin K^1 \right\} < \delta.$$

In view of Lemma 6, this implies the fact that the families $\{u^\varepsilon(x - \varepsilon^{-1}\bar{b}t, t)\}$ and $\{u^0(x + \varepsilon^{-1} \int_0^t \eta(\frac{s}{\varepsilon^2})ds, t)\}$ have the same limit law in V^1. The convergence in law of $u^0(x + \varepsilon^{-1} \int_0^t \eta(\frac{s}{\varepsilon^2})ds, t)$ towards $u^0(x + \Lambda w_t, t)$ follows from Lemma 3 and continuity of the mapping Φ defined above.

References

1. Kleptsyna, M., Piatnitski, A. (1997) Homogenization of random parabolic operators. Homogenization and application to material sciences, GAKKUTO Internat. Ser. Math. Sci. Appl. **9**, 241–255, Gakkōtosho, Tokyo
2. Campillo, F., Kleptsyna, M., Piatnitski, A. (1999) Homogenization of random parabolic operators with large potential. Proc. Seventh Vilnius Conference. Probability Theory and Mathematical Statistics, B. Grigelionis et al. (Eds.) TEV, Vilnius, Lithuania / VSP, Zeist, The Netherlands
3. Bensoussan, A., Lions, J.-L., Papanicolaou, G. (1978) Asymptotic Analysis for Periodic Structures. North-Holland Publ., Amsterdam
4. Jikov V., Kozlov, S., Oleinik, O. (1994) Homogenization of Differential Operators and Integral Functionals. Springer-Verlag, Berlin
5. Kozlov, S.M. (1984) Reducibility of quasiperiodic operators and averaging. Trans. Moscow Math. Soc. **2**, 101–126
6. Kozlov, S.M., Piatnitski, A.L. (1991) Averaging on a background of vanishing viscosity. Math. USSR Sbornik. **70**(1), 241-261
7. Pardoux, E. (1999) Homogenization of linear and semilinear second order parabolic PDEs with periodic coefficients: a probabilistic approach. J. Funct. Anal. **167**, 498–520

8. Avellaneda, M., Majda, A.J. (1990) Mathematical models with exact renormalization for turbulent transport. Comm. Math. Phys. **131**, 381–429
9. Avellaneda, M., Majda, A.J. (1994) Some examples with features of renormalization for turbulent transport. Philos. Trans. Roy. Soc., London Ser. A. **346**(1679), 205–233
10. Fannjiang, A,. Papanicolaou, G. (1994) Convection enhanced diffusion for periodic flows. SIAM J. Appl. Math. **54**, 333–408
11. Fannjiang, A,. Papanicolaou, G. (1996) Diffusion in Turbulence. Prob. Theor. Rel. Fields. **105**, 279-334
12. Zhikov, V. (1997) Diffusion in an Incompressible Random Flow. Functional Analysis Appl. **31**(3), 1–11
13. Garnier, J. (1997) Homogenization in a periodic and time-dependent potential. SIAM J. Appl. Math. **57**(1), 95–111
14. Viot, M. (1976) Solutions Faibles d'Equations Dérivées Stochastiques Non Linéaires. Thèse. Université Paris VI
15. Bouc, R., Pardoux, E. (1984) Asymptotic analysis of PDEs with wide-band noise disturbance expansion of the moments. Stochastic Analysis and Applications. **2**(4), 369–422
16. Liptser, R.SH., Shiryaev, A.N. (1989) Theory of Martingales. Kluwer Academic Publishers
17. Trudinger, N.S. (1968) Pointwise estimates and quasilinear parabolic equations. Comm. Pure Appl. Math., **21**, 205–226
18. Aronson, D.G. (1967) Bounds for the fundamental solutions of a parabolic equation. Bull. Am. Math. Soc., **73**, 890–896
19. Ladyzhenskaia, O.A., Solonnikov, V.A., Ural'tseva, N.N. (1968) Linear and Quasilinear Equations of Parabolic Type. Amer. Math. Soc., Rhode Island
20. Billingsley, P. (1968) Convergence of Probability Measures. John Wiley&Sons, New York
21. Da Prato, G., Zabczyk, J. (1992) Stochastic Equations in Infinite Dimensions. Cambridge Univ. Press

Remarks on Γ-Convergence of Penalized Functionals of Ginzburg-Landau Type in One Dimension

Andrija Raguž

Department of Mathematics, Faculty of Science, Bijenička 30, 10 000 Zagreb, Croatia

Abstract. In this note we study the Ginzburg-Landau functional

$$I^{\varepsilon}(v) := \int_{\Omega} (\varepsilon^2 v''(s)^2 + W(v'(s)) + a(s)(v(s) - g(s))^2) ds$$

for $v \in \mathrm{H}^2_{per}(\Omega)$. $\Omega \subseteq \mathbf{R}$ is a bounded open set, $a \in \mathrm{L}^{\infty}(\Omega)$, $a \geq \alpha > 0$ and $g \in \mathrm{C}^1(\overline{\Omega})$, $|g'| < 1$. W is non-negative continuous function such that $W(\xi) = 0$ iff $\xi \in \{-1, 1\}$.

In view of the approach of Alberti and Müller in [1], we formulate the relaxation and minimization problem related to the functional I^{ε} and we discuss the choice of relaxation and blowup procedure adjusted to capture two characteristic small scales associated to the minimizing sequences.

Also, we prove Γ-convergence result for the integrands, and we highlight the idea of proof for Γ-convergence for integral functionals induced by the chosen relaxation.

Contents

1 Introduction

A natural modification of the energy studied in [1] is obtained by replacing the term av^2 with the penalization $a(v-g)^2$, where g is a given function, so that now we deal with the functionals I^{ε} of the type

$$I^{\varepsilon}(v) := \int_{\Omega} (\varepsilon^2 v''(s)^2 + W(v'(s)) + a(s)(v(s) - g(s))^2) ds \,.$$

By analogy with the case $g = 0$, Alberti and Müller expected that when $|g'| \leq 1$, the minimizers v_{ε} of I^{ε} get closer and closer to u as $\varepsilon \longrightarrow 0$, while the derivatives v_ε' take values closer and closer to ± 1, a behavior which is induced by the term $\varepsilon^2 v''^2$ and the fact that W vanishes at ± 1. Both the local volume fractions of the + and − 'phase' and the typical lengthscale of microstructure are expected to depend on $g'(s)$, with the lengthscale approaching ∞ (measured in units of $\varepsilon^{\frac{1}{3}}$) as $|g'(s)|$ approaches 1. A totally different behavior is expected when $|g'(s)| > 1$, since such a function cannot be approximated by sawtooth functions with derivative ± 1.

There are different ways to treat this problem, and some of them are sketched in [8]. In this paper we focus on the method recently developed by Alberti and Müller, and which relies on the notion of the so-called Young measure on micro-patterns. Such an object can capture fast oscillations over two distinct scales in a sense that we keep track of rapidly oscillating minimizing sequences through Young measure generated by 'blowups' of the original sequence. The main advantage of such an approach is that it provides us with rigorous framework for generating a minimizer at the limit, while, for example, formal asymptotic expansions can give only a good guess of what the minimizer should look like. Here, Γ-convergence is a natural type of convergence for functionals I^ε, since it is essentially equivalent to the convergence of minimizers.

To avoid some degenerate cases we choose $v \in H^2_{per}(\Omega)$, $a \in \mathrm{L}^\infty(\Omega)$, $a(s) \geq \alpha > 0$ for almost every $s \in \Omega$, $W \in \mathrm{C}(\mathbf{R})$, $W \geq 0$, $W(\xi) = 0$ iff $\xi \in \{-1, 1\}$, and we assume that there exists a constant $C > 0$ such that for every $\xi \in \mathbf{R}$ it holds $W(\xi) \geq C|\xi|$. Furthermore, we take $g \in \mathrm{C}^1(\overline{\Omega})$, $\mathrm{Lip(g)} < 1$. In this setting, however, the Γ-limit of I^ε is not of much interest, as all minimizing sequences v^ε converge to zero strongly in $\mathrm{L}^2(\Omega)$. It should be noted that this result immediately implies that a relaxation in the space L^2 or H^1 does not serve our purpose, and we are compelled to look for a relaxation over more general spaces. It turns out that a natural space is a space of parameterized measures or Young measures, as it guarantees well posedness of the relaxed minimization problem (cf. [13] for some simpler examples).

We are primarily interested in studying small scales of oscillations of minimizing sequences, and we need to find a rescaling which will alow us to 'see' all scales. Hence, we will firstly rescale I^ε, and then define it on the space of Young measures with values in natural complete metric space K. Most of our notation is inherited from [1]: Ω denotes a bounded open interval endowed with Lebesgue measure, while the letter s is reserved for a slow variable running through Ω. v denotes real function on Ω, which is periodically extended outside Ω. The space of patterns is denoted by K, and it is the set of all Borel measurable functions $x : \mathbf{R} \longrightarrow [-\infty, +\infty]$, with the agreement that such x is always a function of (fast) variable t. K can be made compact and metrizable topological space by defining a natural topology on K with respect to weak*-topology on $\mathrm{L}^\infty(\mathbf{R}; [-1, 1])$ induced by the mapping $x \mapsto \frac{2}{\pi} \arctan(x)$ (see [1] for details). By $\mathcal{M}(K)$ we denote the set of all bounded Radon measures on K. As usual, we will identify $\mathcal{M}(K)$ with the dual of $C(K)$ by the duality pairing

$$\langle \mu, g \rangle := \int_K g d\mu , \quad g \in C(K) , \quad \mu \in \mathcal{M}(K) .$$

The orbit of $x \in K$ is the set $\mathcal{O}(x)$ of all translations of x, and this set is compact iff x is h-periodic for some finite h. Such a function generates a probability measure ε_x on K which is defined by

$$\langle \varepsilon_x, g \rangle := \frac{1}{h} \int_0^h g(T_\tau x) d\tau , \quad g \in C(K) ,$$

where T_τ is a translation operator acting on K defined by $T_\tau x(t) := x(t-\tau)$, and is extended to a mapping on $C(K)$ by $T_\tau g(x) := g(T_\tau x)$. $\mathcal{I}(K)$ stands for the set of all probability measures on K which are invariant with respect to the action of operator $T_\tau^{\#}$, which is defined as

$$\langle T_\tau^{\#} \mu, g \rangle := \langle \mu, T_\tau^{-1} g \rangle ,$$

and it can be shown is weakly* closed and convex. It is important to note that such a set could be also called a set of 'translatory' invariant measures, which is somewhat misleading since invariance with respect to translations on space $\mathcal{M}(K)$ in the classical sense means something different. However, we refer to ε_x as to elementary invariant measure, and we say that $\mathcal{I}(K)$ is a set of invariant measures. A classical result assures that ε_x is the unique invariant measure supported on $\mathcal{O}(x)$. As usual, $H^2_{per}(\Omega)$ denotes the set of all $H^2_{loc}(\mathbf{R})$ functions, extended by periodicity outside Ω. $BV(\Omega; \mathbf{R})$ is the set of all real functions with bounded variation on Ω, and it is endowed with a natural norm. Finally, following [1], by Sx we denote the set of all discontinuities for some $x \in K$.

2 Some preliminaries

Before we proceed, we give a precise meaning to the notions mentioned in section 1.

Definition 1. **(Γ-convergence)** Let us consider a metric space X and a family of functionals $F_\varepsilon : X \longrightarrow [0, +\infty]$. Such a family Γ-converges to a functional F, which is written as $F_\varepsilon \xrightarrow{\Gamma} F$, if the following holds:

(i) For any sequence x_ε of points in X such that $x_\varepsilon \xrightarrow{X} x$ it follows that $\liminf_\varepsilon F_\varepsilon(x_\varepsilon) \geq F(x)$.
(ii) For any point y in X there exists a sequence y_ε such that $y_\varepsilon \xrightarrow{X} y$ and $\limsup_\varepsilon F_\varepsilon(y_\varepsilon) \leq F(y)$.

We say that a family F_ε continuously converges to a functional F if in (i) we can write equality as well, and we write $F_\varepsilon \xrightarrow{C} F$.

Definition 2 (Relaxation). Let us consider a metric space X and a functional $J : X \longrightarrow [0, +\infty]$. **Relaxed functional** $\overline{J}$ is any lower semicontinuous functional acting on the space $\overline{X}$, where $X \hookrightarrow \overline{X}$, $\overline{X}$ is compact and has the property

(i) $(\forall\, x \in X) \quad J(x) = \overline{J}(x)$.
(ii) $\inf_{x \in X} J(x) = \min_{\overline{x} \in \overline{X}} \overline{J}(\overline{x})$.

Definition 3 (Young Measures). Let $\varphi : \Omega \longrightarrow K$ be a Borel measurable function. A mapping $\delta_\varphi : \Omega \longrightarrow \mathcal{M}(K)$ defined by

$$\delta_\varphi(s) := \delta_{\varphi(s)}$$

is said to be **the elementary Young measure** associated to φ. The set of all elementary Young measures is denoted by $EYM(\Omega; K)$.

By $YM(\Omega; K)$ we denote the set of all **Young measures**, i. e. the completion of $EYM(\Omega; K)$ in the natural weak* topology, namely the topology on the dual of $\mathrm{L}^1(\Omega; \mathrm{C}(K))$ induced by natural duality pairing

$$\langle \nu, \Phi \rangle := \int_\Omega \langle \nu_s, \Phi_s \rangle ds \ , \quad \Phi_s \in \mathrm{C}(K) \ .$$

Although the apparatus of Young measures has been deployed in recent years by a number of researchers, it seems that there is still no unifying approach to their elementary properties.

We will use a version of the fundamental theorem for Young measures adopted from a short survey in [1], section 2 (see [3] for other comments and results).

Theorem 1. *For every sequence of measurable maps $u^k : \Omega \longrightarrow K$ there exists a subsequence (not relabeled) which generates the Young measure ν, i.e.,*

$$\delta_{u^k} \overset{*}{\rightharpoonup} \nu \ .$$

Moreover, ν has the following properties:

(i) If $f : \Omega \times \mathbf{R} \longrightarrow \mathbf{R}$ is a Carathéodory function satisfying $|f(s, x)| \leq h(s)$ for some $h \in \mathrm{L}^1(\Omega)$, then

$$\lim_{k \longrightarrow +\infty} \int_\Omega f(s, u^k(s)) ds = \int_\Omega \langle \nu_s, f_s \rangle ds \ .$$

(ii) The sequence of maps (u^k) converges in measure to some $u : \Omega \longrightarrow K$ if and only if ν is the elementary Young measure associated to u.

3 Formulation of a two-scale problem

Definition 4 (Blowup). Let us consider a function $v : \Omega \longrightarrow \mathbf{R}$. The ε-blowup of function v is denoted by $R^\varepsilon v : \Omega \longrightarrow K$ and is defined at a point $s \in \Omega$ by the mapping $R_s^\varepsilon v : \mathbf{R} \longrightarrow \mathbf{R}$, $R_s^\varepsilon v(t) := \varepsilon^{-\frac{1}{3}} v(s + \varepsilon^{\frac{1}{3}} t)$.

Such a transformation has been used in [1] to study the effects mentioned in the introduction, and we now turn our attention to a precise formulation of the procedure developed in the mentioned paper.

- Characterize the class of all Young measures $\nu \in YM(\Omega; K)$ which are generated by sequences of ε-blowups of functions v^ε.
- Rewrite the rescaled functionals $\varepsilon^{-\frac{2}{3}} I^\varepsilon(v)$ as $\int_\Omega f_s^\varepsilon(R_s^\varepsilon v) ds$ for a suitable choice of $R_s^\varepsilon v$ and f_s^ε.
- Identify the Γ-limit f_s of f_s^ε for almost every $s \in \Omega$.
- Identify the Γ-limit of appropriately defined relaxed functionals on the space $YM(\Omega; K)$.
- Determine the minimizer for the relaxed functional at the limit and prove its uniqueness.

In our paper we focus on the first three steps of the program. To begin with, we use the following blowup, rather than the one originally introduced (see section 3 for details):

$$R_s^\varepsilon v(t) := \varepsilon^{-\frac{1}{3}} \left(v(s + \varepsilon^{\frac{1}{3}} t) - g(s) - g'(s) \varepsilon^{\frac{1}{3}} t \right) .$$

Firstly, we claim that the Young measure ν generated by a sequence of blowups is invariant with respect to translations.

Proposition 1. *If $\delta_{R^\varepsilon v^\varepsilon} \xrightarrow{\ *\ } \nu$ in $YM(\Omega; K)$, then $\nu_s \in \mathcal{I}(K)$ for a.e. $s \in \Omega$.*

Proof. Let us define $u^\varepsilon(s) := R_s^\varepsilon v^\varepsilon$ and calculate

$$\begin{aligned} T_\tau^{-1}(u^\varepsilon(s))(t) - u^\varepsilon(s + \varepsilon^{\frac{1}{3}} \tau)(t) &= u^\varepsilon(s)(t + \tau) - R^\varepsilon_{s+\varepsilon^{\frac{1}{3}}\tau} v^\varepsilon(t) \\ &= -\, \varepsilon^{-\frac{1}{3}} g(s) + g'(s + \varepsilon^{\frac{1}{3}} \tau) t \\ &\quad + \varepsilon^{-\frac{1}{3}} g(s + \varepsilon^{\frac{1}{3}} \tau) - g'(s)(t + \tau) . \end{aligned}$$

Now an application of Remark 2.6 and Lemma 2.7 in [1] yields the invariance since g' is continuous. □

We proceed by noting that, since v is periodic on Ω, the functional I^ε : $\mathrm{H}^2_{per}(\Omega) \longrightarrow [0,+\infty]$ defined as

$$I^\varepsilon(v) := \int_\Omega \Big(\varepsilon^2 v''(s)^2 + W(v'(s)) + a(s)(v(s)-g(s))^2\Big)ds$$

can now be written in the form

$$\varepsilon^{-\frac{2}{3}} I^\varepsilon(v) = \int_\Omega f_s^\varepsilon(x_s)ds \ ,$$

where

$$x_s := R_s^\varepsilon v \ , \quad f_s^\varepsilon(x_s) = \gamma_s^\varepsilon(x_s) + h_s^\varepsilon(x_s) \ ,$$

and

$$\gamma_s^\varepsilon(x_s) := \frac{1}{2r}\int_{-r}^{r} \Big(\varepsilon^{\frac{2}{3}} x_s''(t)^2 + \varepsilon^{-\frac{2}{3}} W(x_s'(t) - g'(s))\Big)dt \ ,$$

$$h_s^\varepsilon(x_s) := \frac{1}{2r}\int_{-r}^{r} a_s^\varepsilon(t)\big(x_s(t) - R_s^\varepsilon g(t)\big)^2 dt \ , \quad a_s^\varepsilon(t) := a(s+\varepsilon^{\frac{1}{3}} t) \ .$$

Since g' is continuous, we deduce

$$a_s^\varepsilon R_s^\varepsilon g \longrightarrow 0 \quad \text{a.e. } s \in \mathbf{R} \ ,$$

and therefore

$$h_s^\varepsilon \xrightarrow{\ C\ } h_s \ ,$$

where

$$h_s(x) := a(s)\frac{1}{2r}\int_{-r}^{r} x(t)^2 dt \ .$$

Here we note that the invariance property and continuous convergence from above is still preserved (by Rademacher's Theorem) also in the case where g is Lipschitz-continuous function. Next part of our consideration is to prove that the integrands f_s^ε Γ-converge in the space K, assuring therefore that requirements of Step 3 are fulfilled also in this case. Choice of our rescaling reduces such a claim to an application of a classical Γ-convergence result, namely the famous Theorem of L. Modica and S. Mortola (cf. [7]).

Definition 5. Let $I \subset \mathbf{R}$ be a fixed interval. A function x belonging to K is said to be of the class $\mathcal{S}^s(I)$ if the following holds: x is continuous on I, x is piecewise affine on I, and $x' \in \{g'(s)-1, g'(s)+1\}$ on I. If x is extended from I to $\mathbf{R}$ by periodicity, and satisfies $x \in \mathcal{S}^s(J)$ for any interval J, we say that x belongs to $\mathcal{S}^s_{per}(I)$.

Theorem 2 (Modica-Mortola). *Let us choose $\Omega \subseteq \mathbf{R}^d$ bounded and open set, and let us define for $\varepsilon > 0$ a functional $G_\varepsilon : \mathrm{L}^1(\Omega) \longrightarrow [0,+\infty]$ by*

$$G_\varepsilon(v) := \begin{cases} \int_\Omega \varepsilon|\nabla v|^2 + \frac{1}{\varepsilon}W(v), & \textit{if } v \in \mathrm{H}^1(\Omega),\ |v| \le 1 \\ +\infty & \textit{otherwise.} \end{cases}$$

Then the family (G_ε) Γ*-converges to a functional* $G : \mathrm{L}^1(\Omega) \longrightarrow [0,+\infty]$ *defined by*

$$G(v) := \begin{cases} A_0|Dv|, & \text{if } y \in \mathrm{BV}(\Omega;\{-1,1\}) \\ +\infty & \text{otherwise,} \end{cases}$$

where $A_0 := 2\int_{-1}^{1}\sqrt{W(\xi)}d\xi$.

Corollary 1. *For almost every* $s \in \Omega$ *it holds* $f_s^\varepsilon \xrightarrow{\Gamma} f_s$, *where* $f_s^\varepsilon, f_s : K \longrightarrow [0,+\infty]$ *are defined as*

$$f_s^\varepsilon(x) := \begin{cases} \frac{1}{2r}\int_{-r}^{r}\varepsilon x''^2 + \frac{1}{\varepsilon}W(x' - g'(s)) + a_s^\varepsilon x^2, & \text{if } x \in \mathrm{H}^2\langle -r,r\rangle \\ +\infty & \text{otherwise,} \end{cases}$$

$$f_s(x) := \begin{cases} \frac{A_0}{2r}|Sx' \cap \langle -r,r\rangle| + a(s)\frac{1}{2r}\int_{-r}^{r}x^2, & \text{if } x \in \mathcal{S}\langle -r,r\rangle \\ +\infty & \text{otherwise.} \end{cases}$$

Proof. Let us define W_s by

$$W_s(\xi) := W(\xi - g'(s)) .$$

It is easily seen that the statement of the corollary is merely a reformulation of one-dimensional version of the Modica-Mortola theorem, taking into account that a constant A_0 is preserved by the choice of function W_s and class $\mathcal{S}^s$, and that the sum of a Γ-convergent and a continuously convergent sequence Γ-converges to the sum of corresponding limits. □

4 The choice of blowup

The reason why it is natural to consider such a particular blowup is not plausible and requires a further comment. First attempt would be to check if the very same blowup introduced in [1] leads to Γ-convergence of both integrands f_s^ε and functionals F_g^ε in such a way that the corresponding limits f_s and F_g admit a unique periodic minimizer, i. e., the Young measure generated by blowups of minimizers of F_g^ε (see section 4) is supported on the orbit of a single periodic function $y \in K$ which takes values in $\mathbf{R}$, rather than in the whole range of the pattern space. It is easy to verify that such an attempt fails since we can easily construct a minimizing sequence whose Young measure is concentrated at the function taking both values $-\infty$ and $+\infty$ on a set of strictly positive measure. In this case, when trying to rewrite I^ε in terms of f_s^ε, the penalizing effect appears in the lower order term alone, and therefore it is reasonable to look for another blowup which will allow penalization to influence the principal term of f_s^ε. Therefore, we turn our attention to the construction of a blowup which itself contains the penalizing term g. When doing so, an additional objective is to find a class of penalizing terms as large as possible, as it was emphasized in the introduction. Here we will briefly explain some of the difficulties in that respect. Let us consider (for the time being) a function $g \in C^1(\Omega)$ and let us define the following blowup:

$$R_s^\varepsilon v(t) := \varepsilon^{-\frac{1}{3}}\left(v(s+\varepsilon^{\frac{1}{3}}t) - g(s)\right).$$

Now I^ε can be written in the form

$$\varepsilon^{-\frac{2}{3}} I^\varepsilon(v) = \int_\Omega f_s^\varepsilon(x_s)ds ,$$

where

$$f_s^\varepsilon(x_s) = \gamma_s^\varepsilon(x_s) + h_s^\varepsilon(x_s) ,$$

$$h_s^\varepsilon(x_s) := \frac{1}{2r}\int_{-r}^{r} a_s^\varepsilon(t)\left(x_s(t) + \varepsilon^{-\frac{1}{3}}g(s) - \varepsilon^{-\frac{1}{3}}g(s+\varepsilon^{\frac{1}{3}}t)\right)^2 dt ,$$

and

$$\gamma_s^\varepsilon(x_s) := \frac{1}{2r}\int_{-r}^{r}\left(\varepsilon^{\frac{2}{3}}x_s''(t)^2 + \varepsilon^{-\frac{2}{3}}W(x_s'(t))\right)dt , \quad x_s := R_s^\varepsilon v .$$

The continuity of g' implies

$$h_s^\varepsilon \xrightarrow{C} h_s ,$$

where

$$h_s(x) := a(s)\frac{1}{2r}\int_{-r}^{r}\left(x(t) - tg'(s)\right)^2 dt ,$$

so that, in fact,

$$f_s^\varepsilon \xrightarrow{\Gamma} f_s .$$

Thus, while trying to prove a result analogous to the main result in [1], subsection 3.5, in our setting we have so far verified condition (a). Conditions (b) and (c) follow from general theory. Indeed, the assumptions of Theorem 2.12 in [1] are fulfilled, and it remains to be checked that we are able to find a suitable family of transformations on K so that we can apply abstract results studied in sections 4 and 5. To this end, we define

$$G_g := \{T_{\tau_g} : \tau \in \mathbf{R}\} , \quad G := G_g \cup G_{-g} ,$$

where T_{τ_g} is a transformation on K acting as follows

$$T_{\tau_g}(x)(t) := x(t-\tau) + g'(s)t .$$

It turns out that G is a group with respect to composition,and now we are attempting to prove the invariance of ν. Consider $v_\varepsilon : \Omega \longrightarrow K$, a sequence of measurable functions extended by periodicity to $\mathbf{R}$, and a Young measure

ν generated by mappings $R^\varepsilon v^\varepsilon : \mathbf{R} \longrightarrow K$. Then we check if ν_s is invariant under the action of G for a.e. $s \in \mathbf{R}$. It is enough to prove that

$$T^\#_{\tau_g} \nu_s = \nu_s$$

for almost every $s \in \mathbf{R}$. Let us define $u^\varepsilon : \mathbf{R} \longrightarrow K$ by $u^\varepsilon(s) := R^\varepsilon_s v^\varepsilon$. Since G acts continuously on $YM(\Omega; K)$, we immediately deduce that the Young measure generated by the sequence $s \mapsto T_{\tau_g}(u^\varepsilon(s))$ is exactly $(T^\#_{\tau_g}\nu_s)_{s\in\Omega}$, supposing that

$$\delta_{u^\varepsilon} \xrightarrow{\ *\ } \nu \ .$$

Then we compute

$$\begin{aligned} T_{\tau_g}(u^\varepsilon(s))(t) &= u^\varepsilon(s)(t-\tau) + g'(s)t \\ &= u^\varepsilon(s - \varepsilon^{\frac{1}{3}}\tau)(t) + \varepsilon^{-\frac{1}{3}} g(s - \varepsilon^{\frac{1}{3}}\tau) \\ &\quad -\varepsilon^{-\frac{1}{3}} g(s)(t-\tau) + g'(s)t \ , \end{aligned}$$

and it is obvious that there is no way of getting around the singularity coming from last three terms, since we know that mappings $s \mapsto u^\varepsilon(s)$ and $s \mapsto u^\varepsilon(s + \varepsilon^{\frac{1}{3}}\tau)$ generate the same Young measure, and we have no invariance in the limit. On the other hand, if we let transformations in G to depend on s, it is not even clear how G acts on K. More precisely, it makes full sense to define the action of $T_{\tau_{g(s)}}$ on (every) x_s belonging to K, but not on arbitrary $x \in K$. Therefore, it would be natural to consider the following definition: An element $\nu \in YM(\Omega; K)$ is 'invariant' iff it holds

$$T^\#_{\tau_g} \nu = \nu \ ,$$

where $(T^\#_{\tau_g}\nu)_s := T^\#_{\tau_{g(s)}}\nu_s$, and, as usually, for $\varphi \in \mathrm{C}(K)$ we define

$$\langle T^\#_{\tau_{g(s)}}\nu_s, \varphi\rangle := \langle \nu_s, (T_{\tau_{g(s)}})^{-1}\varphi\rangle \ .$$

Provided that we want to pursue the matter in this direction, we should also define

$$\mathcal{I}(\Omega; K) := \{\nu \in YM(\Omega; K) : T^\#_{\tau_{g(s)}}\nu_s = \nu_s \quad \text{a.e. } s \in \mathbf{R}\} \ ,$$

and it is quite clear that these notions require restatement of a large number of nontrivial results obtained originally in [1]. Next attempt would be to try to define blowups which yield translation invariant measures ν_s in a usual sense. Previous calculations suggest that the penalizer should appear in the principal term of f^ε_s, which, combined with the fact that blowup must preserve the gradient of functions in a pattern space, leads to choice

$$R^\varepsilon_s v(t) := \varepsilon^{-\frac{1}{3}}\left(v(s + \varepsilon^{\frac{1}{3}}t) - g(s + \varepsilon^{\frac{1}{3}}t)\right) .$$

Main results in [1] can be restated now, as calculations which insure conditions (a), (b) and (d) in [1], section 3, present no difficulty. Having in mind that the form of this particular blowup by itself requires $g \in \mathrm{H}^2(\Omega)$, we rewrite the term $g(s+\varepsilon^{\frac{1}{3}}t)$ as a first order correction of $g(s)$, namely, we refine previous blowup to get

$$R_s^\varepsilon v(t) := \varepsilon^{-\frac{1}{3}}\left(v(s+\varepsilon^{\frac{1}{3}}t) - g(s) - g'(s)\varepsilon^{\frac{1}{3}}t\right),$$

and so we deal with f_s^ε defined in section 3.

5 Convergence of relaxed functionals

In order to complete Step 4, or, equivalently, to state a version of the main result in [1], let us recall that in our setting the notion of f_s-uniform approximability (cf. [1] , section 5) plays a crucial role, and by means of a following result a convergence result for relaxed functionals in the case $g=0$ can be proved.

Lemma 1. *Let $\mathcal{D}_g$ be a subset of $YM(\Omega;K)$ characterized as follows: $\nu \in \mathcal{D}_g$ if and only if we can find a partition $(I_j)_{j=1}^\infty$ of Ω up to a set of measure zero such that the restrictions $\nu : I_j \longrightarrow \mathcal{M}(K)$ are of the form $\nu_s = \varepsilon_{x_j^s}$ for almost every s in I_j, where $x_j^s \in \mathcal{S}^s_{per}(I_{h_j})$ and $I_{h_j} = \langle 0, h_j \rangle$. Then the set $\mathcal{D}_0$ is F_0-dense in $YM(\Omega;K)$, i.e., for every element $\nu \in YM(\Omega;K)$ such that $F_0(\nu) < +\infty$ there exists a sequence (ν_k) in $\mathcal{D}_0$ with the properties*

(i) $\nu_k \xrightarrow{*} \nu$
(ii) $\limsup_{k\longrightarrow\infty} F_0(\nu_k) \le F_0(\nu)$.

The proof of the claim is delicate and essentially guarantees the upper bound inequality for limits in the definition of Γ-convergence in the case of relaxed functionals. Details can be found in sections 3, 4 and 5 of [1].

Remark 1. At this point we note that the set $\mathcal{D}_g$ may not be F_g-dense in $YM(\Omega;K)$. For instance, if g' is injective, for chosen $\overline{s} \in \Omega$ the mapping

$$s \mapsto f_s(x^{\overline{s}}), \quad x^{\overline{s}} \in \mathcal{S}^{\overline{s}},$$

does not satisfy condition (c) in [1], section 3. To avoid such a difficulty, we approximate g by a sequence (g_k) of piecewise affine Lipschitz functions, we introduce notation

$$R_s^{\varepsilon,k} v(t) := \varepsilon^{-\frac{1}{3}}\left(v(s+\varepsilon^{\frac{1}{3}}t) - g_k(s) - g_k'(s)\varepsilon^{\frac{1}{3}}t\right)$$

for the blowup induced by g_k, and by $\mathcal{S}^{s,k}(I)$ we denote the class of all continuous piecewise affine functions on some interval I with derivative in the set $\{1+g_k'(s), -1+g_k'(s)\}$. In accordance with the preceding notation, we introduce integrands $f_s^{\varepsilon,k}$ and f_s^k acting on K so that $f_s^{\varepsilon,k} \xrightarrow{\Gamma} f_s^k$ for any $k \in \mathbf{N}$ as $\varepsilon \longrightarrow 0$. Another natural set, namely, $\cup_{k=1}^{+\infty} \mathcal{D}_{g_k}$ in general is not F_g-dense, and thus we are not able to apply the construction explained in [1] directly.

In view of the above remarks, we are able to prove the following result by combining Cantor diagonal argument and approximation by sequence (g_k): If we define the functionals $F_g^\varepsilon, F_g : YM(\Omega; K) \longrightarrow [0, +\infty]$ by

$$F_g^\varepsilon(\nu) := \begin{cases} \int_\Omega \langle \nu_s, f_s^\varepsilon \rangle ds, & \text{if } \nu_s = \delta_{R_s^\varepsilon v} \text{ for some } v \in \mathrm{H}^2_{per}(\Omega) \\ +\infty & \text{otherwise,} \end{cases}$$

$$F_g(\nu) := \begin{cases} \int_\Omega \langle \nu_s, f_s \rangle ds, & \text{if } \nu_s \in \mathcal{I}(K) \text{ for a. e. } s \in \Omega \\ +\infty & \text{otherwise,} \end{cases}$$

then $F_g^\varepsilon \xrightarrow{\Gamma} F_g$. In the forthcoming paper [9], we shall discuss the proof of the convergence in detail. As in [1], we expect that a unique minimizer for F_g is a non-homogeneous Young measure which is in almost every point an elementary invariant measure.

Acknowledgments. The author benefited greatly from discussions with Prof. Müller during his stay at the Max Planck Institute for Mathematics in the Sciences, Leipzig, Germany (1999/2000). Also, the author would like to thank Prof. Antonić for valuable comments during preparation of this paper. This work is supported in part by the Croatian Ministry of Science and Technology through project 037015–Oscillatory solutions of partial differential equations, and by DAAD foundation (324 A/99/08103).

References

1. Alberti, G., Müller, S., (2001) A new approach to variational problems with multiple scales. Comm. Pure Appl. Math. **54**, 761–825
2. DeSimone, A., Kohn, R.V., Müller, S., Otto, F. (2000) Magnetic microstructures, a paradigm of multiscale problems. In: Ball J.M., Hunt J.C.R. (Eds.) Proc. ICIAM 99 (Edinburgh, 1999). Oxford University Press, 175–190
3. Ball, J.M. (1989) A version of the fundamental theorem for Young measures, in PDE's and Continuum Models of Phase Transitions. In: Rascle, M. et al. (Eds.), Lecture Notes in Physics **344**, Springer, 207-215
4. Ball, J.M., James, R.D. (1987) Fine phase mixtures as minimizers of energy, Arch. Rational Mech. Anal. **100**, 13–52
5. Evans, L.C., Gariepy, R.F. (1992) Lecture notes on measure theory and fine properties of functions. CRC Press, Boca Raton
6. Kohn, R.V., Müller, S. (1992) Branching of twins near an austensite-twinned-martensite interface. Philosophical Magazine A **66**, 697–715
7. Modica, L., Mortola, S. (1977) Un esempio di Γ-convergenca, Bull. Un. Mat. Ital. (5) **14-B**, 285-299
8. Müller, S. (1993) Singular perturbations as a selection criterion for minimizing sequences, Calc. Var. **1**, 169–204

9. Raguž, A. (2002) preprint
10. Tartar, L. (1979) Compensated compactness and applications to partial differential equations. Nonlinear analysis and mechanics, Heriot-Watt symposium IV. Research notes in mathematics, Pitman, 136–192
11. Tartar, L. (1987) The appearance of oscillations in optimization problems. In: Knops, R.J., Lacey, A.A. (Eds.) Non-classical continuum mechanics. Proceedings of the London mathematical society symposium, Durham, July 1986, London mathematical society lecture notes series 122, Cambridge University Press, 129–150
12. Tartar, L. (1995) Beyond Young measures, Meccanica **30**, 505–526
13. Young, L.C. (1980) Lectures on the calculus of variations and optimal control theory. Chelsea

Homogenization of a Free Boundary Problem: the no-Slip Condition

Ben Schweizer

Institut für Angewandte Mathematik, Universität Heidelberg, INF 294,
D-69120 Heidelberg, Germany

Abstract. We present results on the homogenization of a fluid problem with a free boundary. We make reference to results published in 2000. There we have used a slip-condition along the walls in order to give a rigorous derivation of the limit equations. One expects that the same limit equations hold also for a no-slip condition. This contribution analyzes the specific difficulties that appear in the case of a no-slip condition; we present some results that indicate how homogenization could be proved in this case.

1 Setting of the fiber problem

The behavior of water in a porous medium is a typical object of interest in the theory of homogenization. Our aim is to study this physical system in the case that a third phase (air) is present. We have to deal with a fluid-air interface (a free boundary) within the porous medium.

We use the geometry of the problem and the equations as they were introduced in [2]. In this contribution we are particularly interested in equation (4) below. For small m it has the character of a slip condition, for large m it has the character of a no-slip condition. In [2] we were able to give homogenization results in the case $m \leq 1$.

The condition $m \leq 1$ was cause for discussions. One would expect that the homogenized equations remain unchanged in the case $m > 1$. We want to give here conditions on how this extended result can be achieved. We present the setting of the problem in this section and report results of [2] and [3] in section 2. In section 3 we deal with the no-slip condition and present new calculations concerning this case. It will be necessary to use test-functions that satisfy the no-slip condition. With their help it is possible to derive the equation for the trace of the pressure in terms of the limit η_0. The situation is more involved when it comes to the closure condition that relates η_0 with limits of h^ε, the functions $h_0(x)$ and $h_1(x, y)$. Here one needs a result concerning the product of two unknowns (η^ε with ∇h^ε). We derive relation (25) which is formally equivalent to (24), the one needed to conclude the homogenization result.

In order to simplify the geometry we study the following setting: The porous medium consists in a 3D volume where cylinders of length 1 and cross section of size ε are taken as obstacles (holes). The cylinders are composed as

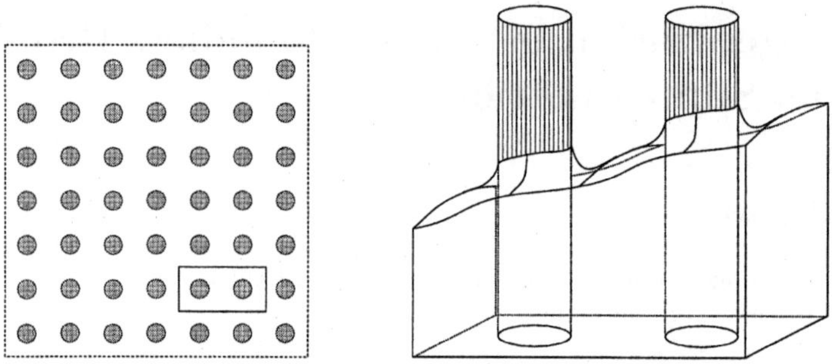

Fig. 1. Domain of h^ε and the local geometry

vertical columns in a periodic fashion and with distance of order ε between neighbouring fibers.

We next give the mathematical description of the problem. We use the following domains:

$$\begin{array}{rcll} R &=& \{(y_1,y_2)|0<y_1,y_2<1\} & \text{the unit cell,}\\ Q &\subset\subset& R, \qquad \partial Q \in C^k & \text{section of the fibers,}\\ F &=& R\setminus Q & \text{microscopic fluid phase,}\\ \Pi &=& \{(x_1,x_2)|0<x_1,x_2<1\}, & \\ \Pi_\varepsilon &=& \Pi\setminus\varepsilon(\mathbb{Z}^2+Q) & \text{the base,}\\ \Omega_\varepsilon &=& \Pi_\varepsilon\times(0,\infty), \qquad \Omega=\Pi\times(0,\infty). & \end{array}$$

We consider a free boundary problem in Ω_ε. The dependent variables are $(v^\varepsilon, p^\varepsilon, h^\varepsilon)$. We use the notation $x = (\tilde{x}, x_3)$ and $\nabla = (\tilde{\nabla}, \partial_{x_3})$. For given functions $h^\varepsilon : \Pi_\varepsilon \to \mathbb{R}$ and $h_0 : \Pi \to \mathbb{R}$ we consider the domains $G_\varepsilon := \Omega_\varepsilon \cap \{x_3 < h^\varepsilon(\tilde{x})\}$ and $G_0 := \Omega \cap \{x_3 < h_0(\tilde{x})\}$.

We study the Stokes equations in this perforated domain. The scaling $\varepsilon^2\Delta$ is made in order to have the mean velocities of order 1. With normal vectors n and tangential vectors τ the equations are

$$-\varepsilon^2\Delta v^\varepsilon(x) + \nabla p^\varepsilon = f \qquad \text{in } G_\varepsilon, \tag{1}$$

$$\nabla\cdot v^\varepsilon = 0 \qquad \text{in } G_\varepsilon, \tag{2}$$

$$v^\varepsilon\cdot n = 0 \qquad \text{on } \partial_l G_\varepsilon, \tag{3}$$

$$\varepsilon^m\tau\cdot\partial_n v^\varepsilon + \gamma\tau\cdot v^\varepsilon = 0 \qquad \text{on } \partial_l G_\varepsilon, \tag{4}$$

$$v^\varepsilon\cdot n = 0 \qquad \text{on } \partial_u G_\varepsilon, \tag{5}$$

$$T(v^\varepsilon,p^\varepsilon)\cdot n = \mathcal{H}(h^\varepsilon)\cdot n \qquad \text{on } \partial_u G_\varepsilon, \tag{6}$$

$$v^\varepsilon = 0 \qquad \text{on } \{x_3=0\}, \tag{7}$$

$$\sin(\text{contact angle}) = \alpha \qquad \text{on } \partial\Pi_\varepsilon. \tag{8}$$

The equations are complemented with the volume condition $\int_{\Pi_\varepsilon} h^\varepsilon = |F|$ to have average height 1. We used $\partial_u G_\varepsilon$ and $\partial_l G_\varepsilon$ for upper and lateral boundary. In equation (6) T stands for the stress-tensor and $\mathcal{H}$ denotes the mean curvature of the surface $\Gamma = graph(h)$. It will be useful to define

$$\eta^\varepsilon(\tilde{x}) := \frac{\nabla h^\varepsilon(\tilde{x})}{\sqrt{1+|\nabla h^\varepsilon(\tilde{x})|^2}} \tag{9}$$

and to use further on $\mathcal{H} = \nabla \cdot \eta^\varepsilon$.

2 Homogenization results

For solutions $(v^\varepsilon, p^\varepsilon, h^\varepsilon)$ we consider trivial extensions of these functions and the two-scale limits (v_0, p_0, h_0) (in fact, for the pressure first a pressure constant $\hat{p}(\varepsilon)$ has to be subtracted). The aim is to derive equations that are satisfied by the limit functions. To derive them we will additionally consider $\eta^\varepsilon(\tilde{x}) \to \eta_0(\tilde{x}, y)$ in the sense of two scale convergence on Π.

The limit problem is a set of equations for averaged velocity $V = \frac{1}{|F|}\int_F v_0$ and averaged pressure P living on a subdomain of Ω, combined with an equation for the averaged height h_0 defined on Π. The limit problem is again a free boundary value problem, the equations (11)–(13) hold on the domain $\Omega_{h_0} := \{(x_1, x_2, x_3) : x_3 < h_0(x_1, x_2)\}$.

$$P(\tilde{x}, h_0(\tilde{x})) = -\tilde{\nabla} \cdot E(\tilde{\nabla} h_0(\tilde{x})), \tag{10}$$

$$\partial_3 P(\tilde{x}, x_3) = f_3, \tag{11}$$

$$(V_1, V_2)(x) = M \cdot (f_1 - \partial_{x_1} P, f_2 - \partial_{x_2} P)(x), \tag{12}$$

$$\nabla \cdot V(x_1, x_2, x_3) = 0, \tag{13}$$

$$V_3(x_1, x_2, 0) = 0, \tag{14}$$

$$(V \cdot n)(\tilde{x}, h_0(\tilde{x})) = 0, \tag{15}$$

together with the normalization $\int_\Pi h_0 = 1$. The 2×2-matrix M is defined via a cell-problem and E is a nonlinear map defined on a neighborhood U of $0 \in \mathbb{R}^2$. $E : U \to \mathbb{R}^2$ is defined as follows.

$$E(\xi) := \frac{1}{|F|}\int_F \eta(y)\, dy, \tag{16}$$

where the function $\eta : F \to \mathbb{R}^2$ solves the following auxiliary problem:

$$\tilde{\nabla} \cdot \eta(y) = \alpha \frac{|\partial Q|}{|F|}, \qquad y \in F, \tag{17}$$

$$\tilde{n} \cdot \eta(y) = \alpha \qquad y \in \partial Q, \tag{18}$$

$$\eta(.) = \frac{\xi + \nabla\varphi(.)}{\sqrt{1 + |\xi + \nabla\varphi(.)|^2}},$$

for some periodic function $\varphi : F \to \mathbb{R}$.

The main result of [2] is: Averages of the functions $v^\varepsilon, p^\varepsilon, h^\varepsilon$ converge for $\varepsilon \to 0$ to the functions V, P, h_0 solving the limit problem (10)–(15).

3 The case of a no-slip condition

All but two steps in the analysis of the described problem in [2] and [3] are essentially independent of the exponent m of the slip condition. The two points where the condition $m \leq 1$ is used are proposition 2.11 and lemma 3.3 of [2].

In this section we analyze under what conditions the two steps can be made independent of m. We start by quoting (parts of) proposition 2.11. It provides the first relation for the pressure on the boundary. One has to be careful with an unbounded pressure constant that appears in the problem. We assume that $p^\varepsilon - \hat{p}(\varepsilon)$ is bounded in L^2 and use $p_{-1} = -\alpha\frac{|\partial Q|}{|F|}$.

Proposition 1 (Equations for P). *There exists the limit*

$$\bar{p}_0 := \lim_{\varepsilon \to 0} [\hat{p}(\varepsilon) - \varepsilon^{-1} p_{-1}]. \tag{19}$$

For $m \leq 1$ the two-scale limit p_0 of $p^\varepsilon - \hat{p}(\varepsilon)$ satisfies

$$p_0(x,y) + \bar{p}_0 = \begin{cases} P(x) & \forall y \in F, \\ 0 & \forall y \in Q, \end{cases} \tag{20}$$

with

$$\mathrm{tr}(P)(\tilde{x}) = -\tilde{\nabla}_x \cdot \left(\frac{1}{|F|} \int_F \eta_0(\tilde{x}, y)\, dy \right). \tag{21}$$

Note that the condition $m \leq 1$ appears in the proposition.

Here the following definition of a trace is used:

Definition 1. We use a smooth function $\chi : \mathbb{R} \to \mathbb{R}$ that is identical to 1 on $(-1,1)$ and has its support on $(-2,2)$. We set $\chi_\delta(z) := \chi(\frac{z}{\delta})$. Let $\psi : \Pi \to \mathbb{R}$ be a smooth function. For an L^1-function $P : G_0 \to \mathbb{R}$ we consider the limit

$$\lim_{\delta \to 0} \int_{G_0} P(\tilde{x}, x_3) \chi'_\delta(x_3 - h_0(\tilde{x})) \psi(\tilde{x})\, d\tilde{x}\, dx_3. \tag{22}$$

If the limit exists then it defines a linear functional $\lambda : \psi \mapsto \lambda(\psi)$. If λ is independent of the cut-off function χ then we write for the limit

$$\lambda(\psi) =: \int_\Pi \mathrm{tr}(P)(\tilde{x}) \psi(\tilde{x})\, d\tilde{x},$$

thus defining the distribution $\mathrm{tr}(P)$. On smooth domains and for continuous functions P it coincides with the usual trace.

Claim. The proposition remains valid for $m > 1$.

The proof uses (as the original) the following ε and ϕ dependent terms $\mathcal{A}_1$ to $\mathcal{A}_6$. The weak form of the equations states that these terms $\mathcal{A}_1$ to $\mathcal{A}_6$ sum up to zero for every smooth test-function ϕ.

$$\mathcal{A}_1 = 2\varepsilon^2 \int_{G_\varepsilon} Dv^\varepsilon : D\phi, \qquad \mathcal{A}_2 = -\int_{G_\varepsilon} p^\varepsilon \nabla \cdot \phi,$$

$$\mathcal{A}_3 = \varepsilon^{2-m}\gamma \int_{\partial_l G_\varepsilon} v^\varepsilon \cdot \phi, \qquad \mathcal{A}_6 = -\int_{G_\varepsilon} f \cdot \phi,$$

$$\mathcal{A}_4 = \int_{\Pi_\varepsilon} \eta^\varepsilon(\tilde{x}) \cdot \tilde{\nabla} \left(\phi(\tilde{x}, h^\varepsilon(\tilde{x})) \cdot n^\varepsilon \sqrt{1 + |\tilde{\nabla} h^\varepsilon|^2} \right) d\tilde{x},$$

$$\mathcal{A}_5 = -\alpha \int_{\partial \Pi_\varepsilon} \phi_3(\tilde{x}, h^\varepsilon(\tilde{x}))\, d\tilde{x}.$$

To show equation (21) we insert

$$\phi(x) = \chi_\delta(x_3 - h_0(\tilde{x}))\psi(\tilde{x})\varphi_\rho(\frac{\tilde{x}}{\varepsilon})e_3$$

with χ_δ as in definition 1 and a periodic function $\varphi_\rho \in C_0^\infty(F, \mathbb{R})$ such that $\varphi_\rho \to \chi_F$ for $\rho \to 0$. With the 2D-exterior normal $\tilde{n}$ on F we require that $\nabla \varphi_\rho$ converges to $-\tilde{n}\delta_{\partial Q}$ as a measure. The function φ_ρ was not needed in the case $m \leq 1$. We calculate

$$-(\hat{p}(\varepsilon) - \varepsilon^{-1}p_{-1}) \int_{\Pi_\varepsilon} \psi(\tilde{x})\, d\tilde{x} \to -\bar{p}_0|F| \int_{\Pi} \psi(\tilde{x})\, d\tilde{x},$$

$$\begin{aligned}
\mathcal{A}_2 + \hat{p}(\varepsilon) \int_{\Pi_\varepsilon} \psi(\tilde{x})\, d\tilde{x} &= -\int_{G_\varepsilon} (p^\varepsilon - \hat{p}(\varepsilon))(x)\partial_3\phi(x)\varphi_\rho(\frac{\tilde{x}}{\varepsilon})\, dx \\
&= -\int_{\Omega} (\bar{p}^\varepsilon - \hat{p}(\varepsilon))(x)\chi_\delta'(x_3 - h_0(\tilde{x}))\psi(\tilde{x})\varphi_\rho(\frac{\tilde{x}}{\varepsilon})\, dx \\
&\overset{\rho\to 0}{\longrightarrow} -\int_{\Omega} (\bar{p}^\varepsilon - \hat{p}(\varepsilon))(x)\chi_\delta'(x_3 - h_0(\tilde{x}))\psi(\tilde{x})\chi_F(\frac{\tilde{x}}{\varepsilon})\, dx \\
&\overset{\varepsilon\to 0}{\longrightarrow} -\int_{G_0} (-\bar{p}_0 + P(x))|F|\chi_\delta'(x_3 - h_0(\tilde{x}))\psi(\tilde{x})\, dx,
\end{aligned}$$

where $\bar{p}^\varepsilon$ is the extension of p^ε by $\hat{p}(\varepsilon)$. We next calculate for $\mathcal{A}_4$:

$$\begin{aligned}
\mathcal{A}_4 - \varepsilon^{-1}p_{-1} \int_{\Pi_\varepsilon} \psi(\tilde{x})\, d\tilde{x} &= \int_{\Pi_\varepsilon} \eta^\varepsilon \cdot \tilde{\nabla}\phi - \varepsilon^{-1}p_{-1} \int_{\Pi_\varepsilon} \psi(\tilde{x})\, d\tilde{x} \\
= \int_{\Pi_\varepsilon} \eta^\varepsilon \cdot \tilde{\nabla}\psi \cdot \varphi_\rho(\frac{\cdot}{\varepsilon}) &+ \frac{1}{\varepsilon}\int_{\Pi_\varepsilon} \eta^\varepsilon \cdot \psi \cdot \tilde{\nabla}\varphi_\rho(\frac{\cdot}{\varepsilon}) - \varepsilon^{-1}p_{-1} \int_{\Pi_\varepsilon} \psi(\tilde{x})\, d\tilde{x} \\
\overset{\rho\to 0}{\longrightarrow} \int_{\Pi_\varepsilon} \eta^\varepsilon \cdot \tilde{\nabla}\psi &- \int_{\partial\Pi_\varepsilon} \eta^\varepsilon \cdot \tilde{n}\psi + \varepsilon^{-1}\alpha\frac{|\partial Q|}{|F|} \int_{\Pi_\varepsilon} \psi(\tilde{x})\, d\tilde{x} \\
\overset{\varepsilon\to 0}{\longrightarrow} \int_{\Pi}\int_{F} \eta_0(\cdot, y)\, dy \cdot \tilde{\nabla}\psi.&
\end{aligned}$$

We used continuity of p^ε and of η^ε. In the last line we have used the contact angle condition $\eta^\varepsilon \cdot \tilde{n} = \alpha$ and the limit

$$-\frac{1}{|\partial Q|}\int_{\partial\Pi_\varepsilon} \psi(\tilde{x})\ dS(\tilde{x}) + \varepsilon^{-1}\frac{1}{|F|}\int_{\Pi_\varepsilon} \psi(\tilde{x})\ d\tilde{x} \to 0 \qquad (23)$$

for $\varepsilon \to 0$. Equation (23) can be shown for smooth compactly supported (or periodic) ψ with the help of a cell-wise Taylor expansion of ψ: The first term in the expansion (corresponding to cell-wise constant ψ) vanishes and the second term in the expansion (corresponding to cell-wise linear ψ) is a sum over all cells $\sum_j l(\nabla\psi)$ with some linear function $l : \mathbb{R}^2 \to \mathbb{R}$. This term converges to the integral of the derivatives and vanishes as well.

The expression $\mathcal{A}_1$ tends to zero and $\mathcal{A}_3$ and $\mathcal{A}_5$ vanish, since ϕ vanishes on the boundary of the obstacles. Summing up the preceding expressions we have shown that the following limits exist. They have the values

$$0 = \lim_{\delta\to 0}\lim_{\varepsilon\to 0}\lim_{\rho\to 0}\sum_i \mathcal{A}_i = -\lim_{\delta\to 0}\int_{G_0} P(x)\chi_\delta'(x_3 - h_0(\tilde{x}))\psi(\tilde{x})\ dx$$
$$+\int_\Pi\int_F \eta_0(\tilde{x},y)\cdot\tilde{\nabla}\psi(\tilde{x})\ dy\ d\tilde{x}.$$

Therefore $\mathrm{tr}(P)$ exists in the sense of Definition 1 and (21) holds.

The no-slip condition appears again in the derivation of the closing condition in lemma 3.3 of [2]. The limit of interest there is

$$\int_{\Pi_\varepsilon} \eta_\varepsilon\cdot\nabla h^\varepsilon \to |F|\int_\Pi \mathrm{tr}(P)(\tilde{x})\ h_0(\tilde{x})\ d\tilde{x} + \alpha|\partial Q|\int_\Pi\int_F \Psi\nabla h_1, \quad (24)$$

where $\Psi : F \to \mathbb{R}^2$ is the periodic function with div $(\Psi) = \frac{1}{|F|}$ in F and $\tilde{n}\cdot\Psi = \frac{1}{|\partial Q|}$ on ∂Q. We next give a version of this formula that can still be shown in the case $m > 1$.

Proposition 2. *The solutions $(v^\varepsilon, p^\varepsilon, h^\varepsilon)$ satisfy in the limit $\varepsilon \to 0$*

$$\int_{\Pi_\varepsilon} \eta_\varepsilon\cdot\nabla(h^\varepsilon\cdot\varphi) - \frac{1}{\varepsilon}p_{-1}\int_{\Pi_\varepsilon}(h^\varepsilon\cdot\varphi) \to \left(\int_F\varphi\right)\int_\Pi \mathrm{tr}(P)(\tilde{x})h_0(\tilde{x})\ d\tilde{x} \quad (25)$$

for every periodic $\varphi \in C_0^\infty(F,\mathbb{R})$.

Proof. The proof uses the test-function

$$\phi(\tilde{x},x_3) = h^\varepsilon(\tilde{x})\chi_\delta(x_3 - h_0(\tilde{x}))\varphi(\frac{\tilde{x}}{\varepsilon})e_3$$

with the cut-off function χ_δ of Definition 1. We make use of

$$\int_{\Pi_\varepsilon} \eta_\varepsilon\cdot\nabla(h^\varepsilon\cdot\varphi) = \mathcal{A}_4 = -\mathcal{A}_2 - \mathcal{A}_1 - \mathcal{A}_3 - \mathcal{A}_5 - \mathcal{A}_6 = -\mathcal{A}_2 + o_1(\delta,\varepsilon)$$

with $\lim_{\delta\to 0}\lim_{\varepsilon\to 0} o_1(\delta,\varepsilon) = 0$. We can now calculate the following limit.

$$\begin{aligned}
&\int_{\Pi_\varepsilon} \eta_\varepsilon \cdot \nabla(h^\varepsilon \cdot \varphi) - \hat{p}(\varepsilon) \int_{\Pi_\varepsilon} h^\varepsilon \cdot \varphi \\
&\quad = -\mathcal{A}_2 + o_1(\delta,\varepsilon) - \hat{p}(\varepsilon) \int_{\Pi_\varepsilon} h^\varepsilon \cdot \varphi \\
&\quad = \int_{G_\varepsilon} (p^\varepsilon(x) - \hat{p}(\varepsilon))(x) h^\varepsilon(\tilde{x}) \chi'_\delta(x_3 - h_0(\tilde{x})) \cdot \varphi(\frac{\tilde{x}}{\varepsilon})\ dx + o_1(\delta,\varepsilon) \\
&\quad = \int_{G_0} (p^\varepsilon(x) - \hat{p}(\varepsilon))(x) h_0(\tilde{x}) \chi'_\delta(x_3 - h_0(\tilde{x})) \cdot \varphi(\frac{\tilde{x}}{\varepsilon})\ dx \\
&\qquad\qquad + o_1(\delta,\varepsilon) + o_2(\varepsilon,\delta) \\
&\quad \to \int_{G_0} p_0(x) h_0(\tilde{x}) \left(\int_F \varphi\right) \chi'_\delta(x_3 - h_0(\tilde{x}))\ d\tilde{x}\ dx_3 + \lim_{\varepsilon\to 0} o_1(\delta,\varepsilon),
\end{aligned}$$

with $o_2(\varepsilon,\delta)$ satisfying $\lim_{\varepsilon\to 0} o_2(\varepsilon,\delta) = 0$ by the uniform convergence of h^ε to h_0. Since the left hand side is independent of δ we conclude

$$\begin{aligned}
&\lim_{\varepsilon\to 0}\left\{\int_{\Pi_\varepsilon} \eta_\varepsilon \cdot \nabla(h^\varepsilon \cdot \varphi) - \hat{p}(\varepsilon) \int_{\Pi_\varepsilon} h^\varepsilon \cdot \varphi\right\} \\
&\quad = \lim_{\delta\to 0} \int_{G_0} p_0(x) \chi'_\delta(x_3 - h_0(\tilde{x})) h_0(\tilde{x}) \left(\int_F \varphi\right)\ dx \\
&\quad = \left(\int_F \varphi\right) \int_\Pi (\operatorname{tr}(P)(\tilde{x}) - \bar{p}_0) h_0(\tilde{x})\ d\tilde{x}
\end{aligned}$$

by definition 1. In the above calculation there is included on both sides the error term

$$\left(-\hat{p}(\varepsilon) + \frac{p_{-1}}{\varepsilon}\right) \int_{\Pi_\varepsilon} h^\varepsilon \cdot \varphi \to -\left(\int_F \varphi\right) \int_\Pi \bar{p}_0 h_0(\tilde{x})\ d\tilde{x}.$$

We have shown (25).

We show now the formal equivalence of (24) and (25): Letting φ tend to the characteristic function $\varphi(y) = \chi_F(y)$ we read off the equations of the lemma using

$$\begin{aligned}
\int_{\Pi_\varepsilon} \eta_\varepsilon \cdot \nabla(h^\varepsilon \cdot \varphi) &\sim \int_{\Pi_\varepsilon} \eta_\varepsilon \cdot \nabla h^\varepsilon - \int_{\partial\Pi_\varepsilon} n \cdot \eta_\varepsilon h^\varepsilon \\
&\sim \int_{\Pi_\varepsilon} \eta_\varepsilon \cdot \nabla h^\varepsilon - \alpha\frac{1}{\varepsilon} \int_{\partial Q} (h_0 + \varepsilon h_1),
\end{aligned}$$

and

$$\frac{1}{\varepsilon} p_{-1} \int_{\Pi_\varepsilon} (h^\varepsilon \cdot \varphi) \sim -\alpha \frac{|\partial Q|}{|\Gamma|} \frac{1}{\varepsilon} \int_F (h_0 + \varepsilon h_1).$$

The terms containing h_0 cancel and the terms containing h_1 coincide with the right hand side

$$\alpha|\partial Q| \int_\Pi \int_F \Psi \nabla h_1$$

of (24). The above reasoning could be made rigorous using detailed regularity of the functions h^ε. Note that in [3] estimates better than Lipschitz are shown for h^ε. A condition in order to conclude would be

$$\left\| h^\varepsilon - h_0 - \varepsilon h_1(.,\frac{\cdot}{\varepsilon}) \right\|_{C^0} = o(\varepsilon).$$

In this form this is not contained in [3].

References

1. G. Allaire (1992) Homogenization and two-scale convergence. SIAM J. Math. Anal. **23**, 1482-1518
2. B. Schweizer (2000) Homogenization of a fluid problem with a free boundary. Comm. Pure Appl. Math. **53**, 1118-1152
3. B. Schweizer (2000) Uniform estimates in two periodic homogenization problems. Comm. Pure Appl. Math. **53**, 1153-1179

Radiative Transfer with Many Spectral Lines

Rainer Wehrse[1,2]

[1] Institut f. Theoret. Astrophysik, Universität Heidelberg, Tiergartenstr. 15, D-69121 Heidelberg, Germany
[2] Interdisziplinäres Zentrum f. Wissenschaftl. Rechnen, Universität Heidelberg, Im Neuenheimer Feld 368, D-69120 Heidelberg, Germany

Abstract. The radiative transfer equation, a partial integro-differential equation, is of particular interest for astronomers since it links the spectral properties of the light received (e.g. on Earth) with the properties of the matter from the place of origin (e.g. a star) to the place of the observer.

One major difficulty in its solution is the strong variability of the extinction coefficient entering the equation (e.g. often by more than 6 dex in a small frequency interval). Furthermore, often contributions from more than 10^8 narrow spectral lines have to be included. This has essentially inhibited up to now the accurate consideration of photon fluxes and pressures in radiation-hydrodynamic modelling.

In this contribution two new algorithms developed in collaboration with B. Baschek and W. v. Waldenfels are introduced that allow the efficient calculation of radiation fields with many lines whenever the detailed spectral information is not required. In the first one the extinction coefficient is represented by a 'generalized opacity distribution function'. In a second method the line positions, strengths and profiles are described by a Poisson point process.

The resulting expressions which are valid both in static and differentially moving media become particularly convenient inside a very optically thick medium ("diffusion limit").

Contents

1 Introduction

For astronomers *photons* are by far the most important source of information about the celestial objects they are interested in. Therefore the analysis of light is of paramount importance in astronomy. Fortunately, it is usually sufficient to consider the time independent *specific intensity* $I_\nu(\mathbf{x}, \mathbf{n}, \nu)$, which is essentially a time average $< \mathbf{EE}^* >$ of the electric field vector making up the light. Usually, the specific intensity is either introduced heuristically (cf. [13,7]) by

$$dE = I_\nu(\mathbf{x}, \mathbf{n}, \nu) d\nu d\omega dF \cos\vartheta dt \tag{1}$$

(dE radiant energy of a beam of solid angle $d\omega$ in direction $\mathbf{n}$ with a frequency spread $\nu....\nu + d\nu$ passing per time interval dt through an area element dF at position $\mathbf{x}$ whose normal forms an angle ϑ with the beam direction) or by means of the photon distribution function $\phi(\mathbf{x}, \mathbf{n}, \nu)$ (cf. [8])

$$I_\nu(\mathbf{x}, \mathbf{n}, \nu) = \frac{h^4 \nu^3}{c^2} \phi(\mathbf{x}, \mathbf{n}, \nu) \tag{2}$$

(h Planck's constant, c speed of light).

Due their huge distances from the Earth the objects are usually not spatially resolved and therefore only their flux

$$F_\nu((\mathbf{x}, \mathbf{n}, \nu) = \int_{4\pi} I_\nu(\mathbf{x}, \mathbf{n}', \nu) \mathbf{n} \cdot d\omega' \tag{3}$$

can be observed.

Even in such a case the spectral resolution may still be extremely high; as an example see Fig. 1 where the observed fluxes $F(\lambda)$ of the cool dwarf stars GL1, GL887 and GL832 are displayed for a small wavelength range (about 0.6% of the range over which the human eye is sensitive). It is seen that the number of spectral lines, i.e. narrow depressions, is extremely high in these objects; higher resolution data (as e.g. shown in Fig. 2) even indicate that many features still contain unresolved line contributions. In fact, the modeling of this small range requires the inclusion of about 10^5 spectral lines (see Fig. 3); for the whole range observable today it implies $\sim 10^8$ lines. The positions, strengths, and shapes of the lines make it possible to determine not only the abundance of the various elements present but also the pressures, temperatures and velocity fields of the outer layers of the star.

If high-resolution data as those of Fig. 2 are to be modelled it seems unavoidable to take each line individually into account. However, often just the total flux or some average value over a certain not too small wavelength interval is required. In such a case the distribution of lines in Fig. 3 suggests a statistical treatment. For media without large-scale velocity fields such a modeling has been possible for several decades by means of *opacity distribution functions* and by means of *opacity sampling methods.* However, for differentially moving media these descriptions failed.

In this paper we review –mainly from an astrophysical point of view– a recent attempt by B. Baschek, W. v. Waldenfels and the author [18,3] to remove this problem by modelling the emergent line spectra and the radiative fluxes deep inside of media statistically by means of a Poisson point process. The aspects of probability theory that are involved are well established in

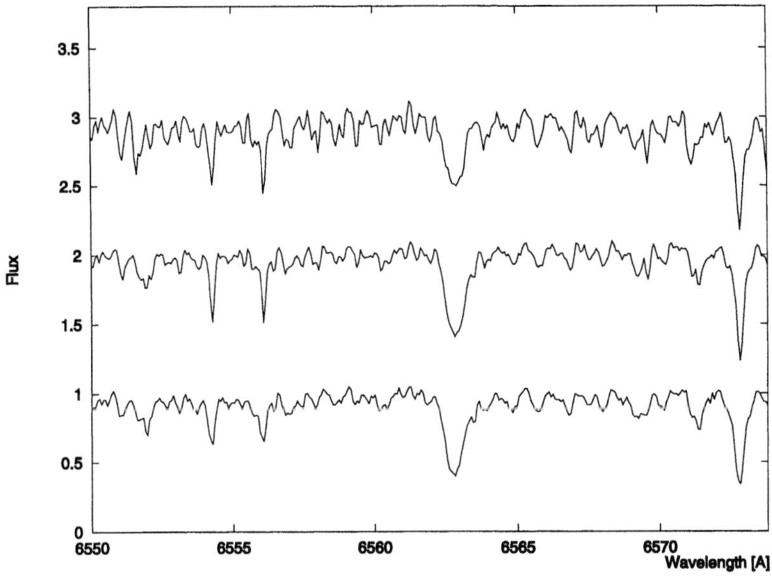

Fig. 1. High resolution spectrum for the cool dwarf stars GL1, GL887, and GL832 (from top to bottom) demonstrating the fast variation of the flux due to many spectral lines (in this case the stronger features are due to hydrogen, titanium, and calcium whereas the weaker ones are predominantly due to titanium oxide). Note that the wavelength range shown is in the red and covers only 0.6% of the sensitivity range of the human eye, which in turn is only a tiny fraction of the range accessible to astronomical instruments

mathematics for long time but –as far as we know– have never been used before in our context.

In the next section we introduce the radiative transfer equation and give some simple solutions. Subsequently (Sect. 3), we discuss the mean value of the specific intensity. In Section 4 a Poisson point process model is presented and the corresponding characteristic function are calculated. The distribution function is then derived by means of Levy's theorem, i.e. by means of a Fourier transformation (Sect. 5), and compared with one obtained by simulations. The following part (Sect. 6) is devoted to the diffusion limit, i.e. to the flux and the momentum transfer from photons to matter ('radiative acceleration') inside an optically thick medium far from the surface. We close with a discussion in which we particularly address presently open problems.

2 The radiative transfer equation and its solution

The specific intensity as measured in a frame comoving with the matter is governed by the *radiative transfer equation* which reads for unpolarized

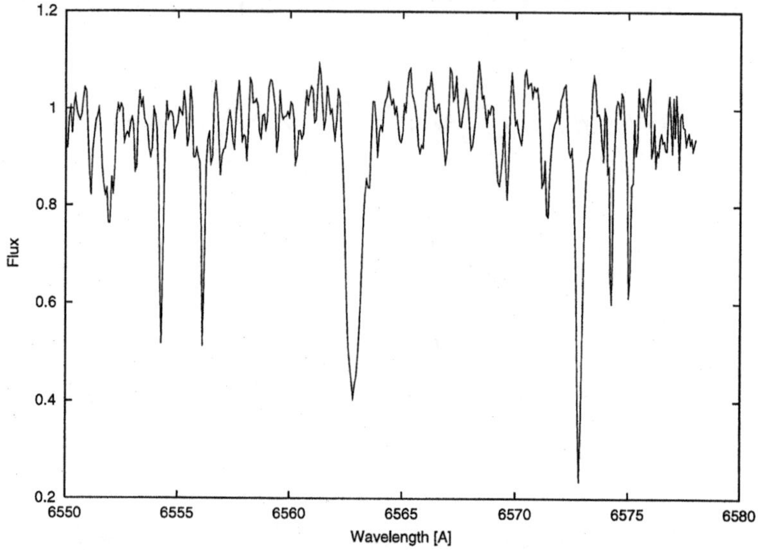

Fig. 2. Very high resolution spectrum of the cool dwarf star GL887 demonstrating that many lines still have substructures due to additional lines not seen in Fig. 1

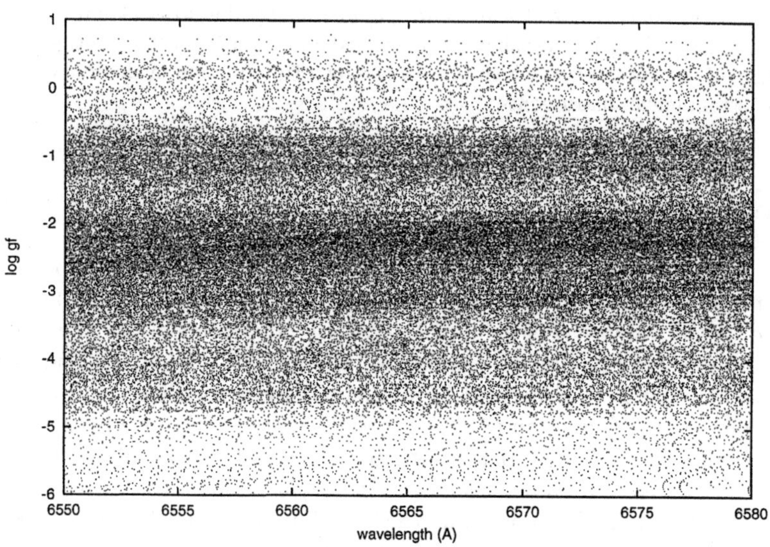

Fig. 3. Distribution of $\sim$ 96000 titanium oxide lines [10] used in the modelling of the spectrum of Fig. 2 in the wavelength-transition probability$\times$statistical weight plane; i.e. the vertical position is a measure of the strength of a line

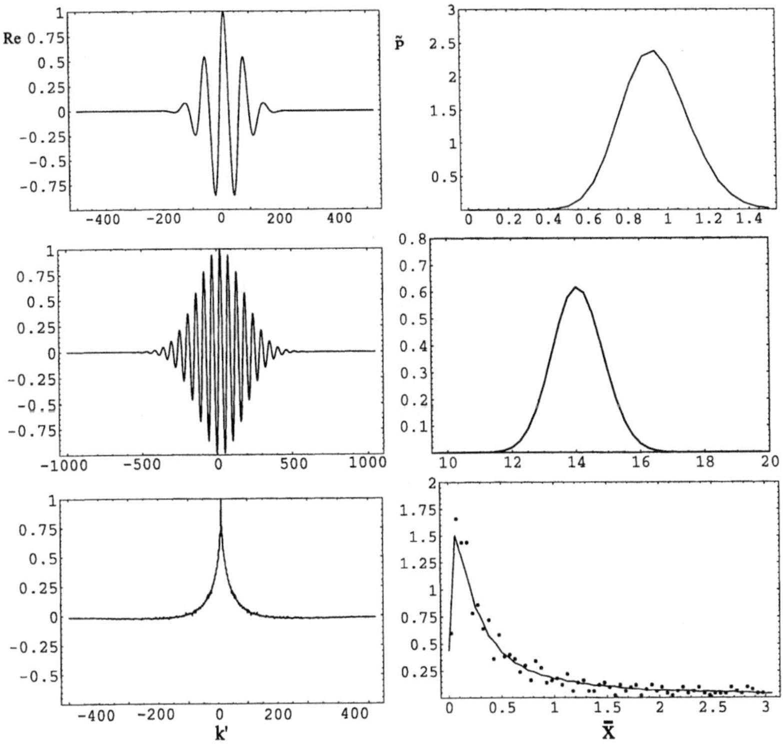

Fig. 4. Generalized opacity distribution functions $\tilde{p}(\bar{x};\xi,\Delta)$ (*right column*) and the real parts of their corresponding expectation values, $\mathrm{Re}(\langle\exp(-\bar{X}(\xi;\Delta)\cdot ik)\rangle)$ (*left column*), for Poisson distributed lines at an arbitrary (logarithmic) wavelength ξ. Shown are examples for three combinations of the line density ϱ and the averaging parameter Δ: (10,100), (100,100), and (1,0.1) (*top to bottom*), where $k = ak'$ with $a = 64$, 512, and 64 (*top to bottom*). For comparison, we have added (as *dots*) in the *lower right* panel values for the generalized opacity distribution function calculated according to the procedure described in Sect. 3. The lines have the same statistical properties in both cases. Note that the small wiggles in the *lower left* panel result from inaccuracies in the $\hat{\xi}$-integration (cf. Eq. 22) as a consequence of a highly oscillatory behavior of the integrand. From [3]

radiation in/from slowly moving stationary media (cf. [15])

$$\mathbf{n}\cdot\nabla I + w\frac{\partial I}{\partial \xi} = \frac{dI}{ds} + w\frac{\partial I}{\partial \xi} = -\chi(I-S) \quad \text{with} \quad w = \mathbf{n}\cdot\nabla(\boldsymbol{\beta}\cdot\mathbf{n})\,. \tag{4}$$

The ∇ operator acts on the spatial variables only. The first term lhs. is the transport term and the second describes the frequency shift due to the Doppler effect. $w = \mathbf{n}\cdot\nabla(\boldsymbol{\beta}\cdot\mathbf{n})$ is the gradient of the velocity $\boldsymbol{\beta}$ (in units of the speed of light c) projected on to the ray direction. w is frequency independent since we are using logarithmic frequencies $\xi = -\log\nu + const.$ $\chi(\mathbf{x},\xi)$ is the extinction coefficient so that the term $-\chi I$ takes care of photons lost from a beam by absorption and scattering. The term χS with $S = S(\mathbf{x},\mathbf{n},\xi)$ being the source function accounts for all photons added to the beam.

Depending on the situation the source function can have various forms (cf. [8]). An approximation often used is that of a two-level atom for which

$$\begin{aligned} S(\mathbf{x},\mathbf{n},\xi) &= \epsilon(\mathbf{x},\xi)B(\mathbf{x},\xi) \\ &+\frac{1-\epsilon(\mathbf{x},\xi)}{4\pi}\int_{4\pi}\int_0^\infty R(\mathbf{x},\xi,\xi',\mathbf{n},\mathbf{n}')I(\mathbf{x},\mathbf{n}',\xi')d\xi'd\omega' \end{aligned} \tag{5}$$

(ϵ de-excitation coefficient, B Planck function, R redistribution function, which represents the 'memory' of the scattering particle). By definition I and S are positive quantities. In this paper we assume in addition $\chi > 0$ and therefore exclude e.g. laser emission.

As appropriate boundary conditions the intensities I_{bc} impinging from the outside on to the medium are used. Note that other choices usually lead to unstable systems.

Although the transfer equation is usually derived by means of energy conservation arguments it is in fact a linearized Boltzmann equation. It has successfully been used in spectroscopy both of laboratory and celestial plasmas. However, its range of proper application is still unknown [11,6]

Since in this contribution we want to concentrate on the consequences of extinction coefficients that are highly variable in the frequency domain, we neglect in the following the spatial dependencies of the extinction coefficient and of the velocity gradient. If a variation is to be taken into account one can use the formulae below for sufficiently short distances and then add the contributions [9]. The source function S is assumed to be given. Then the solution of Eq. 4 along a ray in direction $\mathbf{n}$ can be written for $w \neq 0$ [1,2].

$$I(s,\xi;w) = \mathcal{I}_0(w) + \tfrac{1}{w}\int_{\xi-ws}^{\xi}\exp\left(-\tfrac{1}{w}\int_\eta^\xi \chi(\zeta)\,d\zeta\right)\cdot\chi(\eta)S\left(s-\tfrac{\xi-\eta}{w},\eta\right)d\eta \tag{6}$$

$$\begin{aligned} &= \mathcal{I}_0(w) + S(s,\xi) - S(0,\xi-ws)\exp\left(-\tfrac{1}{w}\int_{\xi-ws}^{\xi}\chi(\zeta)d\zeta\right) \\ &- \int_{\xi-ws}^{\xi}\exp\left(-\tfrac{1}{w}\int_\eta^\xi\chi(\zeta)d\zeta\right)\frac{dS\left(s-\frac{\xi-\eta}{w},\eta\right)}{d\eta}d\eta \end{aligned} \tag{7}$$

where

$$\mathcal{I}_0(w) = I(0,\xi-ws;w)\cdot\exp\left(-\tfrac{1}{w}\int_{\xi-ws}^{\xi}\chi(\zeta)d\zeta\right). \tag{8}$$

The second form is derived from the first one by partial integration. Note that the limit $w \to 0$ can easily be performed and that it leads us back to the well known solution for static media.

3 The expectation value of the specific intensity

If we assume that only the extinction coefficient varies stochastically with ξ we can simply write for the *expectation value of the specific intensity* according to Eqs. 6 to 8

$$\Big\langle I(s,\xi;w)\Big\rangle = \Big\langle \mathcal{I}_0(w)\Big\rangle + \tfrac{1}{w}\int_{\xi-ws}^{\xi}\left\langle \exp\left(-\tfrac{1}{w}\int_{\eta}^{\xi}\chi(\zeta)\,d\zeta\right)\cdot\chi(\eta)\right\rangle S\left(s-\tfrac{\xi-\eta}{w},\eta\right)d\eta \tag{9}$$

$$= \Big\langle \mathcal{I}_0(w)\Big\rangle + S(s,\xi) - S(0,\xi-ws)\left\langle \exp\left(-\tfrac{1}{w}\int_{\xi-ws}^{\xi}\chi(\zeta)d\zeta\right)\right\rangle - \int_{\xi-ws}^{\xi}\left\langle \exp\left(-\tfrac{1}{w}\int_{\eta}^{\xi}\chi(\zeta)d\zeta\right)\right\rangle \frac{dS\left(s-\frac{\xi-\eta}{w},\eta\right)}{d\eta}\,d\eta \tag{10}$$

where

$$\Big\langle \mathcal{I}_0(w)\Big\rangle = \left\langle I(0,\xi-ws;w)\cdot\exp\left(-\tfrac{1}{w}\int_{\xi-ws}^{\xi}\chi(\zeta)d\zeta\right)\right\rangle . \tag{11}$$

Eq. 10 indicates that the expectation value of the emergent specific intensity can be calculated in a straightforward way if the characteristic function

$$\varphi(s) = \left\langle \exp\left(-\tfrac{is}{\Delta}\int_{\xi-\Delta}^{\xi}\chi(\zeta)d\zeta\right)\right\rangle \tag{12}$$

of the probability density function $\tilde{p}$ for the extinction coefficient averaged over a ξ-interval $\Delta = ws$ is known. In the next sections we show how to calculate φ and $\tilde{p}$ by means of a Poisson point process.

As an alternative one can use the ergodic hypothesis and calculate $\bar{X}_i = \int_{\xi_i-\Delta}^{\xi_i}\chi(\zeta)d\zeta/\Delta$ at many positions ξ_i in a ξ-interval of suitable length. The normalized distribution of the $\bar{X}_i$ should then be a good approximation to $\tilde{p}$. Note that this procedure in the limit $w \to 0$ leads back to the way used by astrophysicists for decades in the static case [12].

In terms of the probability density the crucial term in Eq. 10 is then evidently given by

$$\left\langle e^{-\bar{X}(\xi;\Delta)\cdot z}\right\rangle = \int_0^\infty e^{-\bar{x}\cdot z}\tilde{p}(\bar{x};\xi,\Delta)\,d\bar{x}\,. \tag{13}$$

4 The Poisson point process

Although the Poisson point process (cf. [5,4]) is one of the most simple stochastic processes it seems to represent very well spectral line distributions that are known from the laboratory and/or from quantum-mechanical calculations [18]. In addition, the process is very flexible to account for the strengths and shapes of the lines and it can easily be combined with the solution of the radiative transfer solution as shown below.

In order to apply the concept of this process to our problem we have to realize that the extinction coefficient is composed of a continuous contribution χ_c which can be considered to be frequency independent and of contributions of individual lines which we write

$$X(\xi) = \sum_{l=1}^{L} \chi_l(\xi) = \sum_{l=1}^{L} \chi(\hat{\xi}_l, \vartheta_l, \xi - \hat{\xi}_l)\,, \tag{14}$$

Each contribution $\chi(\hat{\xi}_l, \vartheta_l, \xi - \hat{\xi}_l)$ can be assumed to be the product of the strength A_l and the profile function Φ which depends on the frequencies ξ, the wavelength of the line center $\hat{\xi}$, and on the type of the line (as eg. Lorentzian or Gaussian). ϑ_l is used to summarize the line parameters.

$$\chi(\hat{\xi}, \vartheta, \xi - \hat{\xi}) = A \cdot \Phi(\hat{\xi}, \gamma, \xi - \hat{\xi})\,. \tag{15}$$

We now assume that within a given ξ-interval $\xi \ldots \xi + \Delta\xi$ the mean density of lines with properties ϑ (i.e. strengths, shape etc.) is given by $\rho(\hat{\xi}, \vartheta)$ and that the actual number L is given by a Poisson distribution with mean $\langle L \rangle = \rho(\xi, \vartheta)\Delta\xi\Delta\vartheta$, i.e.

$$\mathbb{P}\{L = n\} = \frac{\langle L \rangle^n}{n!}\, \mathrm{e}^{-\langle L \rangle}\,. \tag{16}$$

The centers of the lines $\hat{\xi}$ are assumed to be a sequence of independent, identically distributed random variables: the $(\hat{\xi}_i, \vartheta_i)$ $(i = 1, 2, \ldots)$ form a *Poisson point process.*

The characteristic function can now be calculated in a straightforward way. Since we later need a slightly more complicated expression we give it here for

$$h(\hat{\xi}_l, \vartheta_l, \xi - \hat{\xi}_l) = \Lambda \cdot \chi(\hat{\xi}_l, \vartheta_l, \xi - \hat{\xi}_l) - \frac{1}{w} \int_{\xi - ws}^{\xi} \chi(\hat{\xi}_l, \vartheta_l, \zeta - \hat{\xi}_l)\, d\zeta\,. \tag{17}$$

Since

$$\left\langle \exp\left(\sum_{l=1}^{L} h(\hat{\xi}_l, \vartheta_l, \xi - \hat{\xi}_l) \right) \right\rangle = \sum_{n=0}^{\infty} \frac{\mathrm{e}^{-\varrho(S)}}{n!} \int_S \!\!\ldots\! \int \Big(\varrho(\hat{\xi}_1, \vartheta_1) \ldots \varrho(\hat{\xi}_n, \vartheta_n) \Big)$$

$$\times \exp\left(\sum_{l=1}^{n} h(\hat{\xi}_l, \vartheta_l, \xi - \hat{\xi}_l) \right) d\hat{\xi}_1 d\vartheta_1 \ldots d\hat{\xi}_n d\vartheta_n$$

$$= \mathrm{e}^{-\varrho(S)} \sum_{n=0}^{\infty} \frac{1}{n!} \left[\int_S \varrho(\hat{\xi}, \vartheta) \mathrm{e}^{h(\hat{\xi}, \vartheta, \xi - \hat{\xi})} d\hat{\xi} d\vartheta \right]^n$$

$$= \exp\left(\int\limits_{\mathcal{S}} \varrho(\hat{\xi}, \vartheta) \left\{ \mathrm{e}^{h(\hat{\xi}, \vartheta, \xi - \hat{\xi})} - 1 \right\} d\hat{\xi} d\vartheta \right) \quad (18)$$

with $\mathcal{S}$ being the set of possible ϑ-values, the characteristic function reads for the moving case

$$\varphi(k) = \left\langle \mathrm{e}^{-\bar{X}(\xi;\Delta)\cdot ik} \right\rangle = \exp\left(\int_{\Theta}\int_{-\infty}^{\infty} \varrho(\hat{\xi}, \vartheta) \left\{ \mathrm{e}^{-ik\,U} - 1 \right\} d\hat{\xi}\, d\vartheta \right) \quad (19)$$

with

$$U \equiv U(\hat{\xi}, \xi, \Delta, \gamma, A) = \frac{1}{\Delta} \int_{\xi-\Delta}^{\xi} A\Phi(\hat{\xi}, \gamma, \zeta - \hat{\xi})\, d\zeta\,. \quad (20)$$

and for static media ($w = 0$)

$$\left\langle \mathrm{e}^{-X(\xi)\cdot ik} \right\rangle = \Omega(\xi, \xi; 0) = \left\langle \exp\left(-\sum_{l=1}^{L} \chi(\hat{\xi}_l, \vartheta_l, \xi - \hat{\xi}_l) \cdot ik \right) \right\rangle \quad (21)$$

$$= \exp\left[\int_{\Theta} \int_{-\infty}^{\infty} \varrho(\hat{\xi}, \vartheta) \left(\mathrm{e}^{-\chi(\hat{\xi}, \vartheta, \xi - \hat{\xi})\cdot ik} - 1 \right) d\hat{\xi} d\vartheta \right].$$

5 The opacity distribution function

Using Levy's theorem (see e.g. [5]) and setting $\Lambda = 0$ we can immediately derive the distribution function $\tilde{p}$ of $1/\Delta \int_{\xi-\Delta}^{\xi} \chi(\zeta) d\zeta$ from Eq. 19

$$\tilde{p}(\bar{x}; \xi, \Delta) = \frac{1}{2\pi} \int_{-\infty}^{\infty} \mathrm{e}^{ik\bar{x}} \cdot \left\langle \mathrm{e}^{-\bar{X}(\xi;\Delta)\cdot ik} \right\rangle dk \quad (22)$$

$$= \frac{1}{2\pi} \int_{-\infty}^{\infty} \exp\left[ik\,\bar{x} + \int_{A_1}^{A_2} \int_{-\infty}^{\infty} \varrho(\hat{\xi}, A) \left\{ \mathrm{e}^{-ik\,A\,U} - 1 \right\} d\hat{\xi}\, dA \right] dk$$

$$= \frac{1}{\pi} \int_{0}^{\infty} \exp\left[\int_{A_1}^{A_2} \int_{-\infty}^{\infty} \varrho(\hat{\xi}, A) \left\{ \cos\left(k\,A\,U\right) - 1 \right\} d\hat{\xi}\, dA \right]$$

$$\times \cos\left\{ k\bar{x} - \int_{A_1}^{A_2} \int_{-\infty}^{\infty} \varrho(\hat{\xi}, A) \sin\left(k\,A\,U\right) d\hat{\xi}\, dA \right\} dk\,. \quad (23)$$

Three examples for the real part of the characteristic function and the resulting distribution function are shown in Fig. 4.

6 The diffusion limit

In celestial objects one often has to calculate the radiative flux very deep inside a medium where the extinction coefficient does not vary much over a photon free mean path $1/\chi_c$ and the Planck function can be approximated by a linear function. Then the radiation field is essentially local and one refers to the *diffusion limit.* Whereas for static media already in 1924 Rosseland has derived very convenient expressions for both the monochromatic and the frequency integrated case, no corresponding formulae have been available until very recently [15–17] for the differentially moving case.

In order to derive these expressions we introduce the *spectral thickness*

$$\psi(\xi) = \int_{\xi_1}^{\xi} \chi(\zeta) d\zeta. \tag{24}$$

Since for the conditions stated the boundary conditions do not contribute we find from Eq. 7 for the flux $\mathcal{F}(s_0, \xi; w) = I(s_0, \mathbf{n}, \xi) - I(s_0, -\mathbf{n}, \xi)$ in direction $\mathbf{n}$ at position s_0 and at (logarithmic) frequency ξ in a medium extending from $s = 0$ to $s = \tilde{s}$ and expanding/collapsing with velocity gradient w

$$\begin{aligned}\mathcal{F}(s_0, \xi; w) = &\Big(p(s_0, \xi) - q(s_0, \xi, \mathbf{n})\Big) \exp\left[\frac{-\psi(\xi) + \psi(\xi - w s_0)}{w}\right] \\ &-p(s_0, \xi) \exp\left[\frac{-\psi(\xi) + \psi(\xi - w(\tilde{s} - s_0))}{w}\right] \\ &+q(s_0, \xi, \mathbf{n}) \int_{s_0}^{\tilde{s}} \exp\left[\frac{-\psi(\xi) + \psi(\xi + w(s_0 - \ell))}{w}\right] d\ell \\ &+q(s_0, \xi, \mathbf{n}) \int_{0}^{s_0} \exp\left[\frac{-\psi(\xi) + \psi(\xi - w(s_0 - \ell))}{w}\right] d\ell\,. \end{aligned} \tag{25}$$

Here $p(s, \xi)$ and $q(s, \xi)$ are values of the Planck function (at s and ξ) and of its gradient in direction $\mathbf{n}$, resp.. For a fixed value of ξ we then obtain in the limit s_0, $(\tilde{s} - s_0) \to \infty$

$$\begin{aligned}\mathcal{F}(s_0, \xi; w) &= 2q(s_0, \xi, \mathbf{n}) \int_0^{\infty} \exp\left(-\frac{\psi(\xi) - \psi(\xi - w\ell)}{w}\right) d\ell \\ &= 2q(s_0, \xi, \mathbf{n}) \int_0^{\infty} \exp\left(-\frac{1}{w} \int_{\xi - w\ell}^{\xi} \chi(\zeta) d\zeta\right) d\ell\,. \end{aligned} \tag{26}$$

Equation 26 is basic expression for the flux in the diffusion approximation.

Since in most actual situations the velocity gradients are small and in order to get some more insight into the consequences of spectral lines, edges etc. we expand the exponential term around $w = 0$ up to second order

$$\begin{aligned}\exp\left(-\frac{\psi(\xi) - \psi(\xi - ws)}{w}\right) \simeq \mathrm{e}^{-\psi'(\xi)\cdot s} \Bigg[&1 + \frac{1}{2} s^2 \psi''(\xi)\, w \\ &+ \left(\frac{1}{8} s^4 \psi''(\xi)^2 - \frac{1}{6} s^3 \psi'''(\xi)\right) w^2\Bigg] + O(w^3). \end{aligned} \tag{27}$$

Then we can perform the depth integration analytically and obtain for the "w correction factor" θ to second order in w

$$\theta(s_0,\xi;w) = 1 + \frac{\psi''(\xi)}{\psi'(\xi)^2}\,w + \frac{1}{\psi'(\xi)^3}\left(3\frac{\psi''(\xi)^2}{\psi'(\xi)} - \psi'''(\xi)\right)w^2 \tag{28}$$

Replacing $\psi'(\xi)$ by $\chi(\xi)$ etc., the flux in the diffusion limit becomes

$$\mathcal{F}(s_0,\xi;w) = \mathcal{F}(s_0,\xi)\cdot\left[1 - \frac{\partial}{\partial\xi}\frac{1}{\chi(\xi)}\cdot w + \frac{1}{2}\frac{\partial^2}{\partial\xi^2}\frac{1}{\chi(\xi)^2}\cdot w^2\right]. \tag{29}$$

Since in most cases only the total flux is needed we integrate over frequency to obtain

$$\mathcal{F}_{\rm tot}(s_0;w) = \mathcal{F}_{\rm tot}(s_0)\cdot\left[1 + \eta_1(s_0)\cdot w + \eta_2(s_0)\cdot w^2\right], \tag{30}$$

with

$$\begin{aligned}\eta_1(s_0) &= -\bar{\chi}_{\rm R}(s_0)\int_{-\infty}^{\infty}\frac{1}{\chi(\xi)}\frac{\partial}{\partial\xi}\left(\frac{1}{\chi(\xi)}\right)G(s_0,\xi)\,d\xi \\ &= -\frac{1}{2}\,\bar{\chi}_{\rm R}(s_0)\int_{-\infty}^{\infty}\frac{\partial}{\partial\xi}\left(\frac{1}{\chi(\xi)}\right)^2 G(s_0,\xi)\,d\xi\,, \qquad (31)\\ \eta_2(s_0) &= +\frac{1}{2}\,\bar{\chi}_{\rm R}(s_0)\int_{-\infty}^{\infty}\frac{1}{\chi(\xi)}\frac{\partial^2}{\partial\xi^2}\left(\frac{1}{\chi(\xi)}\right)^2 G(s_0,\xi)\,d\xi\,.\end{aligned}$$

(32)

Here $G(s_0,\xi)$ is the weighting function

$$G(s_0,\xi) = \frac{\partial B(T,\xi)}{\partial T}\Big/\frac{\partial B(T)}{\partial T} \tag{33}$$

with

$$B(T) = \int_{-\infty}^{\infty} B(T,\xi)e^{\xi}d\xi. \tag{34}$$

The 'Rosseland mean opacity' $\bar{\chi}_{\rm R}(s_0)$ defined by

$$\bar{\chi}_{\rm R}(s_0) = \int_{-\infty}^{\infty}\frac{1}{\chi(s_0,\xi)}G(s_0,\xi)d\xi \tag{35}$$

and the flux

$$\mathcal{F}_{\rm tot}(s_0) = 2\mathbf{n}\cdot\nabla T\int_{-\infty}^{\infty}\frac{\partial B(T,\xi)}{\partial T}\frac{1}{\chi(s_0,\xi)}e^{\xi}d\xi = \frac{2\mathbf{n}\cdot\nabla T}{\bar{\chi}_{\rm R}(s_0)}\frac{\partial B(T)}{\partial T} \tag{36}$$

refer to the static case.

Equally, we can define an effective opacity

$$\frac{1}{\bar{\chi}_\beta(s_0;w)} = \frac{1}{\bar{\chi}_{\rm R}(s_0)}\cdot\left[1 + \eta_1(s_0)\cdot w + \eta_2(s_0)\cdot w^2\right], \tag{37}$$

which is of interest since up to now in all radiation-hydrodynamical models precalculated values of $\bar{\chi}_{\mathrm{R}}$ are used without any correction terms and this expression gives the opportunity to include the effects of motion with a minimum of code changes.

In a similar way we find for the radiative acceleration

$$\begin{aligned} a_{\mathrm{rad,tot}}(s_0; w) &= \frac{1}{c}\int_{-\infty}^{\infty} \chi(s_0,\xi)\mathcal{F}(s_0,\xi;w)\exp(\xi)d\xi \\ &= a_{\mathrm{rad,tot}}(s_0)\cdot\left[1+\tilde{\eta}_1(s_0)\cdot w + \tilde{\eta}_2(s_0)\cdot w^2\right] \end{aligned} \tag{38}$$

with

$$a_{\mathrm{rad,tot}}(s_0) = \frac{2}{c}\frac{\partial B}{\partial T}\mathbf{n}\cdot\nabla T \tag{39}$$

being the static value and

$$\tilde{\eta}_1(s_0) = -\int_{-\infty}^{\infty} \frac{\partial}{\partial\xi}\left(\frac{1}{\chi(\xi)}\right) G(s_0,\xi)\,d\xi\,, \tag{40}$$

$$\tilde{\eta}_2(s_0) = \frac{1}{2}\int_{-\infty}^{\infty} \frac{\partial^2}{\partial\xi^2}\left(\frac{1}{\chi(\xi)}\right)^2 G(s_0,\xi)\,d\xi\,. \tag{41}$$

Since the flux in the deep interior of a medium can hardly be measured, it is here even more advantageous than for the intensity at the surface to have expressions for the expectation value. We find

$$\Big\langle F(s_0,\xi;w)\Big\rangle = 2g(s_0,\xi)\int_0^{\infty} \mathrm{e}^{-\chi_c(\xi)\cdot s}\,\Omega(\xi,\xi-ws;w)\,ds \tag{42}$$

with

$$\begin{aligned} \Omega(\xi,\xi-ws;w) = \exp\Bigg(\int_{\Theta}\int_{-\infty}^{\infty} &\varrho(\hat{\xi},\vartheta) \\ \times\Bigg\{\exp&\left(-\frac{1}{w}\int_{\xi-ws}^{\xi}\chi(\hat{\xi},\vartheta,\zeta-\hat{\xi})d\zeta\right)-1\Bigg\}\,d\hat{\xi}d\vartheta\Bigg)\,. \end{aligned} \tag{43}$$

Fig. 5 shows the dependence of the expectation value of the flux for the diffusion approximation on the velocity gradient w. It is seen that the flux is a decreasing function of w, i.e. that lines get increasingly more important for larger velocity gradient. This implies e.g. that in order to transport a given flux the temperature gradient has to be larger in a moving medium than in a static one.

For the radiative acceleration the identity

$$\langle U\cdot V\rangle \equiv \frac{\partial}{\partial\Lambda}\left\langle \mathrm{e}^{\Lambda U}\cdot V\right\rangle\Big|_{\Lambda=0}\,, \tag{44}$$

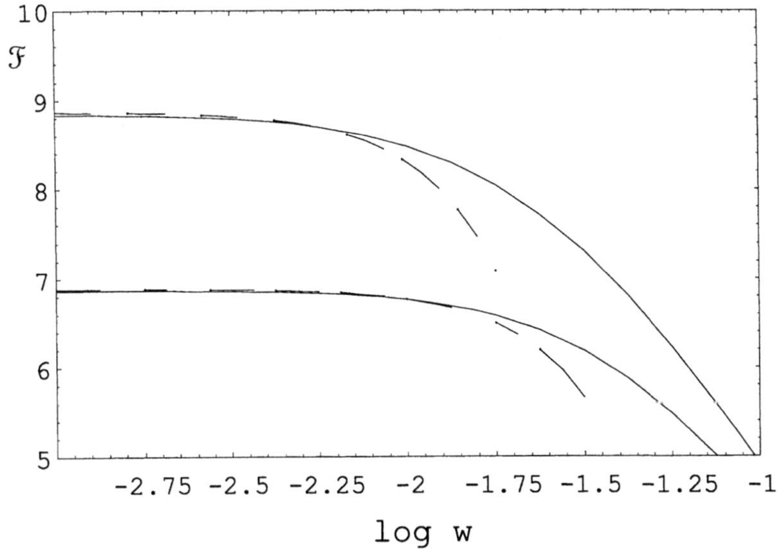

Fig. 5. Variation of the flux inside an optically very thick medium with velocity gradient w. according to Eq. 42 with $2g = 1$ (full curves) and according to the approximation Eq. 50 (broken curves). The upper curves refer to lines of Lorentzian shape with parameters strength $A = 1$, halfwidth $\gamma = 10^{-3}$, density $\rho = 1$ on a continuum $\chi_c = 0.1$. The lower curves have identical parameters except for $\gamma = 10^{-2}$. From [17]

which holds for any two functions U and V, and Eq. 18 have to be used so that

$$\begin{aligned}
\Big\langle a_{\text{rad}}(s_0,\xi;w)\Big\rangle &= \frac{1}{c}\Big\langle \chi(s_0,\xi)\cdot F(s_0,\xi;w)\Big\rangle \\
&= \frac{1}{c}\Bigg[\chi_c(\xi)\cdot\Big\langle F(s_0,\xi;w)\Big\rangle \\
&\quad + \Bigg\langle \sum_{l=1}^{L}\chi(\hat{\xi}_l,\vartheta_l,\xi-\hat{\xi}_l)\cdot F(s_0,\xi;w)\Bigg\rangle\Bigg] \\
&= \frac{2g(s_0,\xi)}{c}\int_0^\infty \mathrm{e}^{-\chi_c(\xi)\cdot s}\,\Omega_{\text{rad}}(\xi,\xi-ws;w)\,ds \qquad (45)
\end{aligned}$$

where

$$\begin{aligned}
\Omega_{\text{rad}}(\xi,\xi-ws;w) &= \Omega(\xi,\xi-ws;w) \\
&\quad\times\Bigg(\chi_c(\xi)+\int_\Theta\int_{-\infty}^{\infty}\varrho(\hat{\xi},\vartheta)\chi(\hat{\xi},\vartheta,\xi-\hat{\xi}) \\
&\quad\times\exp\left(-\frac{1}{w}\int_{\xi-ws}^{\xi}\chi(\hat{\xi},\vartheta,\zeta-\hat{\xi})d\zeta\right)d\hat{\xi}d\vartheta\Bigg)\ . \qquad (46)
\end{aligned}$$

An approximate expression for the flux in the diffusion approximation at small w requires the expansion of Ω around $w = 0$ up to second order

$$\Omega(\xi, \xi - ws; w) = \Omega(\xi, \xi; 0) \cdot \left[1 + \omega_1 \cdot w + \frac{1}{2}\omega_2 \cdot w^2\right] \tag{47}$$

with $\Omega(\xi, \xi; 0)$ given by Eq. (21) and

$$\begin{aligned}\omega_1 &= \frac{1}{\Omega(\xi,\xi;0)} \cdot \left.\frac{\partial \Omega(\xi, \xi - ws; w)}{\partial w}\right|_{(\xi,\xi;0)} \\ &= \frac{s^2}{2} \int_\Theta \int_{-\infty}^{\infty} \varrho(\hat{\xi}, \vartheta)\, \chi'(\hat{\xi}, \vartheta, \xi - \hat{\xi})\, \mathrm{e}^{-\chi(\hat{\xi},\vartheta,\xi-\hat{\xi})s} d\hat{\xi} d\vartheta\,, \end{aligned} \tag{48}$$

$$\begin{aligned}\omega_2 &= \frac{1}{\Omega(\xi,\xi;0)} \cdot \left.\frac{\partial^2 \Omega(\xi, \xi - ws; w)}{\partial w^2}\right|_{(\xi,\xi;0)} \\ &= \int_\Theta \int_{-\infty}^{\infty} \varrho(\hat{\xi}, \vartheta) \left[\frac{s^4}{4}[\chi'(\hat{\xi}, \vartheta, \xi - \hat{\xi})]^2 \right. \\ &\qquad \left. - \frac{s^3}{3}\chi''(\hat{\xi}, \vartheta, \xi - \hat{\xi})\right] \mathrm{e}^{-\chi(\hat{\xi},\vartheta,\xi-\hat{\xi})s} d\hat{\xi} d\vartheta \\ &\quad + \frac{s^4}{4}\left(\int_\Theta \int_{-\infty}^{\infty} \varrho(\hat{\xi}, \vartheta)\, \chi'(\hat{\xi}, \vartheta, \xi - \hat{\xi}) \times \mathrm{e}^{-\chi(\hat{\xi},\vartheta,\xi-\hat{\xi})s} d\hat{\xi} d\vartheta\right)^2 .\end{aligned} \tag{49}$$

Here χ' and χ'' denote the first and second derivative of $\chi(\hat{\xi}, \vartheta, \xi - \hat{\xi})$ with respect to $\hat{\xi}$. We stress that $\Omega(\xi, \xi; 0)$ as well as ω_1 and ω_2 depend not only on ξ but also on s and hence cannot be taken out of the respective integrals in the expression for the monochromatic flux so that

$$\begin{aligned}\left\langle F(s_0, \xi; w)\right\rangle &= 2g(s_0, \xi) \int_0^\infty \mathrm{e}^{-\chi_c(\xi)s} \\ &\quad \times \Omega(\xi, \xi; 0) \cdot \left[1 + \omega_1 \cdot w + \frac{1}{2}\omega_2 \cdot w^2\right] ds \end{aligned} \tag{50}$$

cannot be simplified further. In a completely analogous way one finds for the radiative acceleration

$$\begin{aligned}\left\langle a_{\mathrm{rad}}(s_0, \xi; w)\right\rangle &= \frac{2g(s_0, \xi)}{c} \int_0^\infty \mathrm{e}^{-\chi_c(\xi)\cdot s} \\ &\quad \times \Omega_{\mathrm{rad}}(\xi, \xi; 0) \cdot \left[1 + \tilde{\omega}_1 \cdot w + \frac{1}{2}\tilde{\omega}_2 \cdot w^2\right] ds\,, \end{aligned} \tag{51}$$

where

$$\begin{aligned}\Omega_{\mathrm{rad}}(\xi, \xi; 0) &= \Omega(\xi, \xi; 0) \cdot \Big(\chi_c(\xi) \\ &\quad + \int_\Theta \int_{-\infty}^{\infty} \varrho(\hat{\xi}, \vartheta) \chi(\hat{\xi}, \vartheta, \xi - \hat{\xi}) \cdot \mathrm{e}^{-\chi(\hat{\xi},\vartheta,\xi-\hat{\xi})\cdot s} d\hat{\xi} d\vartheta\Big)\,, \end{aligned} \tag{52}$$

$$\tilde{\omega}_1 = \omega_1 + \frac{\Omega(\xi,\xi;0)}{\Omega_{\rm rad}(\xi,\xi;0)} \cdot \hat{\omega}_1 \,, \tag{53}$$

and

$$\tilde{\omega}_2 = \omega_2 + \frac{\Omega(\xi,\xi;0)}{\Omega_{\rm rad}(\xi,\xi;0)} \cdot (2\omega_1\hat{\omega}_1 + \hat{\omega}_2) \tag{54}$$

with

$$\hat{\omega}_1 = \frac{s^2}{2}\int_\Theta \int_{-\infty}^{\infty} \varrho(\hat{\xi},\vartheta)\,\chi(\hat{\xi},\vartheta,\xi-\hat{\xi})\,\chi'(\hat{\xi},\vartheta,\xi-\hat{\xi})\,\mathrm{e}^{-\chi(\hat{\xi},\vartheta,\xi-\hat{\xi})s}d\hat{\xi}d\vartheta \tag{55}$$

and

$$\hat{\omega}_2 = \int_\Theta \int_{-\infty}^{\infty} \varrho(\hat{\xi},\vartheta)\chi(\hat{\xi},\vartheta,\xi-\hat{\xi}) \times \left[\frac{s^4}{4}[\chi'(\hat{\xi},\vartheta,\xi-\hat{\xi})]^2 - \frac{s^3}{3}\chi''(\hat{\xi},\vartheta,\xi-\hat{\xi})\right] \mathrm{e}^{-\chi(\hat{\xi},\vartheta,\xi-\hat{\xi})s}d\hat{\xi}d\vartheta \,. \tag{56}$$

Analogously to the expression for the flux, also the quantities $\Omega_{\rm rad}$, $\tilde{\omega}_1$ and $\tilde{\omega}_2$ in the monochromatic acceleration, depend on ξ *and* on s.

7 Discussion

In this paper we have reviewed the statistical treatment of radiation fields with many spectral lines by means of a generalized opacity distribution function and by means of a Poisson point process with special emphasis on the diffusion limit. The expressions derived have all been based on an analytical solution of the transfer equation that has been found recently. The density distribution (and therefore the geometry) has been assumed to be smooth and simple. In addition, we have assumed that the source function is given, i.e. we have essentially neglected scattering.

In spite of these simplifications our expressions are suitable for many applications (as e.g. for Mira stars, novae and accretion disks in astrophysics) and can reduce by orders of magnitude the computation times and memory requirements of radiation-hydrodynamical modeling; in fact, they make for the first time a modeling of such objects possible with realistic radiation fields.

In view of the progress in observational accuracy and sensitivity it is evident from this review that it also is highly desirable to develop in addition algorithms that allow to take spatial inhomogeneities and varying scattering fractions ϵ (see Sect. 2) statistically into account. It would in particular be advantageous if a differential equation for the mean values of the radiation field could be derived.

Acknowledgments. I would like to express my sincere thanks to B. Baschek and W. v. Waldenfels for their collaboration: without them most results presented here would not have emerged. Very helpful discussions with A. Mikelic and G. Shaviv are cordially acknowledged. I also thank M. Scholz for critically reading the manuscript and M.S. Bessell for securing and let me use the M dwarf spectra. This work has been supported by the Deutsche Forschungsgemeinschaft (SFB 359, project C2).

References

1. Baschek B., Efimov G.V., von Waldenfels W., Wehrse R. (1997) A&A **317**, 630
2. Baschek B., Grüber C., von Waldenfels W., Wehrse R. (1997) A&A **320**, 920
3. Baschek B., von Waldenfels W., Wehrse R. (2001) A&A **371**, 1084
4. Feller W. (1966) An Introduction to Probability Theory and Its Applications, Vol. II. John Wiley & Sons, New York
5. Fisz M. (1978) Wahrscheinlichkeitsrechnung und mathematische Statistik. VEB Deutscher Verlag der Wissenschaften, Berlin
6. Mandel, L., Wolf, E. (1995) Optical coherence and quantum optics. Cambridge Univ. Press, Cambridge
7. Mihalas D. (1978) Stellar Atmospheres. Freeman, San Francisco
8. Oxenius, J. (1986) Kinetic theory of particles and photons. Springer, Heidelberg
9. Peraiah A. (1984) Discrete space theory of radiative transfer. In: Kalkofen W. (ed.) Methods in radiative transfer. Cambridge University Press, Cambridge, p. 281
10. Plez, B. (1998) A&A **330**, 1109
11. Rutily, B. (1999) In: Transfert de rayonnement en astrophysique, Ph. Stee, ed., Observatoire de Nice, p. 5
12. Strom S.E., Kurucz R.L. (1966) JQSRT 6:591
13. Unsöld, A. (1958) Physik der Sternatmosphären. Springer, Heidelberg
14. Wehrse R., Baschek B. (1999) Physics Reports **311**, 187
15. Wehrse R., Baschek B., von Waldenfels W. (2000) A&A **359**, 780
16. Wehrse R., Baschek B., von Waldenfels W. (2000) A&A **359**, 788
17. Wehrse R., Baschek B., von Waldenfels W. (2002) A&A submitted
18. Wehrse R., von Waldenfels W., Baschek B. (1998) JQSRT **60**, 963

Selected abstracts

Relaxation of multi–well energies in linearised elasticity

Neven Balenović

Department of Mathematics, University of Zagreb
Bijenička 30, Zagreb, Croatia

Abstract. We study the relaxation of multi–well non–convex energy functionals in the context of linearised elasticity. These functionals are known to be lacking minimisers in the usual sense, due to the fact that minimising sequences tend to oscillate rapidly aiming to satisfy imposed restrictions, thus exhibiting microstructures. One of mathematical tools known to be able to give some information on microstructures are H-measures (L. Tartar 1990 and independently P. Gérard 1991). Following the ideas of Kohn, Ball & James and Tartar, we deploy the pseudodifferential calculus in order to express the relaxed functional in the terms of minimisation with respect to H–measures.

Homogenization of the Lavrentiev–Bitsadze equation in partially perforated domain

Gregory A. Chechkin

Department of Differential Equations, Faculty of Mechanics and Mathematics
Lomonosov Moscow State University, Moscow 119899 Russia

Abstract. We consider a model homogenization problem for the Lavrentiev–Bitsadze equation

$$u^\varepsilon_{tt} + \ (\mathrm{sign}\, t)\, u^\varepsilon_{xx} = f(x,t)$$

in a partially perforated domain $\Omega_\varepsilon \in \mathbb{R}^2$ with a Neumann boundary condition on the boundary of holes. The perforated part of the domain lies in the upper half–plane $t > 0$ and has a locally periodic structure. The parameter ε characterizes the cell of periodicity. We study the asymptotic behavior of the solution of such a problem as $\varepsilon \to 0$. We prove the existence and the uniqueness theorem for the posed problem. Then, we construct the homogenized problem and on the base of the a priori estimates of the solution we prove the weak convergence of the solutions of this problem to the solution of the homogenized problem.

This is a joint work with Vladimir A. Kondratiev.

Micromagnetics : Equilibria and Relaxation

Irene Fonseca

Department of Mathematical Sciences and Center for Nonlinear Analysis
Carnegie Mellon University, Pittsburgh, PA 15213, USA

Abstract. Minima of the energy for large magnetic bodies with vanishing induced magnetic fields

$$E(m) := \int_{\Omega} [\varphi(m) - \langle h_e ; m \rangle] \, dx + \frac{1}{2} \int_{\mathbb{R}^3} |h_m|^2 \, dx$$

are completely characterized in terms of the anisotropic energy density φ and the applied external magnetic field $h_e \in \mathbb{R}^3$. More generally, one considers an energy functional for a large ferromagnetic body of the form

$$F(m) := \int_{\mathbb{R}^N} f(x, \chi_{\Omega}(x) m(x), u(x), \nabla u(x)) \, dx$$

where $(\chi_{\Omega} m, \nabla u)$ satisfies Maxwell's equations, i.e. $u \in H^1(\mathbb{R}^N)$ is the unique solution of $\Delta u + \operatorname{div}(\chi_{\Omega} m) = 0$ in $\mathbb{R}^N$. It is shown that if f is a Carathéodory function satisfying very mild growth conditions then the relaxation of F with respect to $L^{\infty} - w*$ convergence in $L^{\infty}(\Omega; \overline{B(0,1)})$ is given by

$$\mathcal{F}(m) = \int_{\Omega} Q_M f(x, m(x), u(x), \nabla u(x)) \, dx + \int_{\mathbb{R}^N \setminus \Omega} f(x, 0, u(x), \nabla u(x)) \, dx$$

where $Q_M f$ is the quasiconvex envelope of f relative to the underlying partial differential equations. This class of integrands includes those of the type

$$f(x, m, u, h) = \varphi(x, m, u) + \psi(x, u, h)$$

with $\psi(x, u, \cdot)$ non convex.

The first part of this work was undertaken in collaboration with Bernard Dacorogna and the relaxation results were obtained jointly with Giovanni Leoni.

The first order term in a homogenized boundary condition for the heat equation

Stephan Luckhaus

Math. Institut Univers. Leipzig
Augustusplatz 10/11 D-04109 Leipzig Germany

Abstract. The paper presented is a joint work with J. Filo, Bratislava. The boundary data investigated are Dirichlet and Neumann data rapidly alternating on the boundary. Since on general smooth boundaries periodicity can be defined only modulo deformations, it makes sense to look for a generalization which most naturally is almost periodicity. Here almost periodicity has to be understood in the sense of capacity. Convergence of the first order term in the expansion can then be derived from the uniqueness of the solution to the blow up problem in halfspace.

Coarse-grid correction by discrete homogenization for elliptic problems

Nicolas Neuss

IWR, Universität Heidelberg,
Im Neuenheimer Feld 368, 69120 Heidelberg, Germany

Abstract. Given a diffusion or elasticity problem with periodically oscillating coefficients, homogenization theory tells us that its solution is approximated by the solution of the so-called homogenized problem. Transfering this into a discrete setting we obtain an approximation property which is suited for proving the convergence of a two-grid iteration wherein the coarse-grid correction is computed by solving a discretely homogenized problem. Robust convergence of this method is proved both theoretically and numerically.

Characteristic Lengths and Times Associated with Processes in the Root Zone

P.A.C. Raats

Wageningen University Paaskamp 16, 9301 KL Roden, The Netherlands

Abstract. Uptake of water from soils by individual plant roots can be described by a nonlinear diffusion equation and appropriate initial and boundary conditions. The model involves one or more characteristic lengths characterizing the root/soil geometry and two characteristic times characterizing, respectively, the diffusive transport and the ratio of supply in the soil and demand by the plant roots. Generally, at a certain critical time uptake will switch from demand–driven to supply–dependent. The structure of the solutions will be shown, both for the linearized problem and for some nonlinear cases. The natural roles of i) steady rate, ii) long–time and iii) quasi–steady state approximations will be indicated.

In the next part, water balance and flow models at the scale of entire root systems will be explored. The approach will be to extend the Richards equations to include a sink term to represent the uptake of water by plant roots. This sink term should on the one hand be consistent with the spatial/temporal distribution and functioning of individual roots and on the other hand ultimately account for typical observed spatial/temporal patterns of extraction of water. The latter appears to require some constraint reflecting the minimization of energy expenditure by the entire plant in satisfying its water demand.

In the final part, the multitude of flow and transport problems related to the excange of gases, nutrients and contaminants between soils and plant roots will be outlined.

Non–Linear Flow in Porous Media

A Numerical Study of the Navier–Stokes System with Two Pressures

Stefan Rief

University Kaiserslautern, Germany

Abstract. We consider a periodic porous medium and the stationary incompressible Navier-Stokes equations defined on the fluid part of the medium. Depending on the asymptotic behaviour of the Reynolds number, the method of homogenization yields different limit equations as the edge lenght ε of the periodicity cell tends to zero. If the Reynolds number grows *slowly* as $\varepsilon \to 0$, one gets the well–known Stokes system with two pressures which is a linear set of differential equations and can be interpreted as Darcy's law. If the Reynolds number grows *fast*, the so called Navier–Stokes system with two pressures is obtained which can be interpreted as a non-linear extension of Darcy's law. We will discuss some numerical aspects of this system.

Printing: Strauss GmbH, Mörlenbach
Binding: Schäffer, Grünstadt

GPSR Compliance

The European Union's (EU) General Product Safety Regulation (GPSR) is a set of rules that requires consumer products to be safe and our obligations to ensure this.

If you have any concerns about our products, you can contact us on ProductSafety@springernature.com

In case Publisher is established outside the EU, the EU authorized representative is:

Springer Nature Customer Service Center GmbH
Europaplatz 3
69115 Heidelberg, Germany

Batch number: 10704711

Printed by Printforce, the Netherlands